A Social–Scientific Examination of the Dynamics of Communication, Thought, and Selves

Seif Sekalala
Drexel University, USA

A volume in the Advances in Linguistics and
Communication Studies (ALCS) Book Series

IGI Global
PUBLISHER of TIMELY KNOWLEDGE

Published in the United States of America by
    IGI Global
    Information Science Reference (an imprint of IGI Global)
    701 E. Chocolate Avenue
    Hershey PA, USA 17033
    Tel: 717-533-8845
    Fax: 717-533-8661
    E-mail: cust@igi-global.com
    Web site: http://www.igi-global.com

Library of Congress Cataloging-in-Publication Data

Names: Sekalala, Seif, 1985- author.
Title: A social-scientific examination of the dynamics of communication,
  thought, and selves / by Seif Sekalala.
Description: Hershey, PA : Information Science Reference, [2022] | Includes
  bibliographical references and index. | Summary: "In this book, the
  author focused exclusively on the concept of intrapersonal
  communication, explicating on how and why we communicate with our own
  selves, and how and why scholars can help humans improve and harness
  intrapersonal communication in fields such as AI"-- Provided by
  publisher.
Identifiers: LCCN 2022970004 (print) | LCCN 2022970005 (ebook) | ISBN
  9781799875079 (hardcover) | ISBN 9781799875086 (paperback) | ISBN
  9781799875093 (ebook)
Subjects: LCSH: Self-talk. | Self-perception. | Thought and thinking. |
  Artificial intelligence.
Classification: LCC BF697.5.S47 S429 2022 (print) | LCC BF697.5.S47
  (ebook) | DDC 158.1--dc23/eng/20220228
LC record available at https://lccn.loc.gov/2022970004
LC ebook record available at https://lccn.loc.gov/2022970005

This book is published in the IGI Global book series Advances in Linguistics and Communication Studies (ALCS) (ISSN: 2372-109X; eISSN: 2372-1111)

British Cataloguing in Publication Data
A Cataloguing in Publication record for this book is available from the British Library.

For electronic access to this publication, please contact: eresources@igi-global.com.

Advances in Linguistics and Communication Studies (ALCS) Book Series

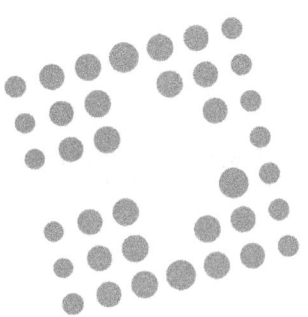

Abigail G. Scheg
Western Governors University, USA

ISSN:2372-109X
EISSN:2372-1111

Mission

The scope of language and communication is constantly changing as society evolves, new modes of communication are developed through technological advancement, and novel words enter our lexicon as the result of cultural change. Understanding how we communicate and use language is crucial in all industries and updated research is necessary in order to promote further knowledge in this field.

The **Advances in Linguistics and Communication Studies (ALCS)** book series presents the latest research in diverse topics relating to language and communication. Interdisciplinary in its coverage, ALCS presents comprehensive research on the use of language and communication in various industries including business, education, government, and healthcare.

Coverage

- Non-Verbal Communication
- Discourse Analysis
- Media and Public Communications
- Language in the Media
- Interpersonal Communication
- Computational Linguistics
- Sociolinguistics
- Cross-Cultural Communication
- Semantics
- Graphic Communications

IGI Global is currently accepting manuscripts for publication within this series. To submit a proposal for a volume in this series, please contact our Acquisition Editors at acquisitions@igi-global.com or visit: https://www.igi-global.com/publish/.

Titles in this Series

For a list of additional titles in this series, please visit: www.igi-global.com/book-series

701 East Chocolate Avenue, Hershey, PA 17033, USA
Tel: 717-533-8845 x100 • Fax: 717-533-8661
E-Mail: cust@igi-global.com • www.igi-global.com

To my mother, Christine Kavutse.
Thank you for everything, and I love you.

Table of Contents

Preface

INTRODUCTION

Sometime around the summer of 2020, I started working on a proposal for a book I had been mulling for several months. Having taught communication studies courses over 10 years in the USA (as well as one semester in China), I had realized that most communication studies scholars/texts habitually gloss over the topic of int*ra*personal communication (i.e., communication from one to the same individual). Thus, I wanted to make a contribution to the study of that under-appreciated topic.

Over one and a half years later, this book is the result of that effort. Whereas it is a truism that int*er*personal-context-based communication is indispensable to most human endeavors in our social and work lives, are our communications with our own *selves* equally important? If "yes," how and/or why?

RELATION TO CONTEMPORARY STUDIES OF COMMUNICATION AND SOCIAL SCIENCE WRIT LARGE

Despite its ancestry from the social sciences writ large, the discipline of communication has—for better or worse—developed its own theoretical-empirical body of knowledge over the past several decades. The most common sub-disciplines covered by courses across undergraduate and graduate levels include interpersonal communication, mass media and public communication (including public speaking), small-group communication, organizational communication, etc. As I note in the introduction, Knapp and Vangelisti's (2009) text—i.e., about interpersonal communication—is emblematic of such coverage.

However, in this book (especially in chapter one), I argue that the discipline's lack of coordination with other social sciences is one of the main reasons for the sub-par growth of theorization and empirical study of intrapersonal communication, to which I also refer as self-communication—or SC. As such, in chapter one, I highlight the efforts that have been made by other social scientists vis-à-vis the study of topics related to SC, including Archer (2007), Hermans and Hermans-Konopka (2010), and symbolic interactionism scholars such as Wiley (1994) and Charon (2007).

CHAPTER OVERVIEW

In any case—regardless of such scholars' previous efforts, how can/should a researcher appropriately study humans' internal communications? My own solution to this quandary was the use of a methodological framework that mostly utilizes autoethnography and narrative analysis, and discourse analysis via a basic meta-review of cognitive- and data-science articles. Using voice-note recordings and transcriptions, I recorded, transcribed, and analyzed my own intrapersonal communications over the course of roughly four months—from November 2020 to March 2021. In addition, I then supplemented that data with analyses of the aforementioned data- and cognitive-science instructional and research literature.

Chapter 1. First, chapter one introduces us to the core theory and practice of the concept in question: what is self-communication (SC), and how do we do it in our day-to-day lives? The opening anecdote demonstrates the misconception that many of us have about communication—particularly via the mode of talking—to ourselves; i.e., that it is *always/often* a symptom of severe mental illness. Instead, based on the available literature from communication studies and other social sciences, I argue that SC is in fact an indispensable part of our human character. Thus, in addition to the sample of research data-analysis I provide in this chapter, readers should closely review the background information I provide about the related studies of SC from various other social science (sub-)disciplines, which can help orient our current renewed focus.

Chapter 2. With that overview of the book's topic(s) in chapter one, I then take a step back in chapter two to attempt a holistic introduction of my authorial-autoethnographic voice to readers, via a serialization of the journal entries and blog posts I have been writing since my high school years. There are multiple valid reasons for this holistic introduction, but two of the more prominent are as follows: 1) Most of the book's data comes from my postulations about general SC and the application of SC to the improvement of our learning-method improvement, which I recorded and transcribed using audio voice-notes. Thus, I believe readers should be given a substantive opportunity to acclimate themselves to the voice of the mind/person providing the upcoming postulations. 2) Secondly, by reviewing those journal and blog entries/posts, I am heeding the advice of many of the scholars I introduce in chapter one, who indeed encourage us to try to understand an individual's psyche based on a holistic consideration of their life-experiences from as early on as possible—i.e., childhood or youth, through adulthood.

Chapter 3. Next, chapter three launches our empirical study of SC in earnest via the topical-thematic study of the voice-note data. Readers are introduced to the four main overarching themes under which I was able to classify most SC instances of the audio voice-note recordings, namely i) *"Types of Self-Communication (SC),"* ii) *"SC And Learning,"* iii) *"Reality And Perception-Processing I: Without Metacognition,"* and iv) *"Reality And Perception-Processing II: With Metacognition."*

I believe the title of the chapter sums up the character of a quintessential SC instance: you find yourself in the kitchen or any other area of the house on a given day—e.g., on a weekend- or other day on which you don't have to go to school or work. And suddenly, you ask yourself: "why did I come to the kitchen?" Perhaps you had planned to clean it, or pick up something to eat, or pick up something that you'd left there accidentally. Regardless, you now find yourself having to recall why exactly you came to the kitchen in the first place.

This chapter also introduces us to an important theme with which I keep grappling throughout the book, namely, the relation or tension between communication with our own selves, versus communication with other members of society. In this context, I highlight two facts, namely: 1) We of course do not live in a social vacuum—rather, we live and interact with other individuals. We thus have to communicate with them constantly. 2) But first, we have to make sense of our world and experiences, and of our relations and communications with these other individuals. A good example of this discussion is in section 3.2.10, as I recount my conversation with an old mentor, and my mind-based reactions to his revelations during and after our conversation.

Chapter 4. We continue our scrutiny of these various SC instances in chapter four. However, in this chapter, I try to specifically hone in on instances that highlight the essence, utility, and versatility of the role of language and symbol systems in our communication. And once again via the chapter's title (about *"tip of the tongue* syndrome)", I try to demonstrate how language and symbol systems can both paradoxically unlock our communication power, and how they can also limit us in many instances of self- and interpersonal communication.

Chapter 5. This chapter continues these discussions, but with the focus mostly shifted back to the examination of the relation or tension between communication with our own selves, versus communication with other individuals of society. Given its strategic midway point in the book, I also attempt to scrutinize the appropriateness of the themes under which I categorize the previous chapters' discussed SC instances.

Chapter 6. Next, I apply the discussion of SC to the secondary focus of the book, namely, the improvement of our learning methods. I demonstrate how SC has helped me as an autodidact attempting to learn software development and data science. The chapter introduces us in earnest to the book's supplementary data-analysis procedure, namely, the qualitative metareview of data- and cognitive-science instructional and scholarly literature.

Chapter 7. In this chapter, we continue our focus on some of the applications of SC by focusing on how it can help us regulate our mental health. The discussion of the relevant SC instances are discussed under one main theme, namely, *thinking-triggered/enhanced equilibrium and distress.* Within that theme, I split the instances further into eight titles, namely: 1) self-checking, 2) temporary memory-disfunction neurosis, 4) meditation, 5) positivity and distress, 6) guilt and intense emotions, 7) "brute-force" learning, and 8) yearning for coaching and self-doubt.

Chapter 8. Finally, we end our focus on potential SC applications by examining its role in the improved design and use of ICT technologies, including AI. After reviewing both the data-sets' findings, I highlight two main solutions that we can use to improve ICT designs and utilities, namely 1) individual (self-directed) mindfulness and individual and collective soul-searching, and 2) consolidation and more mindful use of ICT.

Chapter 9. Having examined all the above core features and potential utilities of SC, I attempt a final holistic review in this chapter. I try to look at SC as more than the sum of its parts, synthesizing the core elements of its characteristics and applications. From data set I (audio voice-note recordings), I conclude that there are five typological categories that can help us study SC, namely: 1) unintentional and deliberate SC, 2) the intra-/inter-personal communication spectrum, 3) language and symbol systems, 4) applications, and 5) other considerations. And from data set II (qualitative metareview of data- and cognitive-science literature), I delineate three main thematic supra-groups, namely: 1) Group one—titled "Learning About Our World"—which consists of the sub-groups of "knowledge" and "epistemology, 2) Group two—titled "Shared Meaning-Making"—which consists of the sub-groups of "Learning and Pedagogy" and "Meaning," 3) And group three—titled "Metacognition"—which consists of the sub-groups of "Empirical Models" and "Cognition and Experimentation."

More importantly, I delineate three main types of all the various SC instances present in our minds, i.e. neutral, positive, and negative. We can thus harness our SC to motivate ourselves as we go through life. In fact, I point out that numerous self-help gurus are often either directly or indirectly discussing SC in their pitches.

Chapter 10. Finally, I conclude the text, with brief recaps of both data-sets, as well as a disclosure of the research limitations, and the opportunities for further research.

TARGET AUDIENCE

Considering my attempt at balancing theory and empiricism with a reader-friendly tone, this book can be helpful to readers of various academic levels from the undergraduate college/university level and beyond. Instructors of various humanities and social-science subjects such as English (e.g., creative writing), sociology, and communication might find it useful. And as I mention in chapter one, part of my motivation for using the interdisciplinary field of cognitive science is to demystify it for students, but I also hope that seasoned experts might gain some new insights about the basic characteristics and utilities of SC-related cognitive-science dynamics.

MAIN CONTRIBUTION AND POTENTIAL IMPACT TO DISCIPLINE

Overall, the study of intrapersonal-/self-communication has not been studied sufficiently to date. In this book, I have tried to provide some substantive evidence of the existence, versatility, and utility of this vital element of personhood. Via primary and secondary research, I have provided a variety of examples of how SC unfolds within and outside of our minds, and how we can harness it better to live more productive and fulfilling lives.

Seif Sekalala
Drexel University, USA

REFERENCES

Archer, M. S. (2007). *Making our Way through the World: Human Reflexivity and Social Mobility.* Cambridge University Press. doi:10.1017/CBO9780511618932

Charon, J. M. (2007). *Symbolic interactionism: An introduction, an interpretation, an integration.* Pearson Education.

Hermans, H. J. M., & Hermans-Konopka, A. (2012). *Dialogical self theory: Positioning and counter-positioning in a globalizing society.* Cambridge Univ. Press.

Knapp, M. L., Vangelisti, A. L., & Caughlin, J. (2013). *Interpersonal communication & human relationships.* Academic Press.

Wiley, N. (1994). *The semiotic self.* Polity.

Acknowledgment

I would like to earnestly thank the staff of IGI Global, who have been wonderful partners throughout the early proposal-preparation and submission, drafting, and finalization stages of this manuscript. I am particularly grateful to: Ms. Melissa Wagner, the editorial director, and Ms. Jan Travers, the IP and contracts director. Thank you so much!

I am also extremely grateful to the scholars who agreed to assist as members of a secondary blind review team (/editorial review board), namely: Dr. Decky Omukoba, the lead reviewer, as well as Drs. Michael Wolfe, Dominique Knutsen, and Eliud Situma. Thank you so much folks, you're truly scholars, and generous friends to boot!

I would also be remiss not to thank Drs. Rachel Reynolds and Douglas Porpora, for putting up with my constant excited updates about the book, and most of all, for your oh-so-generous moral and tangible support over the years. Yours are some of the most prominent of giants' shoulders on which this humble scholar stands. Thank you. And to anyone I am leaving out here—and certainly not on purpose (!)—that has helped smoothen my path to date, thank you.

Chapter 1
The Odd Science Teacher:
Introduction to Intrapersonal Communication

ABSTRACT

This chapter introduces the concept and study of intrapersonal communication. The author defines intrapersonal communication (or self-communication/"SC") and its relationship with interpersonal communication and social-science writ large. He also introduces the two academic (sub-)disciplines/topics—i.e., autoethnography and cognitive-science—that the book utilizes to explicate SC. The chapter also presents the research questions of the book's study, as well as methodological framework with which the author studies SC—including two sample-sets of each of the two main data-types to be utilized, i.e. audio-recorded voice-notes and their attendant transcripts, as well as cognitive-science articles. Thereafter, the author previews the book's research presentation and analysis frameworks.

INTRODUCTION

The Odd Science Teacher

There are two analytical-narrative textual artifacts that can help to orient our study of int**ra**personal or self-communication, as presented in this book. The first one is a story of a primary-school teacher of mine, to whom I'll simply refer as "Mr. 'K,'" and the other is an article by Schwartz & Pines (2019), from the Harvard Business Review.

First, the story of Mr. K: I believe many or most of my schoolmates from elementary and middle school before 1998—at an all-boys' Catholic boarding school in Kampala, Uganda, can recall our very odd science teacher, Mr. K. Mr. K would often walk around our school campus apparently deep in thought, clearly speaking softly to himself while gesticulating dramatically—or even pantomimically, the way one normally does while telling a dramatic story, or arguing with another interlocutor.

DOI: 10.4018/978-1-7998-7507-9.ch001

In other words, Mr. K was behaving in a manner somewhat similar to that of an apparently mentally-ill person. And perhaps he was indeed. But in addition to being one of—if not *the* nicest seventh-grade teacher(s), Mr. K was apparently very smart.

After all, he never carried any books or notes as he lectured and wrote extensive notes for us on the blackboard, and he seemed to have an encyclopedic reserve of facts and factoids, teaching science in an engaging and fun manner! I believe many readers of this book might be able to relate to Mr. K's behavior—i.e., as a characteristic of your own behavior, or that of someone you know—a friend, relative, etc.

And contrary to the false but popular notion that (apparent/visible) intrapersonal communication is a characteristic of insanity (e.g., Zoppi and Legg, 2021), in this book, I argue earnestly that all human beings utilize intrapersonal communication—i.e., communication with our own *selves*—in various contexts. In fact, I also argue that this type of communication is vital to our holistic growth and improvement.

I also try to provide a possible template: 1) for the study of this type of communication, as well as 2) the systematic gradual execution of the above-mentioned personal-professional growth and improvement processes, during and beyond crises such as the current COVID-19 pandemic (as of mid-2021 and beyond). In this regard, I can cite the aforementioned article by Schwartz & Pines (2019) from Harvard Business Review as proof, given the fact that essentially, the process they are describing as vital for leaders' professional growth is in fact (arguably) a method of applied systematic intrapersonal- or self-communication. In this vein, we need to first clearly establish what intrapersonal-/self-communication (SC) is and is not. Throughout the rest of this chapter, I attempt to clearly define intrapersonal-/self-communication (SC), provide the general theoretical—as well as autoethnographic—background from which the study arises, and introduce the research questions and methodological frameworks that I use to answer the questions (including a preview of research findings).

Definition of SC, And Main Purposes Of Study

I: Definition (What Intrapersonal Communication Is), Part I

In this book, the primary definition of intrapersonal communication—informed by the literature-review presented farther below in this chapter—is: ***communication that is clearly self-directed and purposive in the form of abstract and specific thoughts, or verbal/language-based articulations, about the various metaphysical and physical phenomena we actively perceive, recall, or imagine, including our own selves in the past, present, future, and hypothetical timeframes***. That definition is composed of three main parts, namely: 1) A delineation of "*communication that is clearly self-directed and purposive*"; 2) "*in the form of abstract and specific thoughts, or* [the] *verbal/language-based articulations*" contained in said (intrapersonal-)communication, and 3) the objects—i.e., "*...the various metaphysical and physical phenomena we actively perceive or recall, including our own selves (in the past, present, future, and hypothetical timeframes)*," about which we communicate.

One of the most important elements of the above definition is as follows: intrapersonal communication necessarily involves thinking, and many of the thoughts we have in solitude can be defined as instances of intrapersonal communication. In addition, two of the above parts of the primary definition [i.e., 2), and 3)] specify that many of our general and specific mind-based intrapersonal communication thoughts and/or verbalized instances have the characteristic of actively channeling thoughts and communication towards our individual *selves*.

In tandem with the elements of the secondary hypothesized definition explicated below, it should also be noted that intrapersonal-(/self-) communication and cognition can and most often occur(s), with or without our realization of it. In other words, you often do not realize that you are thinking or communicating to/with yourself, even though you are in fact undertaking that action. Some of the few common instances in which we may realize our ongoing intrapersonal-cognition and communication include (among others): talking out loud by yourself (like Mr. K), realization of our extreme happiness or sadness, metacognition, and the act of meditation.

II: Definition (What Intrapersonal Communication is), Part II

The secondary definition of intrapersonal communication is much more expansive. However, readers should note in earnest that I utilize this expansiveness sparingly, given my pursuit of methodological and analytical rigor.

Based on the literature-review and preliminary empirical findings presented in the current and following chapters, we can hypothesize that intra-interpersonal cognition and communication exists on a spectrum. This proposed spectrum is book-ended on two sides by a lack of cognition and communication, and the presence of int*er*personal cognition and communication (please refer to the table below).

Table 1. Intra-interpersonal cognition and communication spectrum (IICC)

A—No Cognition	*B—Cognition 1*	*C—Cognition 2: Intrapersonal Communication*	*D—Cognition 3*	*E—Cognition 4: Interpersonal Communication*
→None-existence → Death → In coma → Unconsciousness (**Important note**: I am leaving it to other scholars to investigate whether, and/or how exactly, unborn and very young children think and communicate to/with themselves.)	→ Dreaming → Thinking clearly or abstractly in general, including via combinations of parts C through E, or other modes. → Feeling—i.e., experiencing emotions such as happiness, sadness, anxiety, etc.	→ Clearly thinking or communicating to yourself, but unaware of the fact → Aware that you're thinking or talking to yourself → Awareness of your own emotions → General metacognition	→ Processing an interlocutor's message in your mind → Forming your sentences before responding during dialogues → Debating in your own mind whether or not to respond to someone during a discussion, or whether or not to ask a question, etc.	→ Communication with other individuals → Cognition in reaction to the communication of our interlocutors

What Intrapersonal Communication is Not

Given the above expansive definitions of intrapersonal communication, a valid and rather important question to ask in this context is: what are the instances of cognition and communication that cannot be defined as intrapersonal communication? And the response to this question is: any reactive cognition and/or communication that is experienced with interlocutors—i.e., between at least two people, and cognition and communication that is triggered in reaction to mediated- and mass-media. One can even argue that to a limited extent, the cognition and communication that occurs during reading is not intrapersonal. In other words, the thoughts and communications we produce in those instances are triggered by others. And often, they are (re-)consumed by these other parties as verbal/nonverbal feedback or reactive thoughts and communications.

Purposes of Study

I decided to write this book after teaching various college communication-studies classes for over 11 years (as of 2021) in the USA (and one semester in China), and in the process, realizing that intrapersonal (/ self-) communication was almost completely ignored, or covered cursorily in most (college) communication-studies courses. The classes that seem to pay the most attention to it are the basic-course classes, usually in the first chapters of the relevant texts, under topics related to the definition of communication. In those sections, intrapersonal communication is listed alongside int*er*personal communication, as well as other general types of communication including mass-media and public communication (including public-speaking), small-group communication, organizational communication, etc. But again, beyond that listing—as well as a basic definition of the phenomenon or concept, not much else is usually mentioned.

And yet as I point out above and in the following sections and chapters, the concept and practice of intrapersonal/self-communication (SC) is vital to the formation and guidance of our various unique *selves* or personalities over the course of our lifetimes. Overall, I submit that the study of SC can vastly improve our understandings and utilities of the human-social and technological phenomena discussed in this book, namely: general cognition, language, reflexivity and socialization, reading comprehension, mental health and illness, even digital-technologies, including AI!

In addition to studying intrapersonal communication in and of itself—i.e., so as to enable a better understanding of it in relation to the above-listed topics and others, this book specifically applies intrapersonal communication to the concept and practice of the improvement of our individual/personal learning methods. Reason: this concept and practice is arguably the main/best conduit through which we can implement the earlier-mentioned goal, i.e., personal-professional growth and improvement.

As an extensive autoethnographic exemplar, I am focusing on my own intrapersonal communication, and in turn, my own learning methods. In large part, the reason for this autoethnographic focus is the fact that it is one of the most apropos methods of executing such a study. As I explain farther below under the introduction of cognitive science, unfortunately, intrapersonal communication does not lend itself easily to many methods of social-scientific study. This characteristic is in contrast to int*er*personal communication, which can be observed via ethnography, experiments, etc.

In fact, another related goal of this study is to demonstrate a new potentially useful method for implementing cognitive-science studies, i.e., via crowd-sourced studies by various individuals, performed via similar systematic autoethnographic studies as the current one (albeit simplified and standardized for ease of use by the general public). Such crowdsourced studies might help cognitive-scientists in their ongoing debates about the utilities and limits of the computational-metaphor-based studies of the human mind. In other words, if human minds indeed function similarly to computers, great; but why not test the hypothesis using as many cognitive self-studies as possible?

Finally, in addition to the objectives stated above, this book is an ode to knowledge for its own sake—especially within each of our individual minds, expressed or not. The book is also an ode to creativity in its various forms, necessarily a product of our complex and sentient human minds. In this context, the study/book is heavily inspired by Douglas R. Hofstadter's "*Goder, Escher, Bach* (1979)." Relatedly, I also try to demonstrate or enact some examples of the above-mentioned nebulous concept and practice of creativity in my own research and writing methods. This attempt will be executed via "*creativity-*

attempt outtakes"—i.e., 1)--serializations of 28 out of the 42 posts from January to November of 2021 of a weekly public-journal on my website (appended to the end of each chapter)*, and 2)--For chapter two only, a brief excerpt from my undergraduate capstone-class creative-writing deliverable, a short novel based on my life. (*<u>Important note</u>: the posts will not be serialized in their original order; rather, they will be arranged in an attempt to match each posts' topics and themes to the appropriate chapter.*)

Moreover, at the end of each chapter, I present a number of discussion-questions and activities that readers of this book—both in scholarly (/clinical) and general-public contexts—can utilize to think critically and/or creatively vis-à-vis SC: what is it? How exactly does it happen to us, and/or how exactly do we do it—both on purpose and by accident/default? Can we delineate between good, bad, neutral—or various combinations of the preceding characteristics—types of SC?

BACKGROUND

Tension Between External And Internal (i.e., Int<u>er</u>personal Versus Int<u>ra</u>personal) Human Communication Studies: The Case For Revisiting Our Social-Scientific Roots

Despite my communication-studies passion and *bonafides* (so to speak), my approach to the study of intrapersonal communication advocates an interdisciplinary approach. My main argument in this context is that one cannot satisfactorily discuss intrapersonal communication as well as the conceptual "*self*," without also focusing on the larger concept of the mind and its dynamics.

However, it should also be noted that a holistic review of various relevant cognitive-science scholarship suggests that the concepts of mind and meaning, as well as symbolic systems, are much more than the sum of their mere parts (e.g., Terrence Deacon, 1997; Putnam, 1988; Kelly et al., 2007; and Hofstadter, 1979). In this regard, Putnam's (1988) view in particular might deserve a special emphasis, as he strongly argues that the computational metaphor of the mind cannot adequately explain human beings' cognitive complexity. And yet, as I detail farther below, that view of the mind, which Thagard (2005) refers to as "CRUM"—i.e., computational-representation understanding of mind—has resulted in a large and insightful body of research.

Regardless, some communication scholars have tried to focus on the topic, and Austin-Lett and Sprague (1976) and Vocate (1994) are two of the most relevant examples. However, as of early 2021, a general database-search—e.g., via Google Scholar—reveals an abysmal number of results (including the above two [sets of] authors), compared to other communication sub-disciplines.

To be fair, regardless of the merits of this book's arguments, human beings are a quintessentially social species. Thus, the discipline of communication-studies should perhaps indeed focus more on the various aspects of our shared meaning-making and purposive interaction. In this regard, Knapp and Vangelisti's (2009) text is emblematic of such efforts. The authors focus on the causal and related dynamics of relationship-building and unravelling, including factors such as environment and culture, effective dialogue, and rituals.

Against the backdrop of the above-explained uneasy coexistence between the two foci in question, I submit that communication-studies scholars might in fact benefit from a holistic and "back-to-the-roots"-style lens, harnessing a variety of social-scientific theories and empirical traditions. I am specifically referring to the theoretical-empirical concepts of human reflexivity (e.g., Archer, 2007), dialogical-self theory (e.g., Hubert Hermans & A. Hermans-Konopka, 2010), and symbolic interactionism (e.g., Wiley, 1994; and Charon, 2007).

Social Reflexivity (e.g., Archer 2007)

Related to various other core concepts of social-science—e.g., human agency and societal-structure, the theory of social-reflexivity posits that humans can, and often do, choose their own unique life-paths, according to their unique desires. Archer categorizes four main types of reflexivity, namely: communicative reflexivity (wherein we seek other individuals' opinions in reaction to our plans), autonomous reflexivity (wherein we cultivate and implement our own plans, regardless of other individuals' opinions), meta-reflexivity (wherein we critique our own plans), and fractured reflexivity (wherein our internal reflections result in indecision and mental turmoil). The research's main conclusion or implication is that our inner dialogues can result in an improvement or worsening of our livelihoods. Overall, Archer's studies make good use of ethnography, as well as extensive qualitative and mixed-methods data-collection and analysis.

Dialogical-Self Theory

Somewhat similar or parallel to the preceding (reflexivity) theory, the proponents of dialogical-self theory argue that in today's world, our *selves* do not necessarily have to conform to taken-for-granted societal norms, or even taken-for-granted personal values. Rather, the current post-modern society's *selves* are dynamic, and constantly (re-)evaluating their positions in relation to others as well as themselves. This dynamism can be effective in various contexts, including conflict resolution between self and other, and within self. Hermans and Hermans-Konopka (2010) particularly provide a rigorous theoretical discussion of the dynamic nature of the self, demonstrating the role of in*tra*personal* communication in self-understanding (*despite referring to the concept differently). The authors also discuss some of the older conceptions of the *self*—e.g., from symbolic interactionism, as a result of language and symbol-systems in general. However, unlike other scholars in this section—i.e., Archer as well as the pragmatism and SI theorists, dialogical self-theory is not distinctive vis-à-vis a substantive use of ethnographic methodology.

Pragmatism and Symbolic Interactionism (SI)

Of all the theorists in this discussion, the proponents of pragmatism and SI might merit the most credit vis-à-vis their rigorous treatment of the concept/phenomenon of the self. In addition to grappling with the human *self*'s precise definition and/or nature, these theorists also examine various dynamics that influence our *selves*' development, unique traits or personalities (especially in the theatre of communication and interaction), and our negotiation of our identities in public and private, among other context. Two particular works can be explicated further in this context, namely Wiley (1994), and Charon (2007).

Wiley (1994) presents a pragmatism-based conception of a human *self* that is essentially three persons within one; i.e., a past ("me"), present ("I"), and future ("you") self. In addition, the self constantly engages in an internal dialogue using symbolic mediums, especially language. This interpretation can help us in our negotiations around identity-politics controversies, as the above characteristics about the *self* also enable us to have empathy for each other.

Relatedly, Charon extends our pragmatism-centric understanding of the self in his explication of the central tenets of SI theory, itself a descendant of pragmatism, namely: 1) social-interaction is key to understanding human behavior, 2) complex intelligence or thinking capacity and self-reference—using symbols and language in particular—is a unique and key human trait, 3) humans do not merely perceive or sense their environment; rather, they actively interpret it 4) instead of the past or future—or perhaps the hypothetical, human action is mainly a result of our present thinking and interactions, and finally 5) humans are active agents in their environments, versus mere subjects of societal structure or other underlying—or overarching—causes or catalysts.

As I mention above, SI theory was built on the foundation of pragmatism, which presumes a number of relevant key principles about human behavior, namely: 1) humans actively interpret their environment, 2) those interpretations are influenced by the utilities we seek, based on our needs and goals, 3) the preceding facts necessarily imply that humans selectively highlight some, not all, aspects of their environment, and 4) ongoing human behavior—not traits or past actions—is what should be studied and interpreted by social-scientists.

Similar to Archer—as discussed farther above, pragmatism and SI theorists are commendable for their use of ethnographic methods in confirming their hypotheses. One scholar that stands out in particular in this context is Erving Goffman, especially his "Asylums" publication (1961).

Gaps or Avenues for Further Study

In totality, all the preceding authors substantively add to our understanding of intrapersonal communication. However, numerous avenues remain for further investigation. For instance, Archer's research does not grapple with some of the key issues of the current research, especially: 1) general intrapersonal communication, not just in relation to practical utilities—i.e., including social-mobility; and 2) relatedly, the basic processes, mechanisms, or parts of intrapersonal communication—e.g., language and other symbol systems, the self, and cognition. Similarly, most authors—save for Archer's parallel general focus on individual self-improvement (especially vis-à-vis career-advancement)—gloss over the basic role of cognition in the development and dynamism of the self. Moreover, scholars have not consistently and rigorously applied the concept of intrapersonal communication vis-à-vis individuals' continuous (/ lifelong) improvement of general, as well as professional skills and abilities.

MAIN FOCUS OF THE CHAPTER

Overview of Book Chapters

- Part I
  - Chapter one: An introduction to the book's topic, i.e. intrapersonal communication: its definition, background of study, related topics, rationale for its study, and a general overview of the book's content and structure.
  - Chapter two: An introduction to the author's authorial-ethnographic voice, via a presentation and basic discussion of journal-entries and blog posts written over a 20-year period.
- Part II
  - Chapter three: Presentation of analysis and interpretation of set 1 data, i.e. voice-notes/ transcripts, with the goal deducing the data's relevance vis-à-vis cognition, reflexivity, and socialization.

- ○ Chapter four: Presentation of analysis and interpretation of set 1 data, i.e. voice-notes/transcripts, with the goal deducing the data's relevance vis-à-vis language and symbol-systems.
- ○ Chapter five: Continuation of analysis and interpretation of set 1 data. Also, an analytical-review of previous chapters' findings in a bid to tentatively confirm potential insights vis-à-vis SC growth and evolution; cognition, reflexivity, and socialization; and language and symbol-systems.
- Part III
  - ○ Chapter six: Presentation of analysis and interpretation of set 1 and set 2 data, i.e. voice-notes/transcripts and meta-review of data- and cognitive-science literature, with the goal deducing the data's relevance vis-à-vis the improvement of personal learning methods.
  - ○ Chapter seven: A revisitation of set 1 data in isolation, i.e. voice-notes/transcripts, with the goal deducing the data's relevance vis-à-vis mental-health.
  - ○ Chapter eight: Presentation of analysis and interpretation of set 1 and set 2 data, i.e. voice-notes/transcripts and meta-review of cognitive-science literature, with the goal deducing the data's relevance vis-à-vis ICT and AI.
- Part IV
  - ○ Chapter nine: An analytical-review of previous chapters' findings in a bid to tentatively confirm potential insights vis-à-vis all the book's topics in question—i.e., as covered in parts two and three.
  - ○ Chapter ten: A holistic recap of the book's discussions, and a reflection on the limitations of the book's research, and potential avenues for further research.

Research Data-Set, Analysis, and Methodological Overview

In tandem with my definitions of intrapersonal/self-communication above, and after a careful consideration of the theoretical-empirical findings of other authors covered therein, the data analysis in this study is composed of two main sets, namely:

I. Discourse-/narrative-analysis of: i)—An extensive systematic collection of autoethnographic meta-analytical intrapersonal communication messages. I recorded this data using an audio-recorder, and I later transcribed it into summary-transcripts. And to a lesser extent ii)—A collection of private and public journal-/blog-entries, written over the course of around 20 years, from high school through 2021 (however, please refer to the caveat in the next paragraph after these research questions below); and

II. In addition to the above empirical data-analysis, I also present a qualitative content-analysis of data- and cognitive-science resources and articles, which I've been collecting between the spring of 2019, through the spring/summer of 2021.

Readers should note that technically, I do not consider chapter two's contents (the serialization of some of my journal and blog posts over the past 20 years) as part of data-set I above. Rather, that chapter's presentations of those private-journal and public-blog entries (since 2002, at the age of 16 while in high-school in Uganda) as well as my posts from two public-journal sites since 2010, are provided in a bid: 1) to orient readers to the book's authorial-ethnographic voice, and 2) perhaps more importantly, to provide a "thick description" (Geertz, 1973) of the author's general autobiographical background, which might be useful to one's understanding of said author's meta-analyses.

In the remaining parts of this chapter, I briefly define each of the three sub-disciplines of my theoretical-methodological framework. Thereafter, I divulge a succinct overview of the immediate autoethnographical-background context of the study, as well as the study's research questions and goals. I also introduce the data collection and analysis methods, and I present a sample of the findings from category I and II above, namely: A) a narrative-analysis on a sample (i.e., 24 out 63, or 38%) of my voice-note recordings, as well as B) brief tabular summaries of a sample of articles from the meta-review data-set. Finally, I present the presentation formats with which I present and analyze the findings of the research throughout the forthcoming chapters.

Introduction to Autoethnography, Cognitive-Science, And Operational Definition of Intrapersonal Communication

The theoretical-methodological framework of the current study is backed by three main social-scientific sub-disciplines, namely 1) autoethnography, 2) cognitive-science*, and 3) intrapersonal communication. (*It should also be noted that the field of cognitive-science is itself interdisciplinary, and necessarily involves STEM fields such as neurology and computer-science, and social-science fields such as linguistics and philosophy, among other disciplines.) In the next parts of this section below, I briefly introduce each of these three knowledge domains.

Brief Introduction to Autoethnography

As explained by the authors referenced below—among many others, autoethnography can be defined as the use of the method of ethnography (i.e., a systematic anthropological study via extensive interaction with a [group of] person[s]), on oneself. In other words, a scholar using autoethnography closely or critically analyzes their experiences, in a bid to extrapolate potentially useful research findings that might be relatable to the general population. Ever since its earnest debut as a social-science method in the 1970s, various authors have defended and demonstrated the efficacy of autoethnography as a valid qualitative research data-gathering and grounded analysis method (e.g., Bochner and Ellis, 2016; and Jones, Ellis, Adams, & Allen-Collinson, n.d.). In consideration of this fact, I will not belabor those points further.

However, contrary to the misguided assumption that the methodology easily renders a researcher prone to narcissism, it can be argued that the main goal of autoethnography is the revelation or uncovering of micro-level social-scientific insights—i.e., as experienced *uniquely* by various individuals. In turn, the macro-level relevance of these insights can be confirmed by extensive (quantitative) studies. In other words, arguably, the most important task for a researcher undertaking the use of the method should be to ask and strive to answer the following question: *how can other people in similar predicaments to mine, benefit from the experiences I am sharing in my scholarship*?

And as I explain farther below, I indeed earnestly hope to help other American and international citizens who might find themselves in the same dilemmas as I have over the past six years. Yet by utilizing autoethnography, I am arguably making myself more vulnerable. How so? By exposing cognitive and even personality traits and quirks, as well as mental health conditions that would otherwise be private. But that is a calculated risk, and I am certain of its beneficence value.

Cognitive-Science

In addition to the methodology of autoethnography and intrapersonal communication, this study utilizes the interdisciplinary field of cognitive-science (CS). Before defining it below, it might be worth noting that part of the reason I chose to use and highlight CS, is a desire to demystify it for myself and other scholars who might be intimidated by it, thus missing out on its numerous helpful utilities. So, what *is* it (CS)? Simply put, CS is the study of the nature, mechanics/dynamics of the (human) mind and its functions, including meaning-making (e.g., Thagard, 2005; and Friedenberg and Silverman, 2006).

It might be helpful for communication and social science scholars to dichotomize two general typologies of study related to our discipline, namely: 1) the study of communication and interaction the way we observe it in society—i.e., as ordinary citizens and as social-scientific scholars, and 2) the mind-based (or internal) antecedents and ongoing causal and related dynamics of said communication and interaction, which are much harder to study, as they cannot be observed as easily as the latter typology.

But despite that difficulty of study—i.e., because of the mind's unobservable character, a lot of progress has in fact been made to date in CS research. In the current study, I attempt to harness that research progress, applying it—along with autoethnography and intrapersonal communication, to a critical reflection on the career-related and general challenges I have experienced in the last five or six years. That attempt will utilize Thagard's (2005) synthesized collection of six main modes or mechanisms by which our minds are believed to process information, according to most cognitive scientists. Thagard (2005) refers to that paradigm or framework as the "Computational-Representational Understanding of Mind (p. 20)," abbreviated as "CRUM." By default, those six modes/mechanisms are the main conduits with which minds can be, and often are, studied. The modes in question—which I utilize in my data analysis farther below—are: 1) logic, 2) rules, 3) concepts, 4) analogies, 5) images, and 6) connections.

Operational Definition of Intrapersonal Communication

The third and final pillar of my theoretical-methodological framework is intrapersonal communication. As I mention farther above, the sub-discipline has not been actively studied compared to other communication sub-disciplines. But despite their older publication-dates, the attempts by Austin-Lett and Sprague (1976), as well as Vocate (1994), are noteworthy efforts. The current study thus builds on their efforts, in the hopes of renewing interest and research in the sub-discipline. Specifically, I utilize the following general and theoretical-methodological considerations and prescriptions:

I.   From Vocate (1994):
   1.   Despite the stark hindrances in our way as social-scientists to study the neurological, bio-physiological/chemical, and other *brain*-based features, we can certainly explore the cultural aspects that influence our intrapersonal communication dynamics,
   2.   The study of intrapersonal communication is more effective if executed via a triangulation of methods,
   3.   The study of intrapersonal communication does not mean that we have to ignore other individuals' perspectives; rather, we need to recognize the role that others play in our internal dialogues, and
   4.   The study of intrapersonal communication should be carried out with human participants from their early years of childhood, through later years of adulthood.

II.   Austin-Lett and Sprague (1976): this text is apparently written mostly for the benefit of (undergrad-uate-college) students and the general public, versus advanced communication and social-science scholars. Among others, the authors' foci raise the following questions:

1.   Individuals should strive to really learn their personalities; how and why do you constantly and predictably think and behave the way you do, and what do others know about your thinking, emotions, and behavior? Relatedly, what are your likes and dislikes, goals, fears?

2.   Do you realize the effects of your perceptions—i.e., your interpretations of your reality, in your various environments (e.g., at home, work, in public places, etc.)?

3.   In an expansion of point number 1 above, do you know how your physical, psychological, and other needs, as well as your values (e.g., bequeathed from family and/or society), influence your actions?

4.   Do you realize the influence of abstract symbol-systems such as language—which are also influenced by culture and other variables—on your self-expression?

5.   Finally, do you realize and appreciate the never-ending nature of intrapersonal communica-tion and growth, and can you confidently share and appreciate or understand your own and others' intrapersonal experiences as stated above, in your social-interactions?

Research Background and Questions

As I mention in the introduction of this chapter, the COVID-19 pandemic has indeed been markedly disruptive for many individuals and families around the world. In that regard, this author is not an excep-tion. I—along with various other family members of mine—contracted the disease, but luckily healed from it quickly. However, for me, the pandemic also coincided with a painful and unfair job loss—i.e., unrelated to the pandemic, an additional episode in a long and depressing job-search saga that I have endured since September of 2015, when I graduated from my Ph.D. program (in communication-studies, from at Drexel university, Philadelphia).

Of course, job-search struggles are not a new or exotic tale in the USA, especially for millennials in general, and for millennial-academics in particular (e.g., Alton, 2016; and McKenna, 2016). Regardless, I humbly submit that my own story might contribute to a better understanding of some of the intersec-tional social dynamics relevant to this societal problem.

A Struggling Black (Ugandan-American) Young Man, A Millennial-Academic in America

I was born in Uganda in 1985, to a Ganda father and a Ugandan-born ethnic-Rwandan (Tutsi) mother, and I arrived in the USA in February 2004 on a student visa. I subsequently completed my Associate's, Bachelor's, and Master's degrees between 2004 to 2010. Around 2009, for reasons I cannot divulge (not for dramatic purposes, but for my own and my family-members' wellbeing), I was granted political asylum. Later, I also earned permanent residency, as well as citizenship via naturalization.

In September 2012, I enrolled in the Ph.D. program, and I was thankfully able to complete it in three and a half years. But since then, I have submitted hundreds of junior-professorship applications in vain. As I mention above, my situation is not unique. However, I can confidently relay at least two unique race-/identity-politics-related events that compounded my distress. The first was my experience in Wenzhou, China, where I had accepted a full-time lecturer position, in which I only lasted one semester.

Why? Unfortunately, Chinese natives in that part of the country—unlike their counterparts in the bigger metropolitan areas such as Beijing and Shanghai—are not used to seeing Black people. Strangers would often gape at me endlessly. Once, a lady in a pharmacy seemed to want to reach out and touch my hair!

Because of that markedly discomforting situation and other factors, I returned in December of 2017, and the job-search resumed. Sometime around April or May of 2018, I received an offer from a university in the central-midwestern region of the US, for a full-time "permanent" non-tenure-track position. Towards the end of my first semester there, a kind colleague—a tenured professor in the department— warned me about the presence of bigoted colleagues in the department, individuals that would probably try to have me removed. Ultimately, that friendly warning was prescient. Around March/April 2020, I was copied on a letter from the chair to the dean and provost, recommending my contract's discontinuation. The main reason given for that recommendation was a contrived reason; a lack of "meritorious" teaching ability, based on student evaluations of my very first semester at the institution. Interestingly, the evaluations for the last semester at that institution turned out to be some of—if not *the* best—I have ever received in my teaching career.

"Up by [My] Own Bootstraps" through Continuous (Re-)Learning

Having endured all the above experiences—and heeding my kind colleague's friendly warning, I decided around late fall 2019 to chart a path forward that would hopefully ensure better employment prospects. I should also note that I happen to have a voracious appetite for learning in general, and I struggle with *impostor syndrome*, which is currently reified by my futile job-search. In other words, for my own peace of mind—not for braggadocio or other undesirable reasons, I need to: 1) clearly understand what I'm teaching, and 2) sound to myself like "I know my stuff!"

First, with a kind mathematician-colleague's help and later on my own, I strengthened my statistics expertise, and I also started collecting and reading/using a number of relevant autodidactic-support resources for data science. I also started various free online software-development learning programs— e.g., FreeCodeCamp and TheOdinProject, and even earned an AI certificate from the University of Helsinki. Later, I also embarked on joint research with a project-management guru, and later earned a CAPM certification.

Moreover, in addition the above foci, I also have a longterm continuous goal—which I am starting to implement now (and will hopefully continue to purse through and beyond my retirement)—to revisit and keep honing my general skills in math, physics, chemistry, and biology. Reasons: 1) I intend to use my continuously-improving knowledge in those fields to pursue any communication-research that might require interdisciplinary expertise in those fields. 2) But again, honestly, as I have already mentioned, the other *raison d'etre* for that goal is the fact that I am simply very inquisitive, and I really enjoy learning in general. In that regard, the first three sets of knowledge on which I am focusing via basic to intermediate books and online resources on are: 1) probability and stochastic process—also related to data-science (e.g., via William Dembski's "The Design Inference"), 2) quantum mechanics—e.g., via "30-Second Quantum-Theory," by Brian Clegg (Ed.), and 3) MIT's online "*OpenCourseWare*" courses such as "Computational Cognitive-Science" and "Statistics for Brain and Cognitive-Science," among others. Finally, in addition to my software-development, statistics, and communication-studies and project-management autodidactic-learning and research, I also try to read a variety of fiction and nonfiction books. As of this study's timeframe, I am reading four books simultaneously, namely (in random order): 1) Barack

Obama's "A Promised Land," 2) Thomas Picketty's "Capital in the 21st Century," 3) Ken Follet's "A Column of Fire," and 4) John Grisham's "Rogue Lawyer."

Resultant Strategy and Research Questions

As many readers might agree, the above learning agenda is rather ambitious, notwithstanding the fact that my learning is self-paced, and I do not have a deadline as such. Regardless, I soon realized that I needed to come up with an effective longterm strategy, for the purpose of improving my learning methods and self-motivation. As a solution to that challenge, I decided to start rigorously keeping track of my thoughts and intrapersonal communication, in a bid to identify some of my own cognitive- and intrapersonal communication-related bottlenecks to effective learning and self-motivation. Subsequently, I came up with the following general/research questions:

RQ 1: *How can I analyze my own thoughts and/or intrapersonal communication, to help me improve my learning methods and self-motivation?*

Relevant method: Careful study via recording and transcribing or curating/assembly, and analyzing via general qualitative- and/or narrative-analysis of: i)—voice-notes/transcripts, and to a limited extent, ii)—via the "Creativity Attempt Outtakes" at the end of each chapter, and the data from the private and public journal-entries I share in chapter two.

RQ 2: *How can I use the techniques of cognitive-science and other relevant academic and applied fields to help improve my learning processes?*

Relevant method: A qualitative content analysis of academic journal-published and other online-derived studies, as well as instructional articles on the study of data-science, and about learning and creativity in general.

I: Chapter 1 Pilot-Study Data Collection, And Preview of Results & Analysis I: Voice-Note Summary-Transcription and Grounded-Theory-Based Analysis

I.1: Introduction

From early November 2020 through the end of March 2021, I systematically recorded voice-notes each week, summarizing my thoughts and intrapersonal communication instances in regard to my learning. For each recording, the default question I was answering was:

"What have I been saying to myself mentally, and/or occasionally, out loud, in reaction to all the things I am learning or reading about?"

By March 27, 2021, I had recorded a total of roughly 153 intrapersonal communication instances, combined in a grand-total of 64 voice-notes, which is also the total number of summary-transcriptions I ended up with. The total recorded time for each summary-transcription unit varies, from a possible minimum of at least two minutes, to a possible maximum of 30 or more minutes. Based on the best-practice principles of the grounded theory analysis paradigm (e.g., Lindlof and Taylor, 2019), I continuously analyzed all the recordings in detail between March and the end of my research analysis stage, around early June (note: the book-submission deadline had been set for the end of July). However, for the purposes of this chapter, I was able to execute a basic qualitative-analysis on a total sample of 24 summary-transcription units, mostly recorded in early January.

I.2: Voice-Note Data-Sample Timeline and Quantity, and Format

Below is a basic breakdown of the current chapter's summary-transcription units' recording timeline and quantity:

- Batch 1.0: Nov 2020 to January 2021: Nine summary-transcription units, and
- Batch 2.1: January 2021: 14 summary-transcription units.
  - Total of both batches combined: 23

Each voice-note summary-transcription is formatted as follows:

Voice-Note No.: [*No. Written Here; Format: Batch X; No. X*]
Date & Time Recorded: [*Answer Here*]
Audio-Length: [*Answer Here Format:* 0 (Mins): 00 (Secs)]
Main Topic/Theme: [*Answer(s) Here*]
Sub-Topic/Theme Summaries:
  - Point 1
  - Point 2
  - Point 3

For instance, below is an example of an actual summary-transcription unit:

Voice-Note No.: Batch 1; No. 3
Date & Time Recorded: Dec 11, 4:10 AM
Audio-Length: 7:19
Main Topic/Theme: Effects of Life's General Events/Features on Learning
Sub-Topic/Theme Summaries:
  - I had the Corona virus
  - I realized that life's general routines and vagaries affect our learning processes
  - Of equal impact: emotions, our other (non-learning-related) thoughts
  - Learning processes are just one part of general cognition
  - Distractions during reading
  - In addition to disciplines such as software-development and math, language-arts, etc., I decided to also analyze the fiction and nonfiction books I'm reading

I.3: List of Main Topics/Themes

From batch 1.0, I delineated a total of nine topics/themes. Below is the list of those topics/themes:

1) Paul Thagard's example of a metaphor to help solve a problem: X-rays to kill cancerous cells
2) Metaphors Vs. Analogies Vs. Similes
3) Effects of Life's General Events/Features on Learning
4) Later Automatic-Recollection of Learned Facts
5) Barack Obama's "A Promised Land" and its relation to the Thaggard text
6) NY Times Article About Game Called "Life"
7) Example of intrapersonal communication via "to-do" lists
8) Various, including: 1) Obama's A Promised Land 2) How do/should we talk to ourselves? Answer via my e-filing methods: "What did I mean by this? Why did I do this [e.g., /i.e., file something within the subfolder that I saved it in]?" 3) "Why did I come to the kitchen?"
9) "What was I doing?" AKA Distraction

And from batch 2.1, I delineated a total of 14 topics/themes. Below is the list of those topics/themes:

1) "Tip of the tongue" syndrome
2) Thinking during meditation
3) 1--Paul Thaggard's Cog-Sci text: images, connections, 2--learning-key/turning-points, and 3--role of emotions in reading and understanding fiction and non-fiction stories
4) "To-do" lists
5) General Types of Intrapersonal-Comm.
6) Other types of intrapersonal-comm.; tip of the tongue and meta-analysis; types of thoughts; distractions in your own mind during conversations or other interactions.
7) Journaling and relation between intra- and interpersonal-comm.; emotions as intrapersonal-comm. instances, including guilt.
8) Emotion of guilt continued; my friend's "ethnography of the mind."
9) Self-doubt about learning; theorization about learning ability; self-communication sans self-reference.
10) Types of self-communication, including metacognition during interpersonal dialogues, longterm-planning and procrastination, and meditation.
11) Spectrum of knowledgeability, from "know nothing" to amateur, through expert; intrapersonal-comm. via/during intense emotions; suspending and confirming understanding; fields of experience.
12) Use of visualization and concept-processing using words while reading fiction and nonfiction, academic, and other genres.
13) Intrapersonal-comm. during ethical dilemmas; hypothesis of "brute-force"-style learning.
14) "Yearning for coaching"; self-doubt in context of "imposter syndrome."

I.4: Exemplars of Potentially-Insightful Findings: Challenge and Solution

In my basic evaluation of the above-explicated summary-transcription units and their corresponding main topics/themes thus far, I was finding it a challenge to identify the summary-transcription units that are more/most important than/of the rest. In other words, out of the sample of 23 units, is it possible to identify—e.g., less than 10 units—that might eventually lead to real insights? At first, the best I could do in that regard was 15 out of 23, a percentage of roughly 65 (%).

But eventually, I was able to overcome the above challenge via two methods: 1) a basic application of Owen's (1984) thematic analysis—especially in consideration of recurrence and repetition of key topics/themes, and 2) a simple method that many of us use in prioritizing the tasks of our "to-do" lists. I asked myself, "which summary-transcription units can I eliminate, and which ones can I prioritize, and in what order? The above elimination methods resulted in the identification of two sets of (arguably) key potential insights, namely: 1) the use of "to-do" lists, "tip of the tongue syndrome," and the role of distractions in triggering intrapersonal communication, and 2) a list of 10 summary-transcription units, ordered by potential-priority.

I.5: Exemplars of Potentially-Insightful Findings, Set 1: Recurrence/Repetition

For the first set of potential insights, I realized that I repetitively discussed the use of "to-do" lists—both in (in)formal written format, and as basic ongoing lists in our minds: E.g., from Voice-Note No.: Batch 2; 2.1.7:

"To-do" lists are a form of us talking to our future selves; "Seif, do not forget to do X, Y, Z." Journaling is somewhat similar; you're also taking to yourself [, recalling what your self did in the past]. However, journaling is also a way for us to preserve our life events for posterity, i.e.,/e.g., after we die; you're both talking to yourself in the journal, and potentially talking to others in the future.

I also realized that the theme of *distractions*—discussed via various arguments, but indeed related to distractions nonetheless—showed up a number of times. E.g., from Voice-Note No.: Batch 2; 2.1.5:

Distractions: getting distracted by random thoughts during class or conversation; but I crossed out this point because I do not think it qualifies as intrapersonal-comm.

+ Experiment: next time with Ethan—my software-engineering coach, "keep a tally of unrelated distractive thoughts."

+ "There are thoughts in general [in our minds], but every time you're thinking, does not necessarily equate to you communicating with yourself!"

Finally, similar to the above themes, "tip of the tongue syndrome" also showed up repeatedly. E.g., from Voice-Note No.: Batch 2; 2.1.1:

Another instance of intrapersonal-comm.: "tip of the tongue" syndrome: "What's that word?! It's on the tip of my tongue!"

+ Editorial note, added during transcription (03/10/21): The word [in this particular case was/] is "zeitgeist!"

I.6: Exemplars of Potentially-Insightful Findings, Set 2: Prioritization

The second set of potential insights is a preliminary priority-list of 10 summary-transcription units (out of the total sample of 23). At this point of my analysis phase, my evaluation and intuition suggests that the units on this list seem promising vis-à-vis the eventual yielding of insights about intrapersonal communication. First, I will present the summarized list below (please refer to appendix A for the full summary-transcriptions of the units listed):

- Voice-Note No.: Batch 2; 2.1.11----------------------------- Potential Priority-No. 1
- Voice-Note No.: Batch 2; 2.1.5-------------------------------Potential Priority-No. 2
- Voice-Note No.: Batch 2; 2.1.7-------------------------------Potential Priority-No. 3
- Voice-Note No.: Batch 2; 2.1.2-------------------------------Potential Priority-No. 4
- Voice-Note No.: Batch 2; 2.1.1-------------------------------Potential Priority-No. 5
- Voice-Note No.: Batch 1; No. 3-------------------------------Potential Priority-No. 6
- Voice-Note No.: Batch 2; 2.1.14-----------------------------Potential Priority-No. 7
- Voice-Note No.: Batch 2; 2.1.15-----------------------------Potential Priority-No. 8
- Voice-Note No.: Batch 1; No. 6-------------------------------Potential Priority-No. 9
- Voice-Note No.: Batch 1; No. 8--------------------------- Potential Priority-No. 10

Out of all units, voice-note number Batch 2; 2.1.11's was particularly rich with what qualitative methodologists refer to as "thick descriptions (e.g., Lindlof & Taylor, 2019)," hence its designation as "Potential Priority-No. 1." Below are examples of this trait:

Voice-Note No.: Batch 2; 2.1.11--*Potential Priority-No. 1*
Date & Time Recorded: Jan 15, 8:51 PM
Audio-Length: 24:08
Main Topic(s)/Theme(s)—Note [to self]: Add after listening to entire recording (!): Spectrum of knowledgeability, from "know nothing" to amateur, through expert; intrapersonal-comm. via/during intense emotions; suspending and confirming understanding; fields of experience.

Sub-Topic/Theme Summaries:

Thick Description Example 1:
- ○ Realization I made about learning processes: knowing a subject **well/thoroughly**—e.g.,/ i.e., as an expert—is different from having a limited amount of knowledge about said subject. [Editorial Note: Depending on where I'm going with the topic, this realization is rather obvious and most likely is not useful.] Related: my upcoming reading of "30-Second Quantum Mechanics Theory." Reading and completing that book does not make me a quantum theory/ mechanics expert!
- ○ Above point continued: before starting your learning-process of subject/topic X, it might be useful to ask yourself, "where on the 'knowledgeability spectrum'—so to speak, of this topic, am I?"

Thick Description Example 2:
- ○ Realization: a lot of intrapersonal-comm. happens as we experience emotions—happiness, sadness, anger, etc. Implication for learning-process improvement: keep the above fact in mind as you read or study; you're more likely to be distracted when emotional, which interferes with your learning process.
- ○ Related qn.: "In an average day [or other time-period, e.g., a few weeks/months, etc.], what percentage of the time do we have intense emotions? Intense sadness, intense happiness,

etc." Another related question in the same context: how much do we communicate with ourselves during these intense emotions?

Thick Description Example 3:

- The importance of what I refer to as "suspending understanding": i.e.,/e.g., while reading a novel or academic book, and you do not fully understand the concepts. My proven hypothesis: keep reading, then try to make sense of the combined content later—e.g., by re-reading, or "connecting the dots," etc.
- That method is harder with STEM subjects, whose topics often build on one another, with the need to master lower-level concepts before continuing onto the advanced concepts. E.g.: need to understand numbers and counting before learning arithmetic—adding and subtracting, etc., and later, algebra, etc.
- Still, "suspension of understanding" is possible even in STEM subjects: e.g., memorizing formulae and using them to solve problems without really understanding the logic behind those formulae. E.g., memorizing the need to divide both sides of a basic algebra equation, e.g., "$2X = 10$" using a common divisor on each side, until the number of the right cannot be divided further. In the above example, $X = 5$. Later, you can learn how the formulae really work; their logic, etc.

II: Chapter 1 Pilot-Study Data Collection, And Preview of Results & Analysis II: Pilot Sample of "Cognitive-Science" Journal & Google-Scholar Qualitative Meta-Review

In compliance with Vocate's (1994) advice above, as well as the recommendations of (qualitative and other) methodologists such as Lindlof and Taylor (2019), researchers should strive to triangulate their research findings, to improve validity and reliability. And as I point out farther above—i.e., under the methodological framework introduction and research questions section, the interdisciplinary field of cognitive-science is one of my methodological-analytical pillars in the current study. Thus, I conceptualized and implemented a secondary data-gathering procedure to enable a basic evaluation of some SC-related helpful techniques that might be of utility to my learning-method-improvement quest. The secondary data in question is composed of two sub-sets of data, namely: 1) instructional articles on the study and practice of data-science, and about learning and creativity in general, and 2) research-journal articles about cognitive-science from two journals about cognitive science, i.e. "*Cognitive Science*" and "*TopiCS In Cognitive Science*."

Unlike the more detailed procedures utilized for the same data in chapters six, eight, and nine, the procedure for this chapter's pilot-study is rather simple, but not necessarily *simplistic*. Over the course of eight or nine months—from May 2020 through early January of 2021, I collected the e-issues of one of the aforementioned cognitive-science journals ("*Cognitive Science*"). That collection also included one issue of a related journal, "*TopiCS in Cognitive-Science*." Unfortunately, I could only make use of the "Open Access" articles, as I was not affiliated with an institution, and thus had no library-membership. In any case, a small number of articles were sufficient for my non-representative sample of articles. Later, I also performed a Google-Scholar search in an incognito Google-Chrome window, utilizing Thagard's (2005) six main categories of cognitive-science theorization and research—i.e., 1) logic, 2) rules, 3) concepts, 4) analogies, 5) images, and 6) connections—to search for articles related to learning-method-improvement. In fact, my search terms were: 1) *logic cognitive science learning*, 2) *rules cognitive sci-*

ence learning 3) *concepts cognitive science learning* 4) *analogies cognitive science learning* 5) *images cognitive science learning*, and 6) *mind connections cognitive science learning*.

For both the above sets of articles, I asked myself the following questions:

1. Regardless of new vocabulary, formulae and notations, methodologies, etc., do I understand the studies' main findings? (Yes/No/Maybe)
2. Can I restate the findings, or interpret their implications, in my own words? (Yes/No/Maybe)
3. What are the main themes or foci of the studies, regardless of their groupings by Thagard's (2005) six categories of computation-representation?
4. On a scale of 1 to 10, how can I rate the helpfulness, relevance, value, etc., of the studies, to me personally—i.e., for my own professional and personal or general growth or improvement?
5. Preliminary result of a tailored adaptation (for purposes of my own learning) of the CASP-checklist-review (designed by the CASP-UK program) for article batch, scale of 1 to 10.

Overall, my basic evaluation of the two samples suggests that Google-Scholar's articles are more helpful to a cognitive science novice like me, who nonetheless desires to make use of its learning-improvement techniques. Below are some details for both sets of articles.

II.1: Results of Basic Qualitative Meta-Review of Pilot Sample's "Cognitive-Science" Journal-Articles

Table 2. Cognitive-science journal-article helpfulness

Batch 1 Sample (Total of 8 Articles)	Question 4: Helpfulness, 1 to 10
Croijmans et al: Expertise Shapes…Wine	5.5
* *Divjak: Exploring and Exploiting Uncertainty*	*6.5*
Segovia-Martin et al: Network Connectivity Dynamics	5
Macdonald et al: Role of Animacy in Children's Interpretation	3
* *Hoerl: Temporal Binding, Causation & Agency*	*8*
Nikolaev: Inflected Novel With Dementia	4
* *Raviv et al: Social Network and Linguistic Structure*	*6*
* *Ruther & Liszkowski: Ontogenetic-Emergence of Cog-Ref*	*6*
Total Scores (Out of 80):	**44/80 (55%)**

* Above-average helpfulness

II.2: Results of Basic Qualitative Meta-Review of Pilot Sample's Google-Scholar Articles

Table 3. Google-Scholar-Article helpfulness by score (1-10, and %)

Article-Group Name	Question 4: Helpfulness,
By Thagard's (2005) Categories & Quantity of Articles (N)	**1 to 10**
1. Logic	
^^ *Isaac et al.: Logic & Complexity in Cognitive Science*	*9*
^^ *Ramscar et al.: Children Value Informativity Over Logic*	*8*
2. Rules	
Bunge: How we use rules to select actions	4
3. Concepts	
^^ *Weidman & Baker: The Cognitive-Science of Learning*	*10*
Halford & McCredden: Cognitive-Science Questions For Cognitive-Development	4
^^ *Booth et al: Evidence for Cognitive Science Principles…in Mathematics*	*9*
4. Analogies	
^^ *Jee et al: Analogical Thinking in Geo-Science Education*	*8*
5. Images	
^^*Dowrick: Self Model Theory: Learning from the future*	*10*
Jakel: Does Cognitive Science Need Kernels?	5.5
^^ *Holmes et al: Can Playing the Computer Game "Tetris" Reduce…Trauma?*	*9*
^^ *Cheng & Tsai: Affordances of Augmented Reality in Science Learning In Science-Learning*	*8*
6. Connections	
Riley et al: Learning From the Body About the Mind	3
Perconti & Plebe: Deep Learning & Cognitive-Science	6.5
Total Scores (Out of 130):	**94/130 (72%)**

^^ Top helpfulness scores

Book-Chapter Research Presentations

General Preview

Having introduced, defined, and briefly previewed the dynamics of self-communication in the first chapter, I attempt an earnest introduction to the autoethnographic-authorial voice that will provide the voice-note data to be analyzed in detail in later chapters. This attempt will be executed via a presentation and basic discussion of a sizable sample of journal and blog entries I have written ever since I was 16, in high school in Uganda. In addition to the above-mentioned goal of familiarizing my readers with the book's autoethnographic-authorial voice, my rationale for using these textual artefacts is the following argument: I submit that they are a quintessential mode of intrapersonal communication, and can thus help us perform a fairly intensive case-study of an arguably "average"/random individual's growth and evolution of SC over a 20-year period. These two chapters—i.e., one and two—are the foundation section of the book.

Thereafter, parts three and four of the book delve into a discussion of the general human and human-made phenomena and systems being investigated in this book, namely, (general) cognition, language, reflexivity and socialization, reading comprehension, mental health and illness, and digital-technology & AI. Interweaved throughout all these topics/chapters, I also focus on the specific goal of how to systematically improve one's own learning methods.

Presentation Format

Regardless of analysis-procedure (as described farther below), I will have to first present the data-set contents of all chapters, unlike the above succinct presentations of the data-set I and II samples for this chapter's pilot-study. This requirement poses a challenge; how should I neatly perform this task—i.e., the presentation of data-sets I & II's contents?

For data-set I (for chapters three to eight), I utilize timeline-aligned divulgence, as well as ad-hoc topical-listing. The topics will be listed under five categories, namely:

1. "Types of Self-communication (SC),"
2. "SC And Learning,"
3. "Reality And Perception-Processing I: Without Metacognition,"
4. "Reality And Perception-Processing II: With Metacognition," and
5. "Discussions of Creativity-Attempt Outtakes."

These themes arose via a basic use of grounded-theory-analysis, data-immersion, and in-vivo and axial coding during the data-transcription, organization and curation stages (the latter—i.e. curation, mostly applies to set-II data), around middle-to-late winter and early spring, 2021 (e.g. as explicated by Lindlof and Taylor, 2019).

But unlike the above format, data-set II (used in chapters six, eight, and nine) will be presented via three main parts, namely: 1) introduction and categorization, 2) significance analysis—via the questions introduced under section 1.8-II above, and 3) thematic analysis. The first two parts will be executed mostly via chapter six (learning) and to a lesser extent, eight (digital-technology and AI), and the last part will be executed via chapter nine (final insights).

A: Conclusive Interpretation of Meaning and Importance AKA, "*So What*?" of Set I Data for Chapters Three to Five and Seven (Presented in Chapter Nine)

A.1

In the book's parts II and III—i.e., in chapters three to eight, I present the results that were derived using a grounded-theoretical qualitative analysis framework. In other words, similar to the analytical process for the sample findings presented farther above, these results were synthesized on an ongoing basis, as the relevant themes and potential insights arose from the data.

In addition, I also use basic versions of Owen's (1984) thematic-analysis method, a prioritization of potentially-insightful voice-note summary-transcripts, and the highlighting of narrative-fragments filled with instances of thick-descriptions. Finally, in compliance with qualitative-analysis best-practices— e.g., as explicated by Harding and Whitehead (2013), and Peterson (2017), I use a triangulation of basic tailored versions of the following techniques: categorization, use of theoretical-conceptualization, in-ductive analysis, use of memos, basic thematic analysis (especially via in-vivo coding and axial-coding processes), and narrative analysis. Throughout the chapters, explications will be provided via excerpts taken directly from the summary-transcriptions.

It should be noted that in the process of detailing his/her findings, a researcher also has to conclusively declare *the overall arguably important meaning* of his/her findings. In other words, after reading said researcher's analyses/findings, readers might react with the question, "great, you've eloquently told us what you found. But what does it *mean*? Why is it important?"

Thus, with the working assumption that my findings have enough prima facie salience, in chapter nine, part of my interpretation of—or my answer to—the above questions is simple: I'm trying to provide good answers to the research questions I posed earlier. However, beyond this simple provision of answers to research-questions, a researcher should arguably strive to: 1) clearly highlight the aforementioned (potential) salience vis-à-vis their research questions, and 2) to adequately expound on the importance of their findings via interpretations based on a systematic use of argumentative rhetoric, logic, and/or theory.

In this vein, in chapter nine's final interpretation of the analysis findings of the upcoming chapters (especially for chapters three to five and seven, as well as chapter two, to a lesser extent), I utilize Linde (1993)'s framework, in addition to Austin-Lett and Sprague (1976), and Vocate (1994), as introduced farther above. Below is a brief explication of Linde's (1993) framework.

A.2

In *Life Stories: The Creation of Coherence*, Linde reports the findings and shares the framework of a study in which she examined the way people talk about themselves over the course of their lifetimes. She defines a "life story" as a discontinuous oral narrative in which the person primarily recounts and evaluates his/her own life, and which contains a reportable event, i.e. an event that can be judged by a listener as one that warrants learning about.

She points out that there are other mediums of re-presenting the history of oneself, including writ-ten biographies and autobiographies, private journals, and psychological profiles. In her study, Linde interviewed 13 "middle-class American speakers" (pg. 52), asking them why and how they had chosen their professions. Her main goal in analyzing this interview data was to trace the coherence systems—the "system[s] of beliefs and relations between beliefs (pg. 163)"—that her interviewees relied on as they recounted their stories.

Before moving on to discuss the main coherence systems she uncovered in her data, Linde briefly highlights some of the ways speaker establish adequate causality in their accounts, and how they repair the disruptions to continuity that might confuse their listeners. She draws on her data to provide examples of narrators that account for causality by citing character (e.g. I became an accountant because as I was always good at math) or "richness of account" (e.g. a confluence of factors that made something possible).

Events that happen without adequate causality can be explained as accidents, and discontinuity can be explained with a variety of strategies including "Irish luck," "temporary discontinuity" (*it's not really surprising I left the priesthood to become an executioner; I had always enjoyed killing squirrels as a kid*), and "self-distancing" (*I did not know any better back then...*) (pg. 151—162). In all, Linde uncovered five coherence systems in her data that are common to upper middle-class American individuals (as of 1993), namely: "versions of Freudian psychology, behaviorism and astrology, as well as some indications of feminism and Catholic confessional accounting (pg. 165)." Coherence systems are ways of explaining causes or phenomena that lie between what we call "common sense," and expert knowledge.

For instance, common sense dictates that if I was good at math as a child, it is no accident that I became an accountant. However, an interviewee who evidently had been exposed to Freudian pop-psychology cited very early toilet-training (starting at six months) by his mother as part of the reason he grew up to become an accountant (pg. 164).

B: Conclusive Interpretation of Set II Data

Unlike those of data-set I as explicated above, the final analysis procedure for data-set II is rather simple, at least "on paper." Specifically, it will be executed via a close examination of the topics and themes that recur in the data-set, and their relation to my own learning processes and/or best-practices.

CHAPTER ONE CONCLUSION

In this chapter, I have introduced the concept of int*ra*personal communication—to which I also refer throughout the book as self-communication or "SC" in short. I have provided a primary and secondary set of definitions of the concept in relation to the concept of int*er*personal communication, and I have explicated the social-scientific academic background of the concept. Finally, I have introduced the methodological framework I will use to present the research in the forthcoming chapters, including a sample of findings uncovered in a pilot-study.

I hope readers' understanding of intrapersonal-/self-communication has been elucidated by this chapter's discussion. Overall, my most forceful argument for the current research is that social-scientific and communication studies are incomplete without rigorous studies of SC. One cannot expect to properly understand the human condition, without systematically studying the dynamics of our internal or self-directed communications. In that vein, I hope this book will spark new efforts in that academic undertaking, the attempt to understand: do individuals communicate with themselves? If "yes," how and/or why do they do so, and what can the dynamics of such a mode of communication help us understand about our minds and societies?

Creativity-Attempt Outtake 1

Post 1; 01/10/21:
Over the years, I've journaled erratically, privately. And from time-to-time, I have written blog posts here-and-there. To me, that fact is regrettable, as **consistent** journaling is a proven way of helping us with many personal intellectual, emotional, and other challenges, not to mention its value in improving our writing skills.

Hence this 2021 challenge. Serendipitously, I am also working on a partly* auto-ethnographic intrapersonal communication and cognitive-science book (*I will also glean results from a literature [qualitative] meta-analysis), scheduled for launch around late 2021 or early 2022. Thus, one of my goals is to synergize these two endeavors—public-journaling, and my auto-ethnographic analysis for that book.

But even beyond June (2021)—the deadline for submitting the book manuscript, I intend to keep up this habit of journaling publicly. Whereas my private journal entries are uninhibited raw reflections on my pains and joys, I believe/hope this public journal will help me hone a literary voice that balances authenticity and professionalism. I also hope to somehow help, inspire, educate, or entertain anyone on the web that will somehow come across this page. Here goes nothing!

Post 2; 01/16/21:
The Challenge Lives; Long Live The Challenge!

Yes, that title is dramatic, and that's part of the point, although I'm not sure about the use/role of that drama for this post. Then again, maybe I do (?); the drama is a good demonstration of the magic–so to speak, the metaphysical and yes, **dramatic** nature of challenges. In this context, a challenge is a self-dare, a goal one sets for themselves, meant to be accomplished within a certain set period of time.

Case in point? The journaling challenge of which this post is a product! As of this moment, I have enrolled myself in let's see…1, 2, 3 three challenges, grand-total, namely, 1) "the 90-day (software-development/) coding challenge," 2) "the 2021 week-end/beginning journaling challenge," and 3) "the reading challenge of the week of 01-16, to 01-23."

Sometime last year, in an essay for a fellowship competition, I discussed the power/utility of figuring out our psychological profiles, as well as the psychological tricks we can use to [jump-]start our brains into accomplishing tasks. For instance, if you love binge-watching shows on Netflix, then you can set yourself a certain number of goals, and make a promise to yourself to binge-watch your favorite Netflix show, only **after** you finish accomplishing those goals!

But one of the interesting features of self-set-challenges is that you do not even have to have a "carrot" to reward yourself with. I guess the "magic" of the self-set-challenge–at least for some of us, comes from the disappointment you'll feel about yourself, if you let yourself down! Well, regardless, here's a verbal/journal-toast to the magical power of challenges, which when combined with that of "to-do" lists, can create productivity super-heroes out of the most ordinary of schmucks, such as Yours Truly!

Post 4; 01/31/21:
It's Always Worth A Try

A couple of weeks ago, I wrote an entry in this public journal about a productivity hack I've discovered in the last few months—i.e., the challenge (to oneself), which I have used for a few months to date, with fairly good results so far. Another productivity trick I can personally vouch for is the famous "to-do" list. A public promise—or a promise to (a) specific someone(s) to whom you owe a deliverable—is yet another effective trick in my experience.

Great, thank goodness for such hacks/techniques. But what does one do, when sometimes (and yes, those times inevitably come up throughout life, here and there), many/most of those tricks are ineffective because of low motivation, which is in turn caused by numerous factors? For let's face it, in such circumstances, to-do lists and self-challenges are rendered weak.

I guess one of the quick solutions for me, for those low-motivation situations, is the use of the other techniques mentioned above, which involve staking your reputation through public promises. For me at least, that is a good motivator; how do I let myself lose face, when all I have to do is simply do X (/"fill in the blank") for my students, or a friend, colleagues, family-member(s), etc.?

But even then, I suppose one can fail in their obligations because of other good reasons outside of our control, as opposed to laziness or lack of motivation. And sometimes, chaotic situations cause us to abandon our fancy plans; in this vein, I really love that Mike Tyson quote, "everybody has a plan, until they get punched in the mouth." Yikes.

One of the answers that come to mind in response to the above dilemmas, is this: it's okay to fail. But there's also a caveat to that maxim: it's okay to fail, **but only if you at least try, or make a good faith effort!**

I believe most folks reading this journal entry can recall those numerous times in which they just did not want to get out of bed, did not feel like writing that paper, etc. At the end of the day, one can only entertain those failures so much, before eventually paying for that mistake—i.e., of easily entertaining failure.

But again, the good news is that you do not have to get up at the crack of dawn every day, or write a perfect paper every time. All you have to do is **try**.

So, here's a promise to myself for this year: I promise to at least try, surely. And try again if I fail or do not perform as well as I wanted to the first time; and try again after that, if I have to, until I achieve whatever it is I am trying to achieve.

Self-Questioning for Creativity-Attempt Outtake 1

State Any Necessary Introductory Remarks, and/or Answer this Question: Does It Make Sense to Me Now—i.e., In The Present—versus Then—i.e., When I Wrote It?

Yes, the posts make sense to me now. The topics and themes discussed/explored are still relevant, after all; please refer to other supporting details under questions 2 to 4 below.

Does it Resonate With Me Now—i.e., In The Present—versus Then—i.e., When I Wrote It?

Yes, the posts still resonate with me. The general common theme among all posts can be described as "resoluteness," or "the resolute intent to try to fulfill personal goals." And indeed, I still agree with all the sentiments expressed!

What Mood and/or State of Mind does the Tone Suggest?

The posts are largely reflective of a calm and determined mind, making an effort in pursuit of self-accountability.

What is/are the Direct and Indirect Relation(/s) to the Current Chapter via Topic(s) and Themes?

The posts in this creativity-attempt outtake (C.A.O) are the first earnest deep dive into my attempt in the book to harness my personal/autoethnographical-analytical voice vis-à-vis a fairly rigorous self-audit via SC.

They are a crash-course-style demonstration of sorts vis-à-vis what written SC looks and "feels" like, so to speak. I also believe their content echoes various aspects of the theories/postulations introduced by the various authors in the chapter, such as Archer's social-reflexivity, as well as pragmatism and symbolic-interactionism.

Any other Evaluations or Comments?

N/A

Activities and Discussion Questions

1. Do you communicate with yourself? If "yes," how? If "no," why not?
2. Do you think talking (versus "communicating" in general) to yourself is weird? Regardless of answer—i.e., "yes"/"no"/"other," why/how not?
3. Draw a line to divide a page into two horizontal parts: then, list the traits that might indicate an unhealthy way of talking to yourself—i.e., on one side of the page, vs. the traits that might indicate a healthy way of talking to yourself—i.e., on the other side of the page.
4. Create a three-column division on a page: this time, on the left side, list the things we can only achieve via interpersonal communication—communication between at least two people; list the things we can only achieve via intrapersonal communication—communication within one person's mind—i.e., quietly thought, or written, or spoken; in the middle column, write the things that both types of communication can achieve.
5. Introduction to journaling activities: Google the search term "journaling activities." Answer the following questions, or do the following sub-activities:
    ◦ Make a list with which to answer this question: which activities are your favorites, and why?
    ◦ Journaling/creative activity 1:
        ▪ Note to class/group instructor/facilitator (*optional*, but highly recommended): Please make arrangements for the group/class to do an activity related to their favorite journaling activity at least once a week. And in place of or in addition to that activity, you can also facilitate creative activities with the class/group that can help individuals study their own SC-processes.
6. Comparison with other types of communication (to also be done for all other upcoming chapters): Look online, or use one of your other communication texts, or consult with your instructor for this activity: look at the content of the chapter, and brainstorm at least two or three concepts from the subdiscipline of interpersonal communication (or other subdisciplines such as small-group-/family-/political-/health communication, etc., or mass-media/persuasion/conflict-resolution, etc.) that you think are related to this chapter. Why do think they are related, and/or what is the relationship? (Discuss as a group/class.)

REFERENCES

Alton, L. (2016, December 22). Millennials are struggling to get jobs - here's why, and what to do about it. *Forbes*. Retrieved December 25, 2021, from https://www.forbes.com/sites/larryalton/2016/12/22/millennials-are-struggling-to-get-jobs-heres-why-and-what-to-do-about-it/?sh=2b2ee6994bb0

Archer, M. S. (2007). *Making our Way through the World: Human Reflexivity and Social Mobility*. Cambridge University Press. doi:10.1017/CBO9780511618932

Austin-Lett, G., & Sprague, J. (1976). *Talk to yourself: Experiencing intrapersonal communication*. Houghton, Mifflin.

Bochner, A. P., & Ellis, C. (2016). *Evocative autoethnography: Writing lives and telling stories*. Left Coast Press. doi:10.4324/9781315545417

Brice, R. (2021, November 24). *CASP Checklists*. CASP - Critical Appraisal Skills Program. https://casp-uk.net/casp-tools-checklists/

Charon, J. M. (2007). *Symbolic interactionism: An introduction, an interpretation, an integration*. Pearson Education.

Deacon, T. W., & International Society for Science and Religion. (2007). *The symbolic species: The co-evolution of language and the brain*. Cambridge: International Society for Science and Religion.

Friedenberg, J., & Silverman, G. (2011). *Cognitive science: An introduction to the study of mind*. SAGE.

Harding, T., & Whitehead, D. (2013) Analyzing data in qualitative research. In Nursing & Midwifery Research: Methods and Appraisal for Evidence-Based Practice (4th ed.). Elsevier.

Hermans, H. J. M., & Hermans-Konopka, A. (2012). *Dialogical self theory: Positioning and counter-positioning in a globalizing society*. Cambridge Univ. Press.

Hofstadter, D. R. (2006). *Gödel, Escher, Bach: An eternal golden braid*. Basic Books.

Holman Jones, S, Ellis, C, Adams, T E, & Allen-Collinson, J. (n.d.). *Autoethnography as the engagement of self/other, self/culture, self/politics, selves/futures*. Left Coast Press.

Kelly, E. F. (2007). Irreducible mind: Toward a psychology for the 21st century. Rowman & Littlefield.

Knapp, M. L., Vangelisti, A. L., & Caughlin, J. (2013). *Interpersonal communication & human relationships*. Academic Press.

Linde, C. (1993). *Life stories: The creation of coherence*. Oxford University Press.

Lindlof, T. R., & Taylor, B. C. (2019). *Qualitative communication research methods*. Academic Press.

McKenna, L. (2016, April 25). Jobs Are Scarce for Ph.D.s. *The Atlantic*. https://www.theatlantic.com/education/archive/2016/04/bad-job-market-phds/479205/

Peterson, J. S. (2019). Presenting a Qualitative Study: A Reviewer's Perspective. *Gifted Child Quarterly*, *63*(3), 147–158. doi:10.1177/0016986219844789

Putnam, H. (1998). *Representation and reality*. MIT.

Schwartz, T., & Pines, E. (2019, April 17). *Harvard Business Review*. Retrieved from https://hbr.org/2019/04/great-leaders-are-thoughtful-and-deliberate-not-impulsive-and-reactive

Thagard, P. (2005). *Mind: Introduction to cognitive science*. MIT Press.

Vocate, D. R. (1994). Intrapersonal communication: Different voices, different minds. Erlbaum.

Wiley, N. (1994). *The semiotic self*. Polity.

Zoppi, L. (n.d.). Is talking to yourself normal? what it means for mental health. *Medical News Today*. Retrieved December 25, 2021, from https://www.medicalnewstoday.com/articles/talking-to-yourself

ADDITIONAL READING

Goffman, E. (1961). *Asylums: Essays on the social situation of mental patients and other inmates*. Anchor Books.

McCool, M. A., Sheffield, T., Garst, K., Ekholm, S. M., Opelt, T., Pannell, K. E., McCool, A. X., Dickey, N., Brown, T., & Hansmann, A. L. (2017). *Passing Cars: The Internal Monologue of a Neurodivergent Trans Girl*. CreateSpace Independent Publishing Platform.

Woodson, C. G. (1998). *The mis-education of the Negro*. Africa World Press.

KEY TERMS AND DEFINITIONS

Autoethnography: The application of the research methodology of ethnography by a researcher, to his/her own life.

Cognition: The mental processing of an individual's experiences, as well as his/her process of learning and sensing.

Cognitive Science: The scientific study of cognition.

Human Communication: The study of human's internal or shared meaning-making processes via sensing, and via the use of symbol-systems and interaction.

Interpersonal Communication: Communication between at least two individuals.

Intrapersonal-(/Self-) Communication: Communication inside an individual's mind, or outside—e.g., spoken or written—but for that same individual's consumption and/or use.

Secondary Intra-Interpersonal Cognition and Communication Spectrum: A tool that can aid our classification of mental and/or communicative states vis-à-vis the presence or absence of cognition, as well as whether or not those states can be defined as either intra-, or interpersonal communication.

Self: An individual's essence of existence in relation to others, as conceived by the mind of this particular individual, and/or as defined by others vis-à-vis that individual's unique characteristics.

Social Reflexivity: A theory synthesized by sociologist Margaret Archer, which suggests four general predispositions by individuals in their reflections vis-à-vis self-improvement plans communicative reflexivity (wherein we seek other individuals' opinions in reaction to our plans), autonomous reflexivity (wherein we cultivate and implement our own plans, regardless of other individuals' opinions), meta-reflexivity (wherein we critique our own plans, which keep us from improving our lives over time), and fractured reflexivity (wherein our internal reflections result in indecision and mental turmoil).

Social Science: The scientific study of human society.

Chapter 2
"Dear Journal…":
An Autoethnographic Case Study of Intrapersonal/Self–Communication Evolution

ABSTRACT

This chapter demonstrates how individuals can execute studies of their own intrapersonal-/self-communication (SC) via reviews of journal and blog entries. The author shares serializations of his journal and blog entries from a time-period of over 20 years, since his junior year of high school—i.e., at the age of 16, through today, at the age of 36 (as of writing this chapter/book). Overall, this chapter demonstrates an example of a process via which an individual's SC voice evolves. In addition to such an evolution, an individual can also cultivate a habit of metacognition—closely examining their own thinking and self-communication processes. As other upcoming chapters will demonstrate, such metacognition and SC-audit processes enable personal and professional growth.

INTRODUCTION AND BACKGROUND

Overview

As I mention towards the end of the previous chapter (one), the goal of the current chapter is to earnestly introduce readers of this book to the autoethnographic-authorial voice that will provide the voice-note data—i.e., the bigger source of the research-findings (versus the cognitive-science articles and emails to self)—to be analyzed in detail in later chapters. To accomplish the above task (i.e., introduction of my autoethnographic-authorial voice), I utilize a general analytical examination of a sample of private journal entries and online blog-posts, which I started writing at the age of 16 in high school in Kampala (the capital), Uganda.

DOI: 10.4018/978-1-7998-7507-9.ch002

Table 1. Author's journal- and blog-entry totals between 2002 to 2021

YR	JAN	FEB	MAR	APR	MAY	JUN	JUL	AUG	SEP	OCT	NOV	DEC	Yr. Totals
2002	18	10	2	8	9	10	15	10	6	6	5	1	99
2003	4	6	7	3	0	0	0	0	0	0	0	0	20
2004	0	2	2	0	0	0	20	2	3	2	5	2	38
2005	2	3	1	0	0	2	3	0	0	0	0	3	14
2006	2	3	1	0	0	0	0	0	0	0	0	0	6
2007	0	0	0	0	0	0	0	0	0	0	0	0	0
2008	0	0	0	0	0	0	0	0	0	0	0	0	0
2009	0		0	0	0	0	0	0		0	0	0	5
2010	0	0	0	0									19
2011									0	0	0	0	13
2012			0		0				0	0	0		10
2013			0		0				0	0	0	0	7
2014	0	0	0		0	0	0	0		0	0	0	4
2015	0	0	0	0	0	0	0	0	0	0	0	0	0
2016	0	0	0	0	0	0	0	0	0	0	0	0	0
2017	0	0	0	0		0	0	0	0	0	0	0	1
2018	0		0		0				0	0	0	0	7
2019													7
2020													14
2021													
Mo.-Total	36	34	20	22	23	23	50	18	20	12	13	14	

Research Question(s):
1) How is my mind processing the experiences my growing body and mind are going through?
2) How is my authorial voice evolving and/or oscillating between private un-self-censored, and blog-style/public self-censored?

The journal-/blog-entries are organized via three main sections, namely A, B, and C. First, section A introduces us to my 16-year-old *self* in earnest; a *self* whose then-journal-entries this current adult (myself) often finds cringe-worthy: at times precocious/wise, at other times a typical raging-hormone teenager, and various other states of being in between those two extremes.

Next, section B briefly provides a summary of a key transition period, i.e. from the life of Seif the teenager in Uganda, to that of Seif, the young-adult in America. This includes a gap year after high school, part of which was spent in England visiting family while frantically planning my trip to the USA, and ending with my arrival in the USA.

Thereafter, section C presents the blog-post entries that I have written haphazardly over the past 11 years since the age of 25 (i.e., in 2010), through the current year, 2021. (It should also be noted that some of the entries of 2021 might be cross-listed under some of the chapter-end "Creativity-Attempt Outtakes.") Alongside my presentation of all these journal and blog excerpts, I will attempt to provide some helpful contextualization and basic analysis, to aid my readers in their conceptualization of the inner workings of my growing/evolving mind.

Besides those three major sections, I also present an excerpt of my undergraduate thesis—a short novel, based on my life—in the "Creativity-Attempt Outtake" section at the end of the chapter, with the full story serialized in appendix "E." Reason: so as to make up for the gap years of 2007 and 2008 vis-à-vis the absence of journaling/blogging. In other words, even though I cannot locate those years' journal/blog entries, that literary text partially demonstrates some themes of the autobiography-related thoughts and self-communications that my mind was processing.

And while 2015 and 2016 are also gap years vis-à-vis journaling/blogging, I unfortunately cannot find—and thus cannot present—any appropriate texts for that time-period in this book. However, readers can look up my Ph.D. dissertation online, which is arguably the most significant text I completed in the time-period. Moreover, I can point out that those two yeas—despite the lack of literary-memorialization—were quite eventful.

Among other events, the major highlights (of 2015 and 2016) were: 1) A visit in the summer of 2015 to my maternal-ancestral native country Rwanda, for a conference and a brief informal capstone research-tour, as I was concluding my PhD dissertation, which is about Rwandan genocide survivors and former refugees; 2) graduation from my Ph.D. program at Drexel university (in Philadelphia, PA) in September 2015; 3) A visit to Uganda in the summer of 2016 after a 12-year absence—i.e., since my 2004 arrival in the USA; and 4) Another trip to Uganda in the fall of 2016 to bring back my mother to the USA, who was afflicted by a then-unknown illness, which was later diagnosed as a gastrointestinal stromal tumor (GIST).

Research Questions and Rationale (i.e., Slightly Revised from the Versions Appended under Table 1 above)

The two research questions I am trying to answer via my analyses of my private journal- and blog-entries/posts are:

1. For my private journal-entries: How is my mind processing the experiences I am going through (from my teenage years through adulthood)?
2. For my public blog posts: How is my SC-authorial voice evolving and oscillating between private un-self-censored, and blog-style/public self-censored?

As the research analysis/discussion explicates in the upcoming chapters, journal entries and to-do lists are indeed important tools for intrapersonal-/self-communication (SC). Moreover, the findings in the coming chapters are presented via a journal-entry style, i.e., by the date and time recorded over the course of five months, from November of 2020, through late March of 2021.

Perhaps most importantly, this chapter—via the journal-/blog-entries examination, can help us paint an idiographic-intensive picture (and hopefully, a vivid one at that) of the growth/evolution of a *self* (my own), from the age of 16 through adulthood (i.e., my current age of 36), the same *self* that provides the postulations of the importance of SC throughout this book. Moreover, as I allude in my introduction in the previous chapter (section 1.2.1), the concept and practice of metacognition is by default an important part of our SC-examinations. And in that vein, research suggests that reflective journaling can help us in that regard (e.g., Ramadhanti et.-al 2020; Alt & Raichel, 2020; and Henter & Indreica, 2014).

Early Insight; Journaling as a Tool for Gauging Mental Health and Illness over Time

When I first came up with the idea of this book project—and when I started its research at the end of 2020, my intention was to present an apt but conservative autoethnography-based study of SC, self-concept, and cognition. However, over the course of my close examination of the journal entries I present in this chapter, I have realized that they offer a rather significant revelation or insight into my character/mind, which might be useful to my readers and other members of the general public.

Apparently, from an early age—i.e., 16, or even my earlier childhood to early youth years (per my recollection), I was already demonstrating the characteristics of clinical anxiety, as well as type-II bipolar-depression or cyclothymia (my mental-health providers seem to be unsure of the precise diagnosis). By turns via my journal-/blog-entries, I perceive the high and low tides of my moods; exuberant and creative one day, and depressive on another day. Or, more commonly, a combination of both moods within one day.

In my discussions of mental health in the current chapter (as well as chapter six, i.e. "Mental Health and Illness"), I will at times use an oblique description-style to describe the severity of my mental-illness over the years. Specifically, I will utilize what I refer to as a "mental-distress scale" of 1 to 10, with 1 being indicative of a calm mind, and 10 the indicator of a severely turbulent mind, which might even include thoughts of self-harm, or a state of borderline panic-attack mode or psychosis. And having briefly introduced the mental illness theme in the current section/chapter, I will substantively grapple with it further in chapter six, which is indeed dedicated to the discussion of mental health and illness in relation to SC.

MAIN FOCUS OF THE CHAPTER

A: Journal-Entries from High School

A.1: Background/Orientation

As I mention previously, this first set of journal-entry excerpts is from early 2002—with the first entry dated January 1 2002, to be precise. So as to aid my readers' understandings or contextualizations of all the upcoming entry-sets (A, B, and C), as well as the upcoming voice-note transcript excerpts, I believe it might be useful for me to divulge a couple of key background-biographical details about my family/upbringing at this juncture.

By far, I believe the biggest influence—for better or worse—on my life was the fact that I was raised by a single mother, despite my father's generous financial- and other resource-support, as well as his moral-support from afar. Another biographical fact that might aid one's understanding of my adolescence and adulthood psyche is my family's socio-economic status, which can luckily be described as "upper-middle class" by any international standards.

My mum had ensured that I got a good primary school educational foundation via a Catholic boarding school, where I discovered early on my exceptional love of the language arts, and was particularly fond of reading novels and writing (English-language) compositions. That love of reading and writing was a great escape from the incessant bullying I was subjected to for being a weakling, and/or "effeminate."

And even though that bullying all but ceased in high school, by my teenage years, the childhood scars were indelible, and my character traits—especially related to my deep thinking and love of creativity in general, and the literary-arts in particular—had ossified. Thus, by my senior year of high school in 2002 (at a semi-international co-educational institution)—surrounded by the typical teenage drama and in the throes of the typical teenage hormone-induced mental turmoil, I was more-than-ready to blow off a considerable amount of creative steam, so to speak.

Two other points are worthy of mention in this context. Like other middle-class kids in Uganda then and now (and I presume in many other countries worldwide), I grew up exposed to a heavy dose of American cultural influences, especially vis-à-vis fashion and entertainment (and to a lesser extent, cuisine as well). This included TV programming (e.g. CNN, super-hero cartoons, and comedies such as *Full House*, *The Fresh Prince of Bel-Air*, etc.); music—especially pop (with Michael Jackson as my favorite childhood artist), R'n'B, and hip-hop. In fact, I am embarrassed to reveal that for a while in high school, I dreamt of being a part-time rapper in the future!

Moreover, very early in my childhood, three of my older siblings had settled in the UK and Germany; and as I started high school in 1999, my immediate elder sister (with whom I had built the closest bond) came to the US for college. Consequently, it was already a foregone conclusion in the back of my mind that I too was destined to live abroad—with the US being the preferred location, thanks to the afore-mentioned cultural influence.

I believe it is important to re-emphasize that the main reason I am explicitly divulging all the above background facts, is that they might aid readers in their understandings of my written/memorialized life-story fragments farther below in this chapter. Moreover, I also believe that readers might attain a better understanding of my general day-to-day thoughts and life-story analyses, presented via the voice-note/transcript data in the upcoming chapters. In other words, similar to any other individual, my psyche and personality have not developed in a vacuum over time. Rather, they have been built via the experiences I have gone through year after year, since my childhood through the present time.

A.2: Entry Subset I: January 2002 (An Introduction to the Mind and SC of Seif, the Teenager at Home in Uganda)

Overall, this first subset of entries is marked by various noteworthy features. First among those features, I am struck by the early presentation of my struggle—which remains to this day—of journaling consistently (i.e., either every day, or otherwise regularly, via deliberately observed intervals). Secondly, I am gratefully amazed by my early propensity for deep thinking. Among others, the entries of January 3 and 26 (2002) below demonstrate this trait.

Another feature that stands out to me is my fervent early/innocent youthful ambitions, as well as my uncertainty—even fear, or more aptly, anxiety—which I was willing to admit to myself and my mother, about my future (preferably, in the USA). Relatedly, my affection for, and yearning for validation from my mother, is rather obvious. And I can unabashedly report that those two character-traits have persisted to this day, despite the proverbial growing pains of the development of various fierce differences of opinions and styles as we both grow older, among other relationship challenges.

January 1, 2002

It's ten minutes past two in the morning. It's the new year. I'm settled in "my bedroom" [i.e., the family recreational room in which we watched TV, which also doubled as my bedroom at night](^)[1] *with my nephew Derek*[2] [who was visiting from the UK, where he lived with his mother, my elder sister](^) *watching TV; a movie, Mrs. Doubtfire.*

In the past years, I never journaled. I just let each and every day go by without writing something about it. I love writing. I would like to be a writer when I grow up. I'd like to be a writer. The thing is:

somehow, I fail to sit down every day to jot down my feelings. You know, my grievances, my happiness, my sadness. I mean, Ken Follet said you do not require so much to sit down and write. But...I just fail.

Oh well, this year I want things to be different. Anyways, at least one thing I'm sure of, is I really want to jot down a few lines about my life daily. I'm an ordinary person. My feelings or actions or aspects about my life are not all that different from anyone else's. Something I know too, once I get down to a fullscap[3] to put something on it with my ball point [pen](^), *I just do not stop. I go on and on and* [on] (^). *I sometimes tend to be a very talkative person. And people tell me this.*

But, who's there to shut you up when you're writing? The paper? The pen? In this way, I can get to pour out the whole ocean of feelings I have bottled up. The more I get to do this, the calmer I'll try to be on the inside. Unfortunately, here I am, always kind of lazy to do it.

Even though Follet said one does not need that much discipline to write, one thing I'm sure of, I lack enough discipline. The little one needs to do something this constructive.

For the past two days, me and my mum have not been fine with each other at all. I spilled tea on her quilt cover. She flipped. Really badly.

I really do not like the way she [flips](^) *out whenever I do something as little.*

She expects me to be perfect. Yet I cannot. <u>I'm only human</u>4.

Oh well, right now, Derek is eating some rice and beans. Cold though. And [drinking](^) *a coke.*

We finished Mrs. Doubtfire and are now watching "Sister-Sister" video-taped episodes. Oh well, it's clocking 3 o'clock now. Very soon, we're going to take to the bedroom and talk till dawn, [and](^) *sleep right around 5 or 6. 6 to be precise. Wake up at 2. A new day begins for me then. Yeah, that's my day basically. Miserably boring. Relaxed though. I'm out for now.*

January 2, 2002

End of my 1st of January. My first day of '02. I'm in the room that's really supposed to be a recreation/TV room[,](^) *next to the dining room. But I have always used it as my bedroom ever since I was a kid. I liked it at first*[,](^) *because the television's found in there* [but later, like any teenager, I craved some privacy, which I couldn't get until everyone went to bed after watching TV](^). *Derek's watching a music video.*

Today has been a good day basically. We woke up at around 3 PM[,](^) *after which mum served us with the oven-baked Irish potato and chicken. We've been on good terms. Nice. What does this year have in store for me?? I wonder. I mean, every* [year/day] (^) *passes by just like this. Really, there's nothing that much I can call a turning point* [that] (^) *I can recall*[,](^) [which](^) *has happened in the past years. Unless of course the time I first flew out to the U.K. in '93. There's quite a lot I'm looking forward to this year of course. O'levels* [i.e., the exams taken after the fourth year of high school, after which students return to complete two years of a more rigorous pre-college curriculum, the A-levels][5]. *Vacation. In two words. In* [And](^) *more: quite a lot.*

Many plans. I just watched a movie on LTV [a Ugandan free-to-air Pentecostal-Christian television channel](^) *though,* [and](^) *I got something they said. "If you want to make God laugh, tell him your plans. I'm sometimes caused to think critically about my life.* [Unclear Word](^)...*both sides. The good one and the bad one. It's not all roses of course. But...I'm okay with it. There's quite a lot I'd like to have. I can do without it though.*

January 3, 2002[6]

It's the end of my day. For the past two days, I've been writing at the beginning of them. Today however, I'm writing at the end of it. I forgot to write yesterday at the beginning of the 3rd. I'm [Unclear Word/Grammar](^)...writing now, at the end of it. Enough of the "guilty explanation" however. This being something I'm just trying to get myself used to, I'm not quite used to it. I forgot to put the small jerry can of water besides mum's bed yesterday as well. Forgetfulness! This disease. It is the one after all that gets [me] (^) to be mediocre in many aspects of my life. I really hope I fulfill these two pledges I made this year. [Editorial note, added 2021: I believe I meant the journaling activity, and putting the small jerrycan of water besides my mum's bed, as she often choked on her saliva in her sleep.](^)

Life...as I said before[,](^) is not all roses. But we have to push ourselves through it. In the good and bad times. It's all part of it, come to think of it. Nowadays, I'm trying to get myself to speak out, in analytical form, the things I've always felt deep within me. As I've always tried to do. I really wish I could get to put out that easily those things within me. Within the deepest parts of me. Those things which are very hard [to](^) analyze and describe. Tell them to whom? My mum. The best person I speak to in this world.

As part of that bid to try and tell "such things" to her, I told her for starters something about the emotions I feel often within me. The different types of emotions. Happiness. Contentedness should I say with my life. For all I've had in my life so far. Anxiety. For what lies ahead of me. If I'll succeed. If I'll make it. If I'll do what's expected of me. What everyone expects of me. But most of all I presume, the <u>challenges</u> *[and](^) responsibility. Having to work to feed myself, to pay my fees and bills* [i.e., college-tuition and living expenses in the USA, as we were already planning ahead for that endeavor—i.e., college in the USA, and we were very skeptical that my father would support it](^).

Guilt. For expectations I haven't lived up to so far. For the luck I have while others toil to survive. This was the first thing I [laid] (^) out for her. [And](^) [s]he (^) relayed back. She similarly has those feelings, and so does everyone, she replied.

At least I got to put that out. I recall sometime I failed to ask what language she used while dating daddy. How weird and impossible I find it that people use Luganda to date!! In simple words, just as I've put it!!! I made it complicated, I failed to ask it. Stopped at the beginning. Perhaps I do complicate a few other feelings I'd like to [relay] (^) and play out...

January 4, 2002; 1:45 AM

I feel veeeery tired. But today hasn't been a bad day on the whole. Went to town with mum. It was boring. Returned home. Sat down to watch TV. By the way, I did remember to take mum's water. I'll go to bed now! Chow! [sic](^)

January 5, 2002; 1:45 AM

Booooring. Oh! Sophie and Sarah on a three way as usual. Apparently had bad news. XX checked on one of her papers of the 1st semester. Actually sent someone to check for her. It said it is going to be redone.

Wherever they checked. Mum wasn't so clear while telling me. Anyways, all this means she won't graduate as [has](^) *been expected. Sophie told mum to give me a lecture from her. I'm fed up* [of the lectures](^)*!!!*

January 6, 2002; 1:45 AM

Derek went back to London. I hate going to the airport. Makes me feel bad. Watching those all those people leave. While I'm going nowhere. Makes me feel jealous. Aunty Grace came too. Lectured me as usual. [I'm](^) *sick* [of all the lectures (?)](^) *Cannot wait to...I do not know. Live alone.*

January 7, 2002

Day wasn't that bad. Went to town. Saw aunty Grace [and](^) *returned her coin purse. Anyways, wasn't too bad after all. Oh, went to the gallery too. I love looking at art works in the gallery.*

January 8, 2002

Called Virginia. Darling. She had noticed my number the day before when I called and hung up. Didn't sound too happy when she [saw/heard](^) *it was me again. Anyways, talked. Damn. I really like that girl.*

January 9, 2002

It's coming to the end of the first week of '02. Boy, the way days go sooo fast. Anyways, nothing really eventful. Same boring day. Routine. Wake up late, do nothing. Anyways, spoke to Resty. Asked for Nike trainers. [Illegible.](^)*...as well. Looking forward to meeting them...out...*

January 10, 2002

It's the 10th of January. Hmm. That magical number 10. I do not know. For some reason, I think it's magical. Today has been a nice day. Has ended nicely so to speak. Reason: I've washed all my dirty clothes. It's not every day that this happens. So...really have reason to be happy. Feel so good. Done some exercise. Feel tired. Steven returned with his daughter. Anyways, I'm out.

January 11, 2002

Nothing really happened. All I found myself doing was not sleeping the whole night. Hoping to write after I felt was just about to sleep (X).

January 12, 2002

Today was a special day. It's not every day that...[incomplete](^) *...*

January 24, 2002

Journaling was more or less the pillar [/flagship initiative](^) *of the really few vague resolutions for this year. I guess I'm not going to be all that successful in life after all if I do not follow the steps I should. Set a clear target. Work to achieve it. Achieve it. Move on.*

I didn't do that. I do not do that. [As in,](^) *I didn't do that at the beginning of this year.*

Oh well, 13 days have passed now without writing in the journal. Going [Illegible] further, to fail about the target thing. Hmmm.(X) *Yesterday, 23rd Jan, XX (XX[7]'s father) died. Committed suicide. Hanged himself to be exact. He had lost it shortly before.* [Besides the above unfortunate event...](^) *Today has been a good day. Very good. I've been on good terms with my mum. Good.*

January 25/26, Saturday, 2002

It's 3 AM in the night. 3 o'clock, and I am still awake. I half love this lifestyle, and half hate it... Stay up till morning... Sleep at daybreak... Woke up at 4 PM. Considering the facts: it symbolizes freedom. I can do whatever I feel like, and no one will say nothing. But...the fact that all I do is sleep, eat, and watch TV... Basically, that's not really normal in my opinion. It's not what people do. Anyways, it's what I live like at my home.

There's some major repairs/construction going on in the house. It was first done outside. [And](^) *the yard is looking great. Now inside. It's kind of chaotic. Shifting stuff from one place to another. But it's really transforming the place. Floor tiles, bathrooms.*

There's nothing I love more than sitting down at the end of my day and writing about it. Oh well, today, as many other days of this holiday, since about the time Stephen came from Germany and began doing repairs, I've felt kind of guilty. Stephen and the other men have been working all day, as I slept throughout, and promptly ate my food when it was time to. Hmmm.

I talked to mum as I always do. And there she was, suddenly, telling me that all I did was to sit in front of her just like a picture. That's all I do. What will I ever do? What can I do? That's what she was telling me. In reply, I told her to mark her words. That someday, I would remind her of saying those words.

My friends, my lovers think I'll fail. It do not please them. My enemies think I'll fail. It does please them. My enemies <u>hope</u> I'll fail. My lovers, friends, <u>fear</u> I'll fail.[8] Enemies; many. Cannot think of all of them though.

Time to retire to bed now. To try and retire. Sleep do not come easy though. For some reason. I'm out.

January 26, 2002[9]

[Stand-Alone Quote At the Top of the Page:](^) *I do not know where this is heading, but I'd sure love to know. There's only one way I'll know. Go on.*

Today in Uganda's history, Y. K. Museveni took over government along with the NRM/A, ousting the Tito Lutwa Okello government. Ending years of dictatorship and wars in Uganda. Historical. I love Museveni, love his ideas. Really love him. Fit to be president.

Today has been a good day, I guess. Woke up late as usual. Spoke to the guy who has been helping XX with the construction. Yeah, I'm not afraid of what he thinks about me anymore. Or what others think. I live for me and not others. Oh well, mum then reminded me the Arnolds [--i.e., my cousins--] (^) were coming.

Few hours later, they were here. Hm, I was really happy to see them, as I am always. Andy, sweet nice little thing. Arnold. Hmmm. How can I describe him? Come to think of it, not simple describing him. Hard thinker [and](^) imaginative. Loves thinking. Good for a kid like him/12 [meaning, at 12 years old](^). Girls really like him. By the way, Monday's his birthday. I love talking to him.

Talking is my art, people tell me. Yeah, I do talk. [It is](^) the only way you can get to really know people.

Interesting the way people stay the same. They just go through development stages. [But](^) it's really not fair to call them development stages.

It's really not fair enough to call them development stages. Coz sometimes[,](^) they do not develop people. [Instead,](^) they transform them while keeping them the same.

I do not know if I am saying what I want to say correctly. But, take an example of myself. I believe I've always been the same, but developing the self within. I think it goes with experiencing more, living one more day.

I do not know if I've written this before, but it's interesting the way people can and will never say all that is on their minds. Because some or most times they do not know [enough](^) to say it.

Sometimes, people cannot describe their feelings. They are not feeling sad. Not empty (X). Just cannot describe it. Other times, they do know it. They just do not know how to say it. I've said this before to Teacher Annet [, my home-room teacher](^).

She listened to me. Carefully. Absorbed. Yes, she does listen to me other times. Always has.

But then she had that look on her face. One of someone thinking hard, dissolved in thought. I love that look on people's faces when I am speaking to them. There she was, my teacher, listening to me.

You know, I kind of love this situation of...having quite a lot stashed away on your mind. You give it an involuntary glance sometimes, and then just keep the sure knowledge [that](^) it's there. Sometimes, one of those things does come [to the forefront of your mind](^), sure. You say it. Or, you memorize it [at the forefront of your mind](^).

I [have](^) written over two hours now. It's high time I went to bed now. Try and sleep. I'm out.

January 29, 2002

Now I know more than ever it's hard to write in a journal. Yesterday, I didn't. However, it was an eventful day. I asked Carol out. Saw an old school. But most of all, [I](^) saw Nurse Amooti, a lady who used to care about me very much. I am tired now. But I have a lot to write. I'll write it sometime. I have to sleep now. Feeling sickly too. Out.

January 31, 2002

I am writing this on the 1st, actually. At 28 past 5. I did not write yesterday however[,](^) because I was shifting from my first room. The room I've used all these years. It was indeed a landmark event, when you really come to think of it.

Mum has just been suggesting that Sinny the little puppy should be sent away. Because it's sick and it will spread it to the other dogs.

A.3: Entry Subset II: February and March 2002 (An Introduction to the Mind and SC of Seif, the High School Teenager in Uganda)

This subset of entries is also marked by a few noteworthy features. First, as the sub-heading above indicates, we are introduced to my mind and self-communications in high school in Uganda. Secondly, we encounter a more intense struggle vis-à-vis self-confidence (the February 8 entry), as well as self-awareness and description of thoughts and emotions (i.e., the February 11 entry).

Moreover, it introduces readers to the aforementioned presentation of characteristics of the two mental-health conditions that would many years later (between 2009 to the present, i.e. 2021) be officially diagnosed as generalized anxiety disorder (i.e., the February 8 entry) and type-II bipolar disorder (i.e., the February 26 entry). Overall, both my recollection and these and other sample-entries suggest an average emotional turmoil score of around 4.5* throughout high school, as well as my undergraduate years in the USA *(*on the aforementioned mental-distress scale of 1 [placid] to 10 [very turbulent, including thoughts of self-harm, or a state of borderline panic-attack mode, or psychosis]*).

The entry-subset also demonstrates a problem that had largely abated in high school, but continued from time-to-time—i.e., getting bullied, given my non-assertive/aggressive character (i.e., the March 5 entry). Finally, the subset also chronicles two common supposedly deviant behavioral phenomena among teenagers, i.e., experimenting with drugs, and clashing with authority figures (i.e., the March 22 entry).

February 2, 2002

Virginia called me to tell me she wasn't going to be at school by 4 PM, that I shouldn't get disappointed. Hm, positive sign. I'll leave her a note. I am [accompanying](^) *Irene* [my neighbor and my mum's goddaughter](^) *to school tomorrow.*

Hope to see Brian too [another nephew who was attending high school in Uganda from the UK, where he lived with his mother, another one of my elder sisters](^). *And other people. Should go to bed now.*
Out…

February 3, 2002

Day was eventful I must say. Went to "Bojas" [i.e., the nickname for "Kabojja Secondary School, another semi-international school](^) *to say hi to Brian. He'll fit in.* [He'd become friends with](^) Hooked up *with some guy called Hassan. From England too.*

Met Paula [my then-girlfriend's older sister, and](^), *Hamil* [who had previously attended my high school](^). *Paula thought I was cute, and I have a nice accent… Oh!* [and](^) *I was with Sheldon the whole time.*

Brian thought Hamil was a "beg friend" [someone who pursues unrequited friendships](^)… *Oh well, should be signing out now. By the way, Richard* [my cousin](^) *came over today. Surprising, really surprising. Married.*
Out…

February 4, Monday, 2002

Today I've returned to school. I had originally planned to come back tomorrow after welcoming back [my elder sister](^) *Sarah* [from the US, where she was visiting my other sister Sophie, who was attending college there](^).

Seems like she missed her flight. I'm back at school. Same old school. Same old people. Not so much to write about though. Hope to write more tomorrow.

February 6, 8:02 AM, 2002

I find this totally outrageous!!! I never journaled yesterday!! Anyways, the day was...just like that. I tried to get myself to learn all the new faces, sadly many though. I wish the school were just like it were [before, with less students[10]](^).

Not many people...nice...fun. It still *is* fun anyway. I'll write again later on.

February 8, 2002 8:00 AM

It's been 2 days now and I'm not writing in my journal!! This is absurd! Anyways, <u>quite</u> a lot has happened in the last 2 days; met a new girl in F.2/S.2 [sophomore year](^). *Weight-lifted, talked to Carlos* [also my nephew—Brian's younger brother, who was attending the junior/primary section of my school.] (^). *I promise, I'm going to continue writing!*

February 8, 2002

I'm finding it kind of hard to recall the day's events. Every day at school though is eventful and <u>scheduled</u>. That's what I love about school. Today, as I've already said, was eventful. I woke up [at the] (^) *usual time.*

Early, didn't keep track of the time though, I've lost count of it lately, having left my watch at home. Anyways, school always gets me to think a lot. For starters, I always get discouraged, and I'm almost sure I'll fail.

Other times, [I have/get inferiority](^) *complexes. He's better than me. He does stuff better. Bluh-bluh.*

Right now though, I'm in preps [i.e., mandatory evening self-supervised learning-review sessions](^). *I have a problem, you know. A real problem. When I look into a book, stuff* [/content](^) *just does not go in* [to my brain](^)! *Somehow. Real problem. Hate to do yo, but I just have to...(finishing the book, I mean). Oh well, I'm not gonna promise again.*

February 11, 2002

Sometimes it's just too hard for me to analyze your [my](^) *day. Coz mixed results of emotions are hard to describe. My* brother [cousin Isaac, with whom we'd adopted the title of "brother" for each other[11]] (^) *just made me so miserable a few minutes ago. He got me to start thinking; all kinds of thoughts.*

He's sad. When I came out of the bathroom, he was crying. He refused to tell what was wrong with him then.

After preps, I went back to dorm and found him lying on the bed, still thinking. He just [Illegible](^) *of what was wrong with him.*

Promised he'll tell me. He's really sad. Someone or something hurt his feelings. I guess part of my hurt because of him is because he hasn't told me what is making him sad.

God knows I'm sad about not writing in my journal for 2 whole days. There's <u>a lot</u> someone must analyze each and every day of their lives.

February 15, 2002

Yesterday was Valentine's day. In this school at least, it's kind of celebrated. There was a lot of excitement here. People were excited. What about? I cannot really tell. Love, perhaps.

But what about love? Anyways, personally, the day was good, even though there wasn't any particular Valentine [i.e., someone in particular](^) *I was celebrating for. My mum perhaps. Lessons stopped at 3 PM.*

We went for P.E [i.e., physical education](^)*thereafter. Then came back to school. I took a nap, then woke up and prepared myself for the evening.*

I wore my new red t-shirt and baggy pants, then set out for class. Nothing special happened. Apart from telling Phiphi that I like her.

February 22, 2002

I've spoken to Aisha today. We had a really long [and](^) *touching—I should say—talk. She told me as usual about her hate for people. I told her too. Burst out actually and told her about my passionate hate for XX. She even showed me bits of her "black book." We gave each other pet names too. She's Tracey. I'm Kade.*

I spoke to Tr. Simon too, earlier today. Yesterday, he told me the reason of his present state of unhappiness these days. His fiancée left him. Allen. For some reason, he trusted me and told me. He told me. The man is so sad. Today he added [that](^) *he's sick. Became sick after he tried calling Allen and she simply said to him, "enough is enough." Well, fate as we say with Tracey is very interesting.*

February 26, 2002

Takes me long to sign it [i.e., to write in the journal](^), *but it does not take me forever. So much has happened within the last few days. My "active" mood has been re-kindled.*

I had it last year [during](^) *second term mostly. Tr. Annet broke to me the news of how she's offering major support for my proposed first novel. And how uncle Pat's a part of it. Yeah, yeah. I* proposed [also drafted](^) *the constitution of CLUFRA* [the acronym for "Club De Français"/the French club, a new club I was suggesting for us to start at the school](^) *too. Tr. Bill has it now. I was suggesting we see Oscar* [the headteacher](^) *to Tr. John. He said it was Bill* [the *defacto* school dean of studies at that time](^) *we had to see first. It's impressive in my opinion. I've done my homework too.*

I lost my planner. It disorganized me, even if I wasn't using it quite that much. We have a biology test tomorrow. On reproduction. Oh, I lost my biology book too. I can say today has ended nicely. Exercised my body a little, worked out my muscles. Had a cool shower. Dressed up nicely. Attended prep. Got a compliment from Nali. Did my homework.

But you cannot have it all. I missed supper, and I'm hungry right now. I do not expect to get something to eat. I've just been speaking to Donald. How I love listening to that boy. He complimented me. "You know…you sound like a journalist…film actor…educated African…" I just have to write tomorrow. I promise myself.

March 5, 2002[12]

God knows I'm not even sure of the date itself. That seems to be the new style anyway. Not being sure. I'm not sure I should act the way I do toward people.

Today, that [derogatory word deleted](^) *annoyed me. He held me by my shirt and began tossing me around.*

Just as yesterday. Only today it really got to me. I just do not know how to even start describing it all. The day has been shit. As I feel like doing now. And I'm hearing a voice of some punk right now. Poking his head through the door. Suckers. They all are. I know it's not fair to say this but…It could be true. I'm out.

March 22, 2002

As I said the last time, too long, but not forever. Last Sunday, which I believe was the 17th of March, I took marijuana for the first time in my life. A milestone, I believe.

Quite a lot has happened since the last time I wrote. So much happens within every 24 hours of activity. It feels so terrible not writing every day. You think about the journal almost every day, feel bad. Think about what has happened. Phrase it in your mind. Not that I do not always phrase other stuff. I'm always phrasing stuff. Tout le temps. Sometimes, I even do it in French.

There are people in this world—school, I should say—who make me sick. Sooo sick. My headteacher is one of those people. Always right. Person who seems to do his job perfectly. Which one would think is to bug people. Always bugging. The kind of person who makes you think. "I'm the adult. The clever [intelligent](^) *person. Your teacher. I tell you. You do as I say."* It's actually hard to find the words to describe him.

I was speaking to Donald the other day. That good friend of mine. And he was telling me the man is only doing his job. I'm his job. I would not love to do that job. Never.

He told me lots more stuff that day. And he was right. He always says stuff that's right. He loves reggae. And the message in reggae is right. He once told me that almost everything in this world irritates him. So the best thing to do is to just "excuse yourself" from it and just live as best as you can. I like his reasoning. Mature, moderate, balanced reasoning. It keeps me going. There are some things in life that keep me going. Like the sound of birds chirping up in the tree. I could go and go and go.(X) [That sentence might have been a distraction-caused error; I might have meant to say/write, "I could go and on and on."](^)

A.4: Entry Subset III: April 2002 (A Brief Exploration of the Thoughts of Seif, the Rebellious High School Love-Sick Teenager)

In this subset, in continuity to the entries above about Virginia, we read the thoughts of a teenage boy who is earnestly participating in the rituals of the mercurial human emotion of romantic love. And in continuation to the March 22 entry above, we read about the repercussions of the teenager's rebellious behavior—i.e., suspension (April 16).

April 9, 2002

I have wanted to update [my journal](^) *for quite a very longtime. Not the ordinary normal every-day need to update from my day. Last Friday, I made a very rare discovery. A rare precious discovery. I discovered someone who's got things in common with me. She wants to be a writer. Actually, she's a writer. She do not believe in school. She wants to do photo-journalism. Little did I know she had a crush on me. She madly loves me. And I think I love her too. I'm so spiritually uplifted.*

April 16, 2002

I returned to school on Sunday from the suspension. In mum's car. With Carlos as well. Carlos cried for a little while after Resty [his and Brian's mother, my eldest sister](^) *and mum had left. He recovered soon afterwards. By the way, Carlos told me of how a P7 (7th grade, last of primary school) girl confessed her love to him.*

It was so interesting. The whole time while he told me about his latest girl adventure, I couldn't stop thinking about Tess. One thing we <u>had</u> in common, I thought, was someone likes us. Difference…mine is a lot more serious. Tess/INK has not yet returned. For heaven's sakes, today's Tuesday! I wonder what she's doing at home.

Now I know what it's like to wait for a lover. I've just realized something. I've been beating myself over not updating about each and every new thing that goes on between us. I just realized however, <u>that's</u> what <u>she</u> does. I do update however. From time to time (de temps en temps).

Gosh I miss my gal!! God knows how much I miss her. I wonder if <u>she</u> feels the same. When I ponder it for some time, I realize she does. I spoke to her cousin today. Young. In love too. With Tracy. I wonder if it will last. Less settled I could say however. She likes him but…not more or the same for that matter as <u>she</u> [Tess/INK](^) *likes me. I wonder if she's so sure of what she wants like Tess is. I miss her.*

I left the house without receiving a word from mum, just as I had entered. I remember vividly <u>that</u> feeling I had as I approached her bedroom. Anxiety. Terrible anxiety. As I always stress, I do not care what the school principal tells her. I know and God knows it's not true. I care [about](^)*her perceptions, and* [her](^) *feelings in the end. I eavesdropped her telling Sophie on the phone she gave up on me. I know she hasn't yet. Not yet. Just not yet. I know Florence had called her prior to my coming. I greeted her; she didn't reply.*

Boy, I miss Tess. Oh! By the way, I hit Zip's nerve. She told [me](^) *I was getting her <u>sssickk</u>. I feel bad about that. But I still miss Tess the more. I wonder…*

I wonder what's written in her journal about me. She told me I'm coded "Kiwi" in it. I miss talking to her. Looking at her. I want to tell her I <u>love</u> her.

B: Summary of Journal Entries Between High School in Uganda and College in the USA

Between the above entry and my arrival in the US, I continued to write irregular journal-entries with similar content as that of the above samples. In other words (among other topics/themes): my boredom and enjoyment—and a variety of other feelings in-between—of high school routines; girls; and my anxiety about, and hope for, the future, etc.

However, the most significant event and/or time-period was the end of high school (specifically, the Ugandan British-style "O'level" stage). Apparently, my final entry in high school is dated November 24, 2002. And one markedly insightful note—what I referred to as the "thought of the day (or 'T.O.D')" states:

One thing I thought about today as I often do [,](^) is this other world of mine up here (my mind). If there's anything that pushes me on, keeps me going, that does.

A huge huge (sic) world it is... An interesting world it is too.

There exist other worlds like this for sure [my then-meaning here is unclear, but I believe I was referring to other individuals' minds](^). *How are they like? I wonder...*

In other words, I was already realizing the nature of my over-active (and arguably, sometimes *creative*) mind.

Thereafter, I spent the whole of 2003—including my brief visit to the UK—striving to achieve my dream of an American college-education. And unfortunately, this summary cannot adequately describe the turbulence of the emotional state in which I was while pursuing that endeavor, especially right before and after the first unsuccessful visa-application in January of 2003 (temporary mental-distress scale score for that time-period: 7.5). Luckily, I received plenty of logistical-, material-, and moral-support from my elder sister and mother. Finally, on the 17th of February 2004, having secured the F1 student-visa, I arrived in the USA (in New Jersey, from where I commuted to Kingsborough Community College in Brooklyn, NY).

C: Blog-Post Entries Since the Age of 25 (i.e., in 2010) Through the Current Year, 2021 (Age: 36)

C.1: Time-Period I; Post-Master's-Degree (Union, NJ)

In this time-period, we revisit the author's mind, six years after his arrival in the United States. One of the major markers of the time-period was my graduation in 2010 from the communication-studies Master's degree program at Kean University (where I had also earned my Bachelor's degree in English-Writing two years prior, in 2008). Immediately thereafter, the chair of the department at that time—i.e., Dr. Christopher Lynch—offered me an adjunct teaching position.

Thus, between the fall of 2010 and the summer of 2012, I taught several sections of the communication-studies basic course (a combination of public-speaking and basic communication-studies theory), as well as a section of "Communication and Conflict Resolution" during the spring of 2011. I was also able to secure a part-time position as an IT specialist at Union County College (at the Plainfield-NJ campus). Moreover, I had also secured political-asylum status in the US; thus, my employment ability was legally unhindered.

But by far, my most significant endeavor during this time-period was the pursuit of admission into a communication-studies Ph.D. program. After two failed attempts, I succeeded in that mission, securing acceptance into Drexel university's "Communication, Culture, and Media" program in the spring (February 1, to be exact) of 2012, with a scheduled start-date of the fall of 2012.

Overall, the entries below do not seem to reflect a mind that is grappling with the above dramatic life-transition events. Instead, my mind seems to be in a constantly pensive and—for lack of a better word—*philosophical* mood. Regardless, the most obvious trait that this entry-set seems to share with all the others is the ongoing oscillation between happy and sad (or creative) moods, a result of my bipolar-II or cyclothymia disorder. "Exhibit A" in this context: the "Reaching Out," "Emotions," and "Apathy" entries, respectively written on July 26 2010, December 3 2010, and February 1 2011, versus the rest of the entries.

Reaching Out; July 26th, 2010

So much seems to happen in the world, if news reports are anything to go by. Spies getting outed. Oil spills. War secret leaks. And on and on. And so much activity, business transactions, other social and economic give and take.

The mind, my mind, which perceives all this, often does so while also bearing in mind that it is doing so from within a vessel that "feels" (for the heart which perceives emotions is also another feature of that vessel) it is above, or apart all this. I view the world from a place of solitude. In social arenas, I am all smiles, pleasant, what have you. I could have a million friends, have a wife that loves me, and still feel like this. Or would I? Do I just think the preceding thoughts out of yearning, for I so desire those things (mutual understanding and connection with others)?

Ever since I was a child, I tried reaching out to others. Siblings, school/class mates, etc. The only person that reached back unconditionally was my mother, and over the years one of my older sisters. But at some point, one just realizes that overall, people just wanna be left alone! Sure, everyone appreciates kindness, love, validation etc. But these things are tiring! I for one find it much easier to just be a loner. But easier is sometimes more painful too, if that makes any sense. And so when I come back home to my little two room basement apartment, with nothing but TV, internet, and a phone, filled with numbers of people I cannot call to chat with (coz they are "busy" or something), that pain and ease combined, kick in.

Where I go from here, only God knows. Ideally, I will make more friends, get a girlfriend/spouse--but I do not want kids. But indeed, only God knows. I can also see myself being a loner for a while, if not forever--what choice do I have?

In the end, only God knows.

Prolificness; November 30th, 2010

Charles Dickens. Leonardo DaVinci. Albert Einstein.

What does it mean to be creative? To be intelligent? These and other such admirable qualities are not the same. But what do they mean, and how does someone achieve them? Are they important anyway?

I believe most of us probably settle to just get by, to do the minimum of whatever it is that we're supposed to do. But what if more people tried each day, or as often as possible, to not only work harder or do their best, but to look for new ways of doing things, to try to force themselves to think different? "Creatures of habits", we mostly are, I guess. There's a cute little parking spot just within the faculty/staff parking area at my job which I always want to park in. If for whatever reason I find it taken, my work-day hasn't gotten off to a good start. Funny, isn't it? We take the same routes to work, buy the same brands, and we have favorite foods.

Of course, it's not necessarily true that excellence or creativity is simply born out of the trying out of different modus operandi. In fact, the complete opposite might be true. If one does the something the same exact way over and over again, they become very good at it. That's how professions are born, no? Plus being creative or intelligent does not necessarily mean that one is going to produce work that's reflective of those qualities. Countless people (e.g. MANY of my students) have TONNES of potential, but just do not want to apply themselves.

In the end, I doubt that we'll ever truly figure out what makes the stuff of genius. I always tell family and friends that yes, even though I'm hardworking, I simply cannot ignore the fact that; call it luck, or blessings, or karma—whatever, it has had at least a small part to play in my success. Sometimes, the right prerequisites, timing, and other factors, along with lots of hard work all just come together to form the perfect storm. And in such cases, good, or even great things happen. Long-shot candidates win office, great books are written, ground-breaking theories are introduced. Haven't you ever read a book that just made the hairs of your neck stand up, a movie that made you cry, or a song that just made your inner spirit soar?

But, what does it really mean to be prolific?

Emotions; December 3rd, 2010

Sometimes, I surprise myself. I am specifically talking about my emotions here.

It occurred to me some time back that perhaps every once in a while, human beings enjoy crying. Women and men. I believe this is connected to the concept of "catharsis". Sometimes, all you need is a good cry, a way for you to let out your sad, depressing emotions, and then you feel better.

I mean, how else do you explain how or why yours truly broke down and started weeping the other day, while recounting a childhood story? Totally unexpected, and you'd probably bowl over laughing if I gave you the specifics. Suffice to say, it really wasn't cry-worthy, in the grand scheme of things. But, like I just said, perhaps I was just in the mood to cry! A relevant tid-bit of information that perhaps had something to do with it; it had been raining cats and dogs the whole day, and the sky looked sooo gloomy! And even though I do not think of myself as one of those folks that get really depressed by the weather, I do not know...perhaps sometimes, it plays a part.

I believe the salience of such episodes is related to the complex world of human emotions. In my case above, the real reason I cried was guilt. When we're happy, we smile or laugh, or just glow within. When sad, we quietly ponder our fates, or perhaps cry, or talk to a friend about it. At times, it might seem we're powerless when it comes to controlling our emotions. A customer service rep pisses you off, and the next thing you know, you're having an out-of-body experience, you can see and hear yourself raging at him/her and wondering, "where did all that come from?" Of course, as with most things in life, we can control and effectively harness our emotions. The trick is figuring out exactly how one wants to, or can do it. For we're all different, with different family roles, vocations, etc. That hotshot CEO can afford to throw tantrums all the time, but not the house maid.

In any case, my little story above ends in a funny way. After telling the childhood story that had made me cry, the person I was talking to said something that just made me start to laugh uncontrollably! The tears weren't even dry on my cheeks, but there I was in stitches!

Apathy; February 1st, 2011

That's what I feel right now. Disengaged, uninterested in anything. I could just sit here all day doing nothing, content to do nothing. Or, I could just sleep forever...

I had planned to upload a blog post titled "Let It Snow, Let It Snow," a tribute to the current snow-storm season that won't let up. But...at the moment, I just feel like sitting here and sulking... I do not want to blame the winter for this. Sometimes, I just get like this. Life can indeed get monotonous and just plain boring sometimes. That's just the way life is.

But, my dear readers, I hope you're reading this while having a good day--fun, different, exciting, all that good stuff.

Me, I'll just keep sulking for now.

"Peace out".

On The "Internal/External Problem-Dichotomy", And "Bad Karma"; August 26, 2011

That title refers to what happens inside of us (our "minds", "hearts", etc.), and what happens around us, or "out there" in the world. Case in point: me and my close friends always pow-pow about our daily personal struggles, triggered by stresses at work, home, etc. Meanwhile, almost every working adult in this world has been in a position in which they have some kind of problem at work, and they have to collectively–with their colleagues, bosses, etc.–discuss or think of ways to solve those problems. We can take it even further, thinking about the way problems are solved at the local community, region, and national level.

THAT is what I think of as the internal/external dichotomy, when we think about "problems" in life. Problems in our own personal lives, and problems "out there", collectively. But a few years ago, I remember telling one of my sisters, that personally, I think even international relations and disputes between nations, are simply a macrocosm of simple and extended family issues. Fights, slights, break up to make up, family events, successions, concerted decision making between parents and children... You do not have to look hard to find comparable events, concepts, systems, etc., in international relations, that mirror the above, and other family dynamics.

It all starts within us. Regardless of how or where we were raised; every single day of our adult lives, we always have to make decisions, deal with our emotions somehow (and there ARE right and wrong ways of doing this), and be considerate of others. Many things can, and often do go wrong while we try to accomplish these tasks successfully. Wrong decisions, dealing with our emotions the wrong way, being inconsiderate of others. A cycle is then created. Because we deal with ourselves and each other in ways that are far from the ideal, we then go on to treat ourselves, and others, even worse than before, and many people "take it out" on others. Bad karma then starts to feed on itself, and on and on we go.

To be sure, many times, an individual does their best to stay calm while others freak out, "to be the adult in the room", to take the high road. But it is really hard doing this, while every other person in the room is going bonkers–and often, about trivial things! That is why I have nothing but respect and admiration for people like President Obama, who seem to be almost unable to have their feathers ruffled.

Note: whereas I'd rather not bring politics into this discussion, Obama is one of the very few people in today's spotlight that are known for having that quality–but I'm sure there are millions of G.O.P potential presidents out there, and dozens of past Republican/Conservative presidents with similar temperaments.

In the end, the remedy to this cycle of "bad karma" is easier said than done. I guess the simplistic way of putting it, is we just have to do our best, to enact better values with ourselves and others, every single day–"one day at a time" indeed. I always think about a belief of mine, and I've probably expressed it to someone before–I will for sure be writing about it again. MOST human beings are GOOD PEOPLE. MOST. They either enact this "goodness" much or some of the time during their lives, or have the PO-TENTIAL to be good, decent human beings. In the end, many of us just choose to be lazy, as far finding that potential for goodness within us, and enacting it during our lifetimes.

C.2: Time-Period II; Ph.D. Program at Drexel University (Philadelphia, PA)

As aforementioned, I had received confirmation of my acceptance into Drexel's communication doctoral program in February of 2012. Coincidentally—in fact, on the same date (February 1), I also received my "green card"—i.e., proof of legal permanent-residence status, having secured political-asylum status two years prior (to boot, my niece Amber was born on that same date!).

And unlike the above entry-set, the one below seems to be more congruent with the time-period markers. This is especially identifiable vis-à-vis my handling of the transition from Union-NJ, where I had lived for seven years, to Philadelphia, PA. The entries also explicitly mention, or seem to imply my struggle to acclimate to the intellectually-grueling routines and tasks of the Ph.D. program—e.g., the February 15, 2013 entry.

April's Very Own; April 22, 2012

I moved to Union, NJ, around 2007. I had been living in Newark for about 1 and 1/2 years, a period that went by almost without incident. "Almost" because, well, Newark being Newark, I did get mugged once. It wasn't that serious. No guns involved, only lost a cheap cell phone and wallet with no money in it–cancelled the credit cards. I did get hurt a little, and the experience was overall harrowing. But again, not that serious.

Perhaps the more remarkable experience about living in Newark, was the mere culture shock of living in an "inner city" area, a place with American-style poverty, bodegas, teenagers who are clearly bored by school and would rather "hang out on corners", the occasional shooting or street fight, name it.

This was the period in which I was beginning to shed my starry-eyed optimism and sheer can-do energy; it was the period in which my idealism came face-to-face, somewhat brutally, with realism. The move to Union provided some respite from the grind of that metamorphosis, but by then, the train was already going downhill.

*** *** ***

So you see, the past 6 years of my life, the "Union" years, are a period in which, well, Seif grew up. I went on to graduate with my bachelors, quickly finished the masters, and started teaching.

I will miss Union, folks. Come late July/early August, yours truly will move on to the City of Brotherly Love, Philly, to embark full-time on his PhD.

But perhaps ten times more than my nostalgia for Union, is that for Kean University. To say that Kean has been good to me would be the understatement of the century. I have cut my teeth here, I have gotten mentors, life-lessons, even some "wisdom", you might say

*** *** ***

Birthdays are the perfect days on which to ruminate about such things, and it was without any fanfare–just the way I like it, that mine came and passed at the beginning of this month. I guess I should be grateful about the fact that I have survived these years on this here earth.

But I am often tempted to think about my life in terms of measuring what I have accomplished in the years that I have been around–bad habit really, I believe, but I still do it.

Perhaps another way to think about it is this: If indeed my work is one of the only few things that give me happiness in this world, then I might as well savor it. No, this does not mean that I become a "workaholic" and turn out even more miserable than I already seem to be. Rather, it means I SAVOR my life's work. And so I will.

In that vein, one of the things I look forward to the most about doctoral school is that immersion in the knowledge that I am drinking in, that constant inquiry and learning; in class, with friends/colleagues, professors, etc.

I have said it before, and I will say it again, folks. We have to ask ourselves what it is in fact that we truly work for in this world. If money, or survival, or food, etc., is all we're searching for, then let us REALLY search for it. Let us become rich, let us collect until we cannot collect anymore. If we only do what we do because it is what we LOVE to do and it is what makes us happy, then, LET'S DO IT!

In Lieu of December's Post, Written On 12/31/12

I believe the earliest memories I have of swimming are from the trips we used to take to see our grandparents in the countryside. For us luckily, "the countryside" just happened to be near Lake Victoria. I'm not sure who exactly used to go along on those trips apart from myself and a couple of my cousins. What I do recall though, is that lake water. I remember the smell of fish too. And seashells. Lots and lots of seashells. Another memory I have is of almost drowning once, when our mum took us to one of the fancier beaches—now that was a rare treat.

Fast forward to today. See, there are not that many activities I can carry out that can truly help me relax, let alone helping me to keep in shape—at least as best I can. If you haven't caught on by now, yes, of course I am talking about swimming—as an exercise.

Every once in a while; regardless of the class I'm teaching, I like telling my students about the concept of "flow" in psychology, as posited by Mihaly Csíkszentmihályi ("CHEEK-SENT-ME-HIGH"). There are some activities in this world that can truly give us joy. Activities in which, once fully immersed and in deep concentration, we are at the peak of our potential. At the risk of oversimplifying or misusing the theory, one can argue that this concept is somewhat similar to the attainment—hard as it might be—of nirvana!

Once I'm in that water, fully immersed or floating, one way or another in the constant motion of the coordinated movements of the swimming strokes, I am alive. Nothing else—nothing else matters at all in this world, in that moment. That is one of purest joys of living I find myself in, most of the time I'm in the swimming pool.

Of course, that feeling is not only limited to swimming. A little while back, I read Eckart Tolle's "A New Earth." Now obviously, Tolle says a lot of wise and poignant things in that book regarding the nature of life, being, the self, etc. But one of the things he mentions towards the end somewhere, is about the power of…wait for it…breathing! That act, basic as it is, without which we cannot live. Imagine for a second how much we take that for granted. Tolle therefore suggests (I might be seriously misquoting him here, so bear with me) just taking the time out to slowly, reflectively breathe. Just close your eyes and SLOWLY breath in and out, savoring each and every inhalation and exhalation, and savoring especially, those 3 or 4 seconds in between your exhalation and inhalation.

What I am suggesting by all this talk about breathing (and I guess meditation, if you will), is that, like swimming, there are indeed a few more activities out there, those that you might enjoy so thoroughly—playing an instrument perhaps, fishing, even gardening, that can put us in a state of "flow." Even an act as simple as breathing; as long as we concentrate fully on it, and savor it. One way or another, there will be a point in your life, in which you are fully immersed—mind, soul and body—in something. When your entire being and your faculties and energy and spirit are channeled into this one thing. In that moment, you are ALIVE. For me; that moment, many times, is when I am swimming.

Goodbye 2012, Hello 2013;
February 15, 2013.

Ah, the story of Seif and his blogging procrastination. Goes something like this: Seif decides to regularly blog, no matter the gyrations of doctoral school. Every beginning of the month that is; hey, who knows, he might even pick up the pace to 2 times a month, right? Ha! Anyways, then the weekly bombardment of readings of Bourdieu, Foucault, Derrida, etc. start, and well…Seif starts to think of blogging as a luxury he can barely afford! Or is it? Perhaps if Seif managed his time better, blogging would keep him sane!!!

Ok. Now that we're done with the venting about blogging, let's do some blogging, shall we?

The winter term is going well so far. Much better than the fall of course. Yes, the rigorousness is still there, but we're now used to the pace. But, I also know not to take that for granted, coz you know how that calm can fool you, before the storm.

So I always think of what to blog at the end of the month, and what crossed my mind over the course of January was the issue of "Learning Lessons." As in, do we learn enough from the mistakes we make on a daily basis so as not to make them again, or so as not to make others similar to them? Personally, I always beat up myself after doing something I should have remembered not to do, or perhaps forgetting to do something I should have remembered to do. It happens to all of us, no? But then I remember one of my mottos (I know it might be a cliché, but it's true nonetheless): everything happens for a reason. I've come to really believe that that is true. And so, after beating myself up over and over again for not learning those lessons I should have learned, I realize that perhaps I was meant to make those mistakes again. The lessons I was meant to learn, I will learn, and the others, well, perhaps I wasn't meant to learn them.

That also reminds me of scientific theories such as the butterfly-effect and its overarching theory (or set of theories, really), chaos theory. Mind you, this has nothing—absolutely nothing, in my opinion, to do with questions of spirituality, the existence of a deity (or of multiple deities), etc. At least not in this context, but you definitely can make a connection between those two ideas, if you please.

Speaking of things happening for reasons; on my way home a couple of hours ago, I see this woman. She is seemingly dressed "normally," in jeans and a hoodie with a purse, clean looking, but she is sitting on a step by the store, crying and telling me—or anyone who looks at her—to help me her. If I had had some dollar bills on me, I really would have given her one or two, I swear. This incident got me thinking about one of the dark sides of urban communities…

One thing I haven't shared before about the city of Philadelphia after living here for a few months, is in regards to its urban feel—its landscape, and its people and their attitudes. Philadelphia—compared to NYC, is a "real" city. As in big, with the usual features you would associate with a big western city. A very good transit system, a fairly big downtown or CBD area—nowhere as dizzying as Manhattan of course—that city swagger and fast pace, name it.

But the one thing that was really jarring about this city is the way its people seemed sooo rude to me, when I first moved here. My perception about the people was something along these lines: WTF is your effin problem??!! I'd go to South Station, ask the ticket ladies which platform is for which train, and they'd effin look me like I'm from Mars!

Meanwhile, the homelessness and scourge of addiction, seems to be really high. I mean it's different when you have a bunch of homeless folk that sit quietly in corners panhandling. But some of the experiences my friends and I have had with the homeless folk here are—again, like I say above—jarring. "In-your-face" is how I can describe it.

Now of course I'm not a nincompoop, I know the gravity and complexity of the issue(s) of homelessness, substance abuse, etc. I get it.

And oh by the way, I've gotten too close [to penury](^) *for comfort myself before, as I might have mentioned here. I almost became homeless once, I've slogged through a few tough months in between jobs, etc. And I also understand not being very amenable to just any job that you come across. I worked in retail for a while, and I know how that feels some days. I get it.*

Not to mention that at the end of the day, luck has a lot to do with it. A lot. Some work hard all their lives and cannot seem to catch a break, whereas some strike it rich easily. But with all that said, do not judge me for feeling just a little too peeved off sometimes, being constantly accosted for change.

I mean, my Catholic school lessons never left me; when I was hungry you gave me to eat, when I was thirsty you gave me to drink, now enter into the home of my father. Indeed.

Yes, we do have to answer the call for charity and kindness in our world. But it occurs to me that in the end, it seems we're all just fallible, complex human beings. Each of us has a self-destructing streak at all times in varying degrees, no matter what.

And so however much we help each other, perhaps we can never truly end suffering and yearning in our lives. Perhaps in a way, we were all born to suffer, and suffer, we shall.

On "High Class Problems" (In Lieu of June's Post);
July 31, 2013

This summer is probably the best I've ever had in all my years living in the U.S. But boy oh boy, is Seif busy these days or what!

Now you see, lately, I've become fond of thinking about and discussing what I like to call "high class problems." One poor schmuck is crying coz he cannot find money to feed his kids, whereas this other guy is complaining about the fact that he has like 5 parties he'll have to attend on the weekend! The humanity!

So you see my friends, that problem right there—Seif being very busy these days; that is a classic high-class problem! The one thing you learn though is you cannot really make fun of someone else's problems, however ridiculous they seem. I mean I've always thought of the English royal family and the fact that no doubt, the queen must endure many moments of duress about the latest controversy that Prince Harry is embroiled in, or the fact that the latest poll says the people want the monarchy abolished, etc.

But you know, sometimes, you have to think of life in a bit of a twisted way to really get it. Here's what I mean. I've had a tendency in the past of being very sanguine in the face of some of the most serious problems you can imagine. And when you think about it, one's physiological and psychological responses to stress—high blood pressure, depression, etc. are the same, or might be even worse for two people in a tight spot. For instance, a businessman worried about the fact that he has to make payments on a business loan that is so far not yielding much profit, and a Seif back in 2008 who was worried he might not have enough to pay his bills and buy groceries!

And yet, we somehow struggle on, eventually making it, or somehow coping and surviving. We do not have any other choice.

And so dear reader, Seif embraces this busy schedule of his more than ever. It does not even feel like work when you're enjoying it, when you feel it's what you were born to do. For me thus, everyday might as well be a vacation. After all, what's not to love about creating, consuming, and sharing knowledge on a daily basis, and getting paid to do it! And getting paid to THINK!!! To think till your head hurts, then think some more! Whoo-hoo!!! Some problems, these are.

Life Is.–Written On 04-21-14; September 10, 2014

So…is this it? After more than 9 (?) months without blogging, is the Mad Professor (the MP) back??? I guess so… ***Smiles, cheesily***

A lot—SO MUCH—has transpired since the last time I blogged. Saying that btw, reminds me of that scene in that Austin Powers movie, when that cute girl tells him he has missed so much over the decades he has been in hibernation, including the fall of the Berlin wall, and the end of apartheid! He-he-he…

Anyhow, like I said, a lot has transpired since last year: Drake released a new album that I played ad infinitum and still cannot get enough of; the MP earned his doctoral candidacy—phew (!!!)—I got to experience Paris for the first time in my life; and, no small development: I got to learn one more very important lesson about romantic relationships.

You see dear reader, what I just did there in that last paragraph is to sap the soul out of a number of experiences that you could write soliloquys and treatises about. But who needs that? Growing, learning, loving, and looking for inspiration sometimes makes us feel like we're not supposed to experience the mundane things of life. And yet, what is life, if not mundane. Those magical moments that we live for are not meant to be experienced every waking moment of every day, folks. That would sap the soul out of them too!

BUT the magic is inevitable many times. And when we come across it, let us not resist it. For instance, in this moment, I cannot help but recall the feelings and thoughts that I experienced while walking through the Detroit Institute of Art a few weeks ago. Being within such close proximity to the works of art of masters you always read about—Degas, Matisse, van Gogh; seeing their brush strokes and oils and canvases, long transformed from those mere materials into mirrors of souls. Being in the presence of an Egyptian mummy from 3000 BCE! I felt goose-bumps dear reader, and I feel them now, just recalling those moments in that museum.

But it's not just such moments that bring out the life from deep within me that I never knew I had—please forgive that sappy cliché of a sentence. But often nowadays, I find myself savoring—the way you savor a sweet, sweet dessert—conversations with perfect strangers, strangers whom I know I'll never meet again, and whom I'm getting nothing from, save for those few minutes of companionship. An old gentleman on a plane seated next to me; a grandma and her grandson in the lobby of the gym; a French woman who's lived in America forever, on a train on my way to Paris (ne t'enquiete pas (!), *she tells me, you'll be fine in Paris*—et elle a complimentais mon Francais!). *And do not get me started on Paris; and Belgium; and Amsterdam.*

Life is good my friends. And life is boring. Life is. It just is. Let us stop trying to make it something. NO, life is NOT "what you make it." It. Just. IS.

C.3: Time-period III; post-Ph.D.-program (Philadelphia, PA; brief stint—i.e., one semester in China; Cherry Hill, NJ; Hays, KS; and again, Philadelphia, PA).

Finally, the entry-set below brings us to the current time-period—i.e., from 2015, through the current date (the fall of 2021, as of writing this). By far, my biggest time- and psychological-marker for this period is my ongoing struggle with securing a permanent job. And compared to the previous two entry-sets, the entries below are arguably the most congruent with the above-mentioned marker (the job-search struggle). In particular, the entries of May 14 2017, as well as that of March 29 2020, demonstrate this characteristic.

No, Not **All of It** Has Been Said And Done;
May 14, 2017

Anecdotal evidence from family and friends tells me that my generation—the Millennials in the age ranges of around (between) 26 and 34, have faced unique and intense challenges vis-à-vis the launching of productive wholesome careers. And indeed, I have read aplenty and have heard from decent sources that this trend is real, and has produced a good number of consequences (young adults moving back in with their parents, "credential creep," etc.). So... What to do, what to do? Supposing you've played by the rules, have gone to school and gotten decent grades, but have fallen victim to underemployment and wage stagnation; what can you do about it?

Of course, plenty of people more intelligent than myself have written about this phenomenon, and have provided good answers (no, I won't provide the links for you here—Google it!). My mission today is far less ambitious. I would like to try and give a pep talk to all my brothers and sisters out there that are going through this crisis.

First, we need to realize that we are lucky, for at least we're still in the prime of our lives, and can thus creatively re-route our energies, and continue on the path to prosperity. Think about it: would not you rather be a young(er) unemployed young man or woman with the ability to learn a new skill or otherwise retool—before you start a family or are otherwise tied down to a place or circumstance, as opposed to being in your late 50s and newly unemployed, without a college degree?

Still, you'll rightly ask me, how oh how, can I get out of this rut? The answers I can then give you might sound cliché, but are worth considering. First of all, perhaps you are not in a rut per se? As in, yes, I get that you didn't toil away in college only to work in retail or other low wage industries for the rest of your life. But we need to really think about our social and vocational/financial lives in more compartmentalized ways.

At what point will be you be satisfied, and why? You might be working at that café and you hate it, yet you've managed to save up some decent monies that you can use later to buy a home or otherwise invest. You might be working part-time, with the burning ambition to move out of your parents' basement, and yet they haven't kicked you out, and are willing to help you take out a loan for you to get that certification that will boost your earnings. See what I mean?

And yes, those other cliché pieces of advice we often hear, are often true. Keep investing in yourself; do your best to make yourself as indispensable as possible at work; beef up on your "soft skills." And finally, to get to the core of why I titled this post as I did: for crying out loud, so as to improve your health (including "happiness") and wealth, do your best to find meaningful ways of actually using creativity in your career and in your life in general.

That concept—creativity—is often invoked but can be hard to pin down vis-à-vis meaning. One good definition, courtesy of Google:

"The use of the imagination or original ideas, especially in the production of an artistic work. Synonyms: inventiveness, imagination, innovation, innovativeness, originality, individuality." So, good sire/lady, for starters:

- Write down your dreams and ideas.
- Rediscover your hobbies.
- Think metaphorically using your favorite sport or other activity. For instance, are you racing through life, or are you in a marathon? Do you win this race by sprinting faster than Usain Bolt, or is it a game of endurance and perseverance?

No, not all of it has been said and done before. You might be the first. Go for it.

06-18-18

I do not know why she's not coming to bed, so I can be able to freely hang out in the living room and kitchen. I'd like to eat, then take my meds, and get ready for bed. But she's remaining steadfastly in the living room.

I do not know if she intends to sleep there tonight and perhaps the next few nights—I guess coz she's so sick of me, the thought of sharing a bed with me is intolerable? Whatever; I do not care. I just want to be able to move and rest freely in the living room and kitchen without her there.

Oh well; I guess I'll just go to the kitchen, make something to eat, come back here, eat it, then prepare to sleep. Story of my life. Spending banal moment after banal moment, wondering what I'm doing on this earth, trying to make sense of my miserable existence.

Day 2 of Challenge 11-11-19

Woke up with a hangover today, which made my day a drag. Not just the hangover, but I was in mood of being really to start the work week. But I did accomplish a couple of things, especially related to planning the week's teaching, and reading the book I have to review for the journal.

But, I also came up with what I think is a good idea, and I'm going to launch it soon. It will be an anonymous blog, on which will I share my journey to become a polymath, a modern-day Da-Vinci!

https://www.youtube.com/watch?v=PVAs3aJHBmA
https://www.youtube.com/watch?v=kEk-BDckjW4
https://www.youtube.com/watch?v=8R-ucUNm64M

03-29-20

I recently realized the priceless value of this activity—journaling—when I came across, or rather, just decided to actually pick up, open, and read my old journal, circa 2005 to 2006. Oh my God, what a re-revelation—so to speak—it was! I read about my first week of college in the USA, my first snow, my first friends and crush—a French-West-African girl, etc. Again, quite a re-revelation.

Oh well, "in other news," the **big**—HUGE, in fact—news from this month, has been my contract revocation by my oh-so-awesome and beautiful and nice and great (SARC) chair, Dr. XX "The Fucking Jerk" XX! The man decided, after all of one semester's teaching evaluations, to give me the boot, as they say. And again, the excuse is the poor student evaluation results from the previous semester.

Anyhow, apparently, XX's prediction came to pass. She'd warned me and Fallon about him, on our visit to her house a few weeks ago. But guess what, dear journal? I am actually ok; NOT bitter, NOT angry, NOT even sad, to tell you the truth. Of course I'm anxious, but then again, I **always** anxious anyway! **Shrugs.** So hey, why should this development worsen my anxiety and depression? Of course, the worst upshot is the HUGE inconvenience and chaos of the logistics of moving, trying to get a new job, etc. But for some reason, and I profusely thank G_d for it, I actually feel mostly ok. The initial shock did hit me hard, seeing that God-awful email addressed to the dean, with the fucking letter I couldn't read because of security settings. But then, somehow, I'm not sure how fast or with what process, that shock fizzled, and I made peace with my new fate. So now, I am sending out job-apps left and right. And I still have a job to do with this semester and the summer on still left. So do it, I will, to the best of my ability.

I do not think there's that much else to report about this past month—March—apart from that development. My usual obsessions and/or intense thoughts, and/or philosophizing, so to speak, still go on daily/hourly/even minutely, so to speak. I ask myself, and God, what my purpose is in life; I desire to execute a TDL—to-do-list—that is as long as my arm; I think up ideas for research studies, and books, and films, etc. And I crave for normalcy; I crave for peace; I crave for the inability to crave.

*I refuse to deceive myself by saying that one day, those obsessions will end. I do know that they will, neither should I really care, I guess/believe. All I should do, is take it one day at a time. See, the thing about clichés is that we disempower their meanings by devaluing their own and our own virtue. "Packed like sardines," loses its oomph. Anyhow, "one day at a time," is one cliché I refuse to castrate. You [know](^) dear journal, I had to actually look up that word—"castrate"—and it still isn't the word I wanted. Anyhow, the point is, I will abide by that motto. And I will continue to pursue discipline and humility. I need to keep looking into mindfulness and Eastern religion-related practices. I need to savor my breaths more, to stop and do some breathing exercises at least five times a day. **Smiles.** Hey, maybe the Muslims were onto something after all!*

05-07-20; Approx. Time? (I guess sometime in the very early morning hours, or at the latest, sometime before 12 PM.) "Kola n'amaanyi, Katonda ajja kuw'emikisa."

I heard the quote above in a song by a Ugandan Luganda hip-hop artist a few years back, and fell in love with it ever since it. In English, it translates as "work hard, and/then God will give you blessings."

Apparently, I missed my monthly 28th journal-writing challenge. In addition to feeling guilty, as if I somehow betrayed myself, I am also blaming myself for failing to be disciplined. From this year onwards,

I want/intend to foster a habit in myself of using discipline, as well as the deliberate and/or mindful action of exercising self-soothing and control—"making peace" with things; annoyances, bad events, etc.

I will either achieve success by implementing that regimen, but regardless, even in poverty and/or failure, etc., I will not blame myself for not having tried my best, or at least just trying! "So help me God!"

CHAPTER CONCLUSION

As I mention previously, the main role of this chapter is to acclimatize readers to the autoethnographic-authorial voice that produces most of the upcoming chapters' postulations in regard to intrapersonal-/self-communication. Thus, among other utilities, I believe the most important role that the chapter can serve is the provision of a holistic understanding of the upcoming postulations, especially the personal-historical context and inspiration from which they arise.

Creativity-Attempt Outtake 2

Beginning of a short novel[13] titled *"Mazimhaka,"* my undergraduate thesis project.

[Note: The full story can be read in appendix E.]

Solomon Mazimhaka could vividly remember one of the worst days of his life [from childhood] (^). He could vividly remember the day when he had realized that his life had been doomed from the very beginning, and there was nothing he could do about it. But how could he explain this to these Ivy League old money patrons?

In fact, almost each and every item in the imposing room, along with the panel of professors just made him more tense. It was one of the opulent conference rooms of the gothic styled main administration building, with a high ceiling, low-hanging chandeliers and meticulous stain glass windows. The walls had a layer of shiny polished mahogany wood in place of paint or wall paper, and floor was covered with white marble tiles. On the floor, between his desk and that of the professors lay a leopard skin. On the wall behind the professors, above the fireplace hung a huge portrait of a stern-looking Governor David Glover, founding benefactor of the university. It was as if the whole room had been arranged to purposefully remind the candidates of their place, albeit subtly.

"Mr. Mazimhaka?" Dr. Lweza's concerned voice startled Solomon. He had been trying to find the right words to answer the question, but words at this point seemed impotent.

He glanced at the cover of his dissertation manuscript. The bold words of the title seemed to be staring [at](^), [or](^)even mocking him: *"Cursed: A Philosophical Reflection on Select Individuals' Perceived Bad Fate."*

The silence was deafening. Patiently, Lweza nudged: "Solomon, it's okay. Try to relax. Just try and remember whatever it was that inspired you to write on this subject."

He still sat motionless, as if in a trance. At first he had panicked, trying to quickly do some damage control. But as he tried more and more to find the words, his mind simply froze. It froze his consciousness and his entire person in the present, and instead vividly revisited another moment in his life, a moment so far away in time.

***

"Solley" (which rhymes with "sorry" [in the British-English pronunciation](^)), as all the boys called him wished he had never been born. Why couldn't they understand? He wondered. Why did they all hate him so much? What had he ever done that had been so terrible for him to deserve such punishment?

The whole school, right from the primary sevens (7th graders) to the primary ones (1st graders) that whole week could not stop talking about the homo that brother Taylor had revealed on Monday morning prayer assembly after a tip-off. All the boys who were recounting the story kept cheekily mimicking brother Taylor's deep voice: "It has come to our attention that Solomon Mazimhaka of P.5 prefers to be kissed by fellow boys!" Anytime someone cut ahead of him in line at the mess, or anytime he dared to contest a dorm cleaning duty turn, Solley was quickly reminded that as a homo, he had no business saying anything to any other normal person. The only day the school had ever experienced such excitement was when the only girl in the school--Jill, the nurse's daughter had arrived.

The prestigious St. Francis Boys' Catholic Boarding Preparatory School at Kkobe Hill had an expansive tract of land overlooking the dead town of Kyengera and the swamps of river Mayanja on the outskirts of Kampala. Run by the Catholic Brothers of the order of St. Francis of Uganda, it was one of the select few elite schools that every old money club patron or politician felt compelled to take their heir.

Self-Questioning For Creativity-Attempt Outtake 2

State Any Necessary Introductory Remarks, and/or Answer this Question: Does it Make Sense to Me Now—i.e., In The Present—versus then—i.e., When I Wrote It?

Unlike the other C.A.Os throughout the book, this C.A.O is markedly different, given its genre as a fiction-writing short-novel-excerpt, albeit based on my (real) life-events. Overall, yes it makes sense, and I am struck by its prescience vis-à-vis the professional/academic path that I envisioned for myself at the time of its writing—i.e., as a future academic (via doctoral scholarship and beyond), which I ultimately managed to achieve. However, I am also ambivalent in regard to the quality of the creative writing (i.e., is it of a professional or otherwise decent caliber?), which is not the case for the most part with my nonfiction writing. In other words, I am much more confident with that (nonfiction) genre of writing.

Does it Resonate with me Now—i.e., In The Present, versus then—i.e., When I Wrote It?

For the most part, yes it does. This particular excerpt grapples with a fictional but real-life-based formative childhood story of humiliation and stigma in reaction to a child's perceived deviant personality and sexual-identity, despite the fact that in my personal experience*, pre-teen children are mostly innocent in regard to these issues (*please note: this statement is a personal, not professional opinion or evaluation). As I mention earlier in the chapter, I was often bullied for being a weakling, and displaying "effeminate" behavior. And unfortunately, those traumatic experiences still affect me—albeit in a limited way—to date, negatively influencing my self-confidence. The excerpt also sets the stage for the rest of the short-novel's story, which is based on my life (as aforementioned). And with that overall story, I try to perform a cathartic exercise using some of the key childhood and early-adulthood events of my life.

What Mood and/or State of Mind does the Tone Suggest?

Unfortunately, the premise of this question is rather unfitting for this particular chapter's C.A.O. However, I suppose one possible answer is thus: the C.A.O demonstrates a dynamic and creativity-aspiring mind, a mind that can use a given person's life-experiences—i.e., the mind's own subject—to create a narrative work of art for the purposes of entertainment and catharsis.

What is/are the Direct and Indirect Relation(/s) to the Current Chapter via Topic(s) and Themes?

The C.A.O is indirectly related to the chapter's topic, i.e., the contents of my journal-entry samples over a period of 19 years. Whereas the journal entries demonstrate a raw form of experience-processing, mind-evolution/dynamism, and oscillation between private and public self, this C.A.O demonstrates an attempt to holistically process the events of a life—i.e., my own—through a creative-fiction lens. In other words or for instance: if/when looked at as a work of narrative art and fused with some fictional elements and events—i.e., some "what-if"-scenarios, what insights or lessons—if any—can we glean from a person's life?

Any other Evaluations or Comments?

I am curious if readers of this book can apply this technique—i.e., the use of their own personal-lives as bases for fictional-stories, and if—as I ponder above (under question-no. 4)—the technique can offer the principals any relevant insights or lessons.

Activities and Discussion Questions

1. Recall the main events that you have lived through in your life so far, regardless of your age. What are the good and the bad? Did you succeed or fail while living through those activities? How/why?
2. Do you ever think about the above topic—i.e. your life's good and bad history, and your future, etc.? If "yes," how often? If "no," why not?
3. Think about your future life:
   i. What do want?
   ii. List at least 5-10 **realistic** and ethical ways of achieving the things you want.
4. Optional: Continue journaling/creative activity 1 as advised at the end of chapter one above.
5. Introduction to journaling activity 2: Keep a journal for the next week—from today (e.g. Monday/Tuesday/Wednesday etc.,) to the same day next week. Write as much as you can about your thoughts, your daily life, etc. Read ahead to next week's review activity to help you write good journal entries.
6. Continue activity 6 from chapter one, "Comparison with other types of communication."
7. Optional: compose a short-story scene—similar to the one in the CAO above and based on your own life, and share it with the group.

REFERENCES

Alt, D., & Raichel, N. (2020). Reflective journaling and metacognitive awareness: Insights from a longitudinal study in higher education. *Reflective Practice, 21*(2), 145–158. doi:10.1080/14623943.2020.1716708

Henter, R., & Indreica, E. S. (2014, May). Reflective journal writing as a metacognitive tool. AFASES International Conference.

Ramadhanti, D., Ghazali, A., Hasanah, M., Harsiati, T., & Yanda, D. (2020). The use of reflective journal as a tool for monitoring of metacognition growth in writing. *International Journal of Emerging Technologies in Learning, 15*(11), 162–187. doi:10.3991/ijet.v15i11.11939

Sekalala, S. (2007, January). *Mazimhaka* [He Who Resolves Disputes] [Unpublished short novel]. Undergraduate Capstone Seminar Thesis.

ADDITIONAL READING

Beers, C. W. (1907). *A mind that found itself.* Longmans, Green and Co. doi:10.1037/10534-000

KEY TERMS AND DEFINITIONS

Blogging: The regular upload of new entries or posts—often written in an informal style—to a blog, a personal or small-organization website designed for that purpose.

Journaling: Regularly writing in a personal journal for purposes of personal self-reflection and record-keeping.

ENDNOTES

[1] *Transcription-Notation Meanings:*
>> *Chapter Two Notation 1 Meaning*: [](^) = Added during transcription, July 2021.
>> *Chapter Two Notation 2 Meaning*: (X) = Stricken-through due to my failure to comprehend my then-meaning.
>> *Chapters Three to Seven Notation 1 Meaning*: (^) = Verbatim quote of voice-note transcript.

[2] Later—around 2007, Derek unexpectedly died one night in his sleep, in his freshman year of college in Michigan, USA, where he had earned a full-ride soccer scholarship; apparently, he had an undiagnosed heart condition.

[3] Today (July 2021), I do not recall what this word means. I believe it might have referred to a sheet of writing paper.

[4] Please note that I added this underlining for emphasis—i.e., for this and other upcoming particular words and phrases—at the time of writing the journal entries.

5   Despite this description, it should be noted that students who have passed O'level exams—without completing the A-levels—from this British-originated system, can join universities in the USA, as I did.

6   Please refer to the section ("2.2.-A.2: Entries I") introduction/overview above for context.

7   Names deleted to protect my family-members' privacy.

8   Note the use of urban-American grammar ("no one will say nothing,") and influence of the hip-hop and R'n'B musical-lyrics style in the 3rd- and 2nd-last paragraphs.

9   Please refer to the section ("2.2.-A.2: Entries I") introduction/overview above for context, vis-à-vis deep thinking.

10   Taibah International School had just been launched a year prior (i.e., in 2000) before my arrival in my junior year (2001). In 2002—my senior year, more students joined us.

11   Isaac is Arnold and Andy's older brother, and we had attended primary school together. He'd joined the (primary) school in 3rd grade, a year after his father—one of my maternal uncles, Charles Rwabahizi—died.

12   Please see 2.2.-A.3 subset introduction above for context.

13   Based on my own life-story, written in my senior-year—2008—of the B.A. English-Writing program (Kean U.).

Chapter 3
"Why Did I Come to the Kitchen?":
Intrapersonal Communication/ Cognition, Reflexivity, and Society – Part I: Early Findings

ABSTRACT

This chapter introduces readers to the findings of the voice-note/transcript data-analysis results. The findings are organized according to four overarching themes, namely 1) "Types of Self-Communication (SC)," 2) "SC and Learning," 3) "Reality and Perception-Processing I: Without Metacognition," and 4) "Reality And Perception-Processing II: With Metacognition." The numerous SC instances garnered prove that individuals are indeed capable of studying their own SC. And in addition to demonstrating the above topics, the author strives to highlight the fact that the processes of SC, as well as the development and maintenance of the human self, are all necessarily influenced by the societies in which we reside.

INTRODUCTION

Format of Chapters Three to Seven

As I mention in chapter one, I began recording voice-notes from around early November of 2020, through late March of 2021, in an attempt to execute a fairly detailed exploration of my own mind's thinking, and intrapersonal-/self-communication (SC) processes. Thus, the next five chapters (i.e., three to seven)—in which I present the book's research findings—can be described using a metaphor of a journey through two parallel metaphysical lanes. One lane traverses through time—i.e., mostly through the above-mentioned five months (but also including flashbacks and reflections about the future) of my own life, as well as the society(/societies) I inhabit, albeit also through the lens of my mind's perception.

DOI: 10.4018/978-1-7998-7507-9.ch003

What was I experiencing and thinking throughout this time period, which included a global event that can be described as a watershed event, i.e., the COVID-19 pandemic?

The other lane traverses haphazardly throughout the virtual and/or brain frontal-lobe-based contents, and habits and/or systems, of my mind. On a day-to-day basis and overall, how does my mind function? How do I think and communicate with/to myself, and why do I do so in that manner?

In addition to attempting to answer these questions in a general fashion, I will also apply the analysis to the specific goal of improving my learning methods. It should also be noted that the findings of both the above metaphysical journey-lanes are consolidated and presented via dates and voice-note numbers (i.e., for the aforementioned voice-notes, recorded November-2020 to March-2021).

In addition to the above-described metaphorical journey, the findings of these five chapters are arranged under four overarching themes, namely, 1) "Types of Self-Communication (SC)," 2) "SC And Learning," 3) "Reality And Perception-Processing I: Without Metacognition," and 4) "Reality And Perception-Processing II: With Metacognition."

These six themes arose from my voice-note- and public-journaling-analysis via a grounded-theory-analysis procedure, i.e., building a familiarity with the data through immersion, and gradual discovery of potential insights via in-vivo and axial-coding analysis around middle-to-late winter and early spring, 2021 (e.g., Lindlof and Taylor, 2019).

Each of the following five chapters—i.e., three to seven—will follow a similar format to the one I use below, i.e., the five overarching themes introduced above, subdivided by topics. The topics are in turn subdivided by dates and voice-note numbers. Finally, readers should note that topics will often be cross-listed/discussed under more than one theme, and many (or even most) topics are cross-listed/discussed in multiple relevant chapters.

About Chapter Three (Cognition, Reflexivity, And Society—Part I)

My main rationale for including the topics (via dates and voice-note numbers) discussed in this chapter is mainly my evaluation of their significance in the context of how our thinking and SC processes are necessarily affected by, and connected to, the societies within which we reside. In other words, the contents of our thoughts and self-communications do not take place in a vacuum. Rather, they are borne by agents partaking in the wider social structure and network of fellow humans.

In fact, as sociologists correctly argue, without socialization as children and adults, we would not be capable of utilizing language and other symbol systems, neither would we be able to publicly and privately conduct ourselves while abiding by shared social standards, mores, customs, etc. (e.g., Little, 2014). Moreover, individuals constantly commune with each other as they partake in shared experiences of natural phenomena—e.g., weather/climate, and disease.

MAIN FOCUS OF THE CHAPTER

Types of Self-Communication (SC)

Nov 30th, 11:25 PM (Voice-Note No.: Batch 1; No. 1); Thagard (2005)'s Discussion of Metaphors as Teaching and Learning Tools— Also Classifiable under the "SC and Learning" Theme

The aforementioned journey began on this date, with what can be described—i.e., using another metaphor—as a dive into the deep-end of my exploration (of my mind's and SC processes). The impetus for this deep-end dive was the marked resonation I felt in response to an example by Paul Thagard (in his "*Mind*" text), of how metaphors can be useful in teaching and learning. First, Thagard presents a seemingly insolvable dilemma: how can physicians and radio therapists use X-rays to kill a tumor without also killing the surrounding cells? He then reveals that a research study revealed that students who were told of a "*dictator's fortress*" story-metaphor were able to answer the above question much faster than students who were not told the story/metaphor.

In the metaphorical story, the key to successfully attacking the dictator's fortress without getting killed by its surrounding landmines, was to attack the fortress using *small groups* of infantry and weapons. Similarly, short bursts of X-rays, as opposed to long uninterrupted bursts, can kill the cancerous cells without significantly destroying the healthy ones. This example is a good demonstration of a concept I kept grappling with throughout my grounded-theory analysis. How can one identify a true *potential insight*? As of early May—the tail-end of my analytical process, I am yet to satisfactorily answer that question. I'm afraid that what might seem insightful to me, might be considered obvious or uninspired by someone else!

Dec 11, 4:10 AM (Voice-Note No.: Batch 1; No. 3); Effects of Life's General Events/ Features on Learning—Also Classifiable under the "SC and Learning" Theme

Like millions of other individuals around the world, I caught the COVID-19 virus. Luckily, apart from the illness' significant but mild symptoms—i.e., flu-like symptoms, as well as loss of taste and smell, I fully recovered, and I did not spread it to my mum, who is elderly and with various comorbidities. I did however infect my sister, whom I was visiting in the Houston, TX metro area (and who takes care of our mother). I also missed out on some valuable quality time with my nephew and niece, as I had to self-isolate within my room in their house.

This experience made me realize an important fact, which I have hitherto only experienced via my general reading of journalistic and social-scientific analyses about educational (under-)achievements and challenges among children of lower socio-economic classes: inevitably, life's general routines and vagaries affect our learning processes. How was I supposed to enjoy my Barack Obama (2020) memoir ("*A Promised Land*") and Thomas Picketty's (2017) "*Capital*," or my John Grisham novel, etc., while dealing with severe nasal and chest congestion, weakness, headaches, etc.?!

Relatedly, I realized that our ongoing emotions, as well as our "non-learning-related" thoughts—i.e., both spontaneous, and/or in reaction to said routines and vagaries—also affect our learning processes. In fact, I realized that learning processes are part and parcel of general cognition, and random thoughts or memories can and often do distract us as we try to read or learn. In turn, these realizations inspired me to include my general fiction and non-fiction reading in my autoethnographic learning-process analyses, to enable a more holistic evaluation, in addition to my analysis of my learning with disciplines such as software-development and math, language-arts, etc.

Dec 15, 6:45 PM (Voice-Note No.: Batch 1; No. 4); Later Automatic-Recollection of Learned Facts—Also Classifiable under the "SC and Learning" Theme

Unlike other longer notes above and below, this note was brief, with a total recorded length of one minute and five seconds. And yet, this (very) brief length belies a potentially significant fact, namely, *"after learning/reading something, the concepts can and … often do come back to you spontaneously later—depending on the context, as you write, discuss, give a lecture, etc. (^)."* As of early May, I find myself grappling with this fact often. How/why does the knowledge we gain over time manifest itself in the forefront of our minds during such activities (writing, discussing, giving a lecture, etc.), yet we often do not realize that we're absorbing significant amounts of knowledge in our day-to-day readings/ learnings? One particularly salient example of this phenomenon is from the context of writing. Every so often, the polar opposite of "tip of the tongue syndrome" fortuitously happens to me as I am writing or speaking. How? I somehow manage to automatically retrieve and use words not only in their right context, but in ways that one can objectively judge as particularly lucid or cogent (unfortunately, other adjectives—e.g., "insightful"—might sound like bragging, which is not my intention)! Even without looking ahead in my transcripts and analysis-sheets, I am certain that this phenomenon will somehow feature again.

Jan 6, 1:06 AM (Voice-Note No.: Batch 1; No. 8); Obama's "A Promised Land;" General Types of Self-Communication Continued; "Why did I come to the Kitchen?"

Before the current era of my life, marked by a constellation of research, teaching, and other tasks, I used to be a much faster reader of fiction and nonfiction, compared to my reading rate today. But not that fast admittedly, as I often pause to savor and ponder what I'm reading. In any case, the new slow-reading style is definitely evident in the nonfiction and fiction books I've been reading since late November, i.e., Thomas Picketty's *"Capital in the 21ˢᵗ Century,"* and Barack Obama's *"A Promised Land."*

On this (Jan 6, Batch 1-8) recording, I reflected on the latter, enjoying the resonance of his disclosure of a character trait that I seem to share with him. In one of the earlier chapters of the memoir, Obama recounts that he *"was a lackluster high-school student. Then in college, he transformed; he would have 'internal self-dialogues,' often thinking deeply about topics such as justice,* [the] *American 'dream'/ democracy, freedom, etc. 'He was living a lot, or was too much in his mind'; his friends told him he needed to loosen up. (^)."* As I mention at the beginning of this paragraph, I happen to share that character trait, as evidenced by some of the same feedback I've been given by family-members and friends over the years. In fact, one particularly memorable instance of such feedback came from my older sister, during the aforementioned visit to her house in Spring, TX. And although I am certain that very many fellow humans around the world have this similar trait, I wonder if it happens to mostly "afflict" those of us with cerebral proclivities (again, I hope I am not coming off as a braggart!).

In this voice-note, I also reflected on another common SC process or habit, also in the context of literacy. In essence, discussing points, facts, arguments, etc., via writing—using both our own and others' evaluations or opinions—is arguably a form of SC, I surmised. Moreover:

Part of my own method for preparing or executing the above tasks [i.e., writing using my own and others' opinions, evaluations, etc.], involves the use of elaborate electronic-filing systems—"folders within folders within folders." Because of this, I'm often forced to try to figure out later: "What did I mean by this? Why did I save this file within the subfolder in which I saved it?" (^)

I also reflected on another similar and rather comical mode of SC, which unfolds via our forgetfulness/distractedness with executing various tasks: *"Why did I come to the kitchen?"*

Jan 7, 6:11 AM (Voice-Note No.: Batch 2; 2.1.3); 1—Paul Thaggard's "Mind" (2005) Text: Images, 2—Learning-Key/Turning-Points, and 3—Role of Emotions in Reading and Understanding Fiction and Non-Fiction Stories—Also Classifiable under the "SC and Learning" Theme

As one of the voice-notes of longer length in all batches, this note featured a number of various topics. Of particular importance was my discussion of Thagard's synthesis/explication of research about images as one of the key ways in which our minds process information. I pointed out that images were significantly more salient to me than the other modes he discusses in his text (i.e., logic, rules, concepts, analogies, and connections). For instance, images were very useful in my understanding of the mathematical concept of fractions. If teachers had not used diagrams to demonstrate what halves, thirds, quarters, and other fractions *looked like*, it would have taken me much longer to understand the concept. But paradoxically, whereas images are indeed very useful to my learning processes in many contexts, they can also present a hindrance to my learning or comprehension in other contexts. For instance, while reading Ken Follet's historical-fiction novels, I often struggle to visualize the villages, towns, and other landscapes, as well as the various human-created artifacts—including the architecture, clothing, etc.—that he describes.

The other two points I discuss in this voice-note are also related to cognitive phenomena that help me considerably in my learning processes. The first phenomenon in this context is the concept of key/turning-points in my learning processes. By key/turning-points, I am referring to particular junctures in books, topics, classes, etc., that somehow holistically crystalize or clarify a big part of the facts or aspects in said learning or reading materials, which were hitherto hindering my learning.

For instance, I encountered one such key/turning-point in chapter six, table 6.10, of the textbook about quantitative research methods for communication studies by Wrench et. al. (2019, p. 188). That table summarizes the various appropriate statistical tests that researchers should use, depending on the variable levels of their data. It breaks down the variable types into three categories, i.e., nominal, ordinal, and interval-ratio; research goals as either measures of differences or relationships among variables (i.e., two groups or variables, or more than two groups or variables); and the corresponding appropriate test. For instance: for measuring relationships between two interval/ratio variables, one should use correlation; and for relationships between more than two interval/ratio variables, one should use regression.

In my opinion, such key/turning-points are vital, especially for individuals such as instructors who have to really understand their subject-matter before teaching it. And in those *"Eureka* (!)" moments when we say to ourselves, "Aha, so that's what that means (!)," we're quintessentially communicating with ourselves indeed.

The last cognitive phenomenon I discuss in this note, which I believe is also useful in our learning process—particularly, in reading-comprehension of fiction and nonfiction—is that of emotions. After all, I highly doubt that many of us would enjoy reading, if it did not trigger the "artificial"—so to speak—emotions such as sadness, happiness, anxiety, etc., that it does, as we read about battle scenes, romantic-relationship development and dissolution, etc.

Jan 8, 4:24 AM (Voice-Note No.: Batch 2; 2.1.4); "To-Do" Lists

Returning to the idea or concept of insightfulness, again, I readily concede that what *I* find insightful might be considered unremarkable or pedestrian by many others. With that said, I have to earnestly note that I for one find the concept and practical use of a *checklist* as quite revolutionary, in a world as complex and unpredictable such as ours. And in this regard, I can count on at least one ideational counterpart-author, namely, Dr. Atul Gawande, who also marvels at the effectiveness of checklists in his 2009 book (Gawande, 2009). But beyond their remarkable utility, what—in essence—is a checklist? My response: a checklist is a form of SC, in which the present self makes an attempt to tell or remind the future self—and often, other individuals as well, to do something(s).

Jan 10, 12:38 AM (Voice-Note No.: *Batch 2; 2.1.5*); General Types of SC

Unlike my certainty vis-à-vis "to-do"(/check-)lists' status as a form of SC, I was much more conflicted about other forms, and the (sub-)topics of this voice-note demonstrate that uncertainty. But first, in an arguably apropos demonstration of the effectiveness of checklists and discipline vis-à-vis routines, I noted editorially that I was behind my goal vis-à-vis recordings. This was a result of my unpredictable life-patterns—i.e., via the volatility of my schedule-adherence with sleep, nutrition, work, and other general activities and tasks. And in large part, this volatility was caused by the absence of a teaching or other *structured* work agenda. After all, no one—e.g., students, a department chair or dean, etc.—was going to berate me for missing my daily recordings or data-analysis procedures!

As for the debatable forms of SC, I wondered in this note whether distractions—i.e., during conversations at home or work, or during class, etc.—were in fact a form of SC. And in this instance, I overruled myself; no, I do not think they (distractions) really count as a form of SC. But I seem to have decided to keep thinking about the question, as in this same note, I also decided to do an experiment on my *self* or mind, the next time I was scheduled to meet with my software-development coach; "*keep a tally of unrelated distractive thoughts (^).*" In any case, my self-debate in this note seems to have substantively contributed to the hypothesis that I later developed and currently present in chapter one's diagram 1.2 (Intra-interpersonal Cognition & Communication Spectrum) via two observations, the first of which was: "*There are thoughts in general [in our minds], but every time you're thinking,* [that instance] *does not necessarily equate to you communicating with yourself(^)*!" (Please see farther below for the second related observation.)

Be that as it may, I realized at least three definitive forms of SC, namely: 1) self-encouragement, 2) self-questioning, and 3) acknowledgement of, and/or exclamation in reaction to corporal states. The first form of SC listed above, self-encouragement, is common in challenging situations. One vivid example I recall is of one of my nephews many years ago in Uganda, as he braced himself and indeed performed a form of self-encouragement, moments before having to swallow bitter anti-malaria pills: "*Come on Carlos* [you can do this!]," he said (and it worked; he swallowed the pills without regurgitating them).

The second form of SC is somewhat antithetical to the latter, as it involves self-doubt, or at least, innocent self-questioning: "*I wonder if I can do that (^)?*" I then explicate the point using a conceptual framework in which I delineate a primary and secondary set of selves. *"I—[/Seif/[Your Name] the primary self, wonder if I—[/Seif/[Your Name]—secondary self-] can do that(^)?"*

The third form of SC listed above—acknowledgement of, and/or exclamation in reaction to corporal states—is a basic and often automatic or reflexive instance of communication, easy to take for granted, but vital in the betterment of our health. This communication-instance is exemplified by the phrase: *"Oh my goodness, my head hurts!"* In other words, often—in reaction to pain or discomfort, or extreme pleasure, etc.—our brains relay these feelings into our sentience, enabling us to adjust to our environments or situations, or get help accordingly, among other utilities.

Finally, I conclude the voice-note with the second of the two aforementioned observations (which evolved into the intra-interpersonal cognition & communication spectrum):

"Any time you're awake, the question is not *if* you communicate with yourself, at least a certain % of the time; the question is *how [or when]*...any time you're awake, *you will* communicate with yourself [in various situations] (^);" edited for clarity."

Jan 10, 12:57 AM (Voice-Note No.: *Batch 2; 2.1.6*); other types of SC; "Tip of the Tongue" and Meta-Analysis; Types of Thoughts; Distractions in your Own Mind during Conversations or Other Interactions

This voice-note, with a recorded length of 13 minutes and thirty-eight seconds, featured a total of six (sub-)topics. First, I discussed the annoying nature of *"tip-of-the-tongue"* syndrome (e.g., Brown, 1991), pointing out that in those instances, *"if you're (your primary self is) referencing your secondary self, as in this instance, then that is SC (^)."* I also pondered the practice of meta-analysis during events such as job-interview or work-presentation rehearsals, observing that a question to yourself such as "how should I answer this question (?)" is clearly an SC instance.

However, I returned to my debate—as introduced under voice-note no. batch 2, 2.1.5 above, continuing to clarify—in my own mind—the difference between SC instances, versus general cognition:

"Every thought is not an instance of SC—i.e., intrapersonal comm., but every thought referencing your secondary self is obviously a thought." That is an important distinction; thinking in general is not necessarily intrapersonal-comm (^).

And I explored the distinction between SC and general cognition further: what about missing a loved one (?), I asked myself. Verdict: *"not necessarily instances of intrapersonal-comm. (^)"*

The next (sub-)topic in this voice-note provides a potentially useful set of examples of general cognition vs. SC. Apparently, I had proceeded with the self-experiment I promised myself earlier (as stated above, under voice-note no. batch 2, 2.1.5):

Two specific instances of distraction from [the] previous day's software-engineering coaching session: 1) Thinking about [an] upcoming job-interview, and 2) Thinking that I had to write an email to someone—"I should remind myself to do that [later]...(^).

I submit that in this text-fragment, instance "1)" is a typical example of general cognition, whereas instance "2)" is one of SC. The voice-note then ends with an application of SC to learning, with a question I had asked myself earlier (I also wrote about it on my public journal page): *"I—[the primary self-] wonder if I—[/Seif/—secondary self-] can write poetry?! (^)"*

Jan 10, 11:21 PM (Voice-Note No.: *Batch 2; 2.1.7*); Journaling and Relation between Intra- and Inter-Personal-Comm.; Emotions as Intrapersonal-Comm. Instances, including Guilt

Here, I returned to the concept and practice of "to-do" lists, comparing them to journal entries. I observed that whereas "to-do" lists are instances of SC intended for our future selves, journaling is an SC instance of recollection from the past. However, I noted one other key distinction: *"… journaling is also a way for us to preserve our life events for posterity, i.e./e.g. after we die; you're both talking to yourself in the journal, and potentially talking to others in the future. (^)."* I also returned to the examination of the essence of SC: *Many times, when we're talking to other people, we're also talking to ourselves; e.g.* [, as demonstrated by] *the common dialogue line in TV/movie dramas: "who are you trying to convince here—me, or yourself? (^)"* And again, I kept fleshing out the hypothesis that I eventually managed to present in chapter one's diagram 1.2:

Intrapersonal-comm. can lead to interpersonal-comm., and vice-versa. In fact, perhaps we're always talking to ourselves, vis-à-vis arranging and screening our thoughts before divulging them? (^)

The internal dialogue seems to continue via a question and answer, from and to my *self*: okay, that (intra-interpersonal communication and cognition hypothesis) might be true indeed; but what about emotions? And whereas I did not provide a definitive or comprehensive answer, I did tackle the particular emotion of guilt. Essentially, a feeling of guilt is an implicit statement that can be expressed as: *"I am a bad person" for having done that bad thing* (^). And thus, I conclude that: *you do not necessarily have to use those or other words; the mere emotion is a de facto intrapersonal-comm.*

Jan 13, 11:34 PM (Voice-Note No.: Batch 2; 2.1.10); Types of Self-Communication, Including Metacognition during Interpersonal Dialogues, Longterm-Planning and Procrastination, and Meditation

Three days after the last voice-note above (Batch 2; 2.1.7), I granted myself an opportune situation with which to somewhat test my new intra-interpersonal communication and cognition (IICC) hypothesis, with the current voice-note. In a phone-conversation, I enjoyed a riveting story by an older African-American gentleman to whom I'll simply refer as Mr. D.E., an old retired mentor of mine from Kean University. Mr. D.E. is in his mid-to-late 70s, and was raised in the Jim Crow-era US-Southern region. In that conversation, I observed my cognition and speaking process.

As Mr. D.E. told me a story about his childhood and his father's careers as a farmer and self-employed cab-driver (as well as their neighbors, who were sharecroppers), *questions kept popping up in my mind in reaction to things he was saying—i.e., follow-up questions* (^). I also realized that outside of contexts such as this book's research, *we meta-analyze our thoughts all the time without realizing it* (^). I reiterated the latter observation with two common instances of SC, namely: *1)--ordinary (non-academic/ research-based) metacognition, 2)--clear and vague emotions, e.g. guilt* (^).

Eventually, the rhetorical pendulum swung again to the opposite direction. Despite my gradually crystalizing ICC hypothesis, I realized that the definition of SC is in fact not a simple "black and white" issue. For instance, *despite to-do lists, I* [often] *experience both clear and **vague** emotions of hesitation, re: thinking about upcoming tasks* (^). This might beg the question: if the "to-do" list is a clear instance of SC, can/should said "clear and **vague** emotions of hesitation" be classified as resultant SC instances, or rather, as instances of general cognition? Other situations—i.e., involving our minds—complicate this question further; one such situation is the practice of meditation: *"If you catch your thoughts wandering as you meditate, is that an instance of intrapersonal-communication? (^)."*

Self-Communication (SC) And Learning

Ultimately, there are a total of four instances noted above—i.e., 3.2.1, 3.2.2, 3.2.3, and 3.2.5—that can be cross-listed under the current theme ("SC and Learning") as well. Otherwise, my analysis of the transcript portion I utilized for this chapter did not suggest any other obvious—or (to me at least,) convincing—instances of voice-notes/transcript-entries relatable to the current theme (of "SC and Learning") in their own right. Instead, I was able to pinpoint and discuss some other relevant instances in this context in the upcoming chapters.

Reality and Perception-Processing I: Without Metacognition

Jan 10, 12:57 AM (Voice-Note No.: Batch 2; 2.1.6); Other Types of SC; Tip of the Tongue and Meta-Analysis; Types of Thoughts; Distractions in your own Mind during Conversations or other Interactions

This note is cross-listed/discussed under the "Types of SC" theme above as well. Its inclusion under the current theme is chiefly due to the non-self-conscious nature of the thoughts being discussed. Unless an individual is working on research similar to this one, many of us do not really care enough to philosophize about the cognitive-linguistic and/or other nature(s) of "tip-of-the-tongue" syndrome; we just know that it's a very annoying feeling[1]! Ditto the instances of SC during job-interview or work-presentation rehearsals, or of strong feelings of missing (a) loved one(s); *"who cares whether/how these are all instances of 'self-communication'?!"* And yet, in the antonymous counterpart of this theme farther below, "Reality And Perception-Processing II: With Metacognition (RP-II)," various individuals—regardless of whether or not they are communication or cognitive-science scholars, express wonderment to themselves or to others about the fascinating nature of some SC instances. Example (of RP-II): our realization of, and wonderment about, the mind's unrelenting nature, including during our meditation sessions (please refer to section "3.5.1," Voice-Note No.: Batch 2; 2.1.2, Jan 6, 10:48 PM, for a relevant example)!

Reality and Perception-Processing II: With Metacognition

Dec 11, 4:10 AM (Voice-Note No.: Batch 1; No. 3); Effects of Life's General Events/Features on Learning

This topic (from its attendant voice-note), is also cross-listed/discussed farther above in this chapter, under the "Types of SC" theme. First, a brief recap of its relevant subtopics or elements might be helpful:

1) realization of life's vagaries on our learning processes, 2) similar impact of emotions, 3) reflection of learning processes as part and parcel of general cognition, and 4) distractions during learning/reading.

I deem the subtopics as relevant to the current theme, given the fact that they involve a considerable amount of what can be referred to as metacognition or mindful cognition or reflection. I made the above realizations deliberately, as opposed to a day-dream manner of thinking, i.e., thinking about something without realizing that you are indeed thinking about said thing, which is how I was able to record them for memorialization.

Jan 6, 10:48 PM (Voice-Note No.: Batch 2; 2.1.2); Thinking during Meditation

I believe the best way for me to convey this note's message might be a wholesale copy-paste from its respective transcript:

---I realized something during or shortly after meditation: it's hard to "quiet" the mind, to avoid thinking.

---Analogy: the way our minds work in that regard—always thinking even when we're trying to meditate, might be similar to a computer opening various random files or performing various other random functions, even though you're taking a break and are not using said computer [but the computer is switched on, vs. "asleep" or shut down] (^).

CHAPTER CONCLUSION

In this chapter, readers are introduced in earnest to the subtopics that I use to discuss the instances of SC that are derived from the bigger data-set of the book's research, i.e., the recorded voice-notes and their attendant transcripts. Those subtopics are grouped under the following categories: 1) "Types of SC," 2) "SC And Learning," 3) "Reality And Perception-Processing I: Without Metacognition," and 4) "Reality And Perception-Processing II: With Metacognition." Various specific instances are discussed under each of the above subtopic categories, proving that principals are indeed capable of studying their own SC (at the very least, in a qualitative fashion).

Creativity-Attempt Outtake 3

Post 5; 02/08/21:
    Of Poetry And Minds

Theme 1: Can *I* Write Poetry?
Another week has come and gone,
and in yet another we toil.
The world hums on;
Tears, happiness, fullness and nothingness.
Time is both our best friend, and our worst enemy,
as are our very minds,
from whence we compose both boring and/or stirring prose,

and boring and/or stirring poetry, alike!
And as the world and time hum on yet,
new minds spring forth,
and some old fall yonder;
another week has come and gone.
–SS.
My Self-Awarded Grade-Range For Above Poem: C+ — B- [☺]

I've been meaning for a while to write one or more entries on my experimental learning website-journal—i.e. "Aspiring Polymath," about poetry. Three themes in particular are on the forefront of my mind about the topic/subject of poetry: 1)–How is it done/learned, and can *I* learn and do it well? 2)–The second theme is very abstract and is hard to articulate, but it is connected to the latter theme about the essence of the art of poetry; and an *art* indeed it is; but is it also a science of sorts? 3)–Is there a clearly identifiable midpoint between poetry and prose? I should point out in earnest by the way, that these thoughts were on my mind long before the celebrated performance by Amanda Gorman at the President Biden's inauguration.

How or why exactly the topic (of poetry) came to my mind when it did, I cannot remember or explain. One truth I'm certain of however, is this: arguably much better than prosaic journal entries, poems—written by and for our amateur selves, can really help us in many respects vis-à-vis collecting our thoughts, meditating and reflecting, and even planning, without the neuroticism that to-do lists can trigger if misused, as they often are.

Also On My Mind, This Week:

- ◦ A rather funny(?)/quirky story: I shared one of my productivity hacks—"the promise, with a self-imposed deadline" (i.e., promising someone[s] to whom you owe a favor or obligation that you will do whatever it is you owe them by a certain date)—with some colleagues, right after using that very hack on them! As in, I made "the promise," then told them about how effective "the promise" tends to be! And oh by the way, it worked; I delivered on my promise to said colleagues, before my self-imposed deadline.
- ◦ I'm continually grappling with the concept of recursion and self-reference, in relation to my upcoming book's research. The question is not if self-reference and meta-cognition/analysis are effective research tools. Rather, the question is **how** can/should we effectively make use of them?
- ◦ Fluidity of mind-based, pre-verbalized concepts: E.g., the/a connection—or lack thereof—between the above two points. I happen to have written the outline of this journal-entry sometime late last week, sans the poetry part (although as I note in the first paragraph above, it's also been on my mind for weeks or months). And the question that arose clearly in the forefront of my mind after writing the above two points on the outline, was the first sentence of the current paragraph/point: is there a connection between the above two points? If "yes," what is it, and how did it arise and help in the composition of the next point(s)? And if "no," how/why do our minds produce such random, ordered, or other (e.g., a combination of random and ordered) thoughts that come into clear focus in said minds?

    ◦   Links between professional failure or success, excellence or giftedness, and regardless, the need for benefactors to give you a helping hand. For many reasons, I tend to intensely dislike cliché advice or boasts about "working hard." Example: >> Qn.: "How can someone succeed in your ["fill in the blank"] industry?" >> Ans.: "By working hard." Like gee, "thanks, captain obvious!" As I mentioned in an older entry, I've realized over time that no matter how gifted or hardworking you are, you'll more-than-likely need a helping hand somehow, somewhere along your professional (and life's general) journey. And I for one will never cease to be grateful to the folks who've helped me get to where I am. I also hope I'll continue to encounter such kind souls. Perhaps most important, I really hope to pay it forward in the future.

Post 6; 02-15-2021:
Of Thoughts, Work, And Pure Existence

Themes/Sub-Topics:
    ◦   Highs and lows of life
    ◦   Use of work to navigate them
    ◦   Ability of thoughts to attend to all our lives' time-periods

To win and to lose, persevere, and hoove,
From platters with 'bundance or crumbs, we choose,
And in toil we march, and for better odds we hope,

…

[Uuurrrgh! I hate this damn "writers' block"! **Sigh** I guess I'll finish this poem–an attempt at iambic-pentameter–later…]

Apparently, I'm starting this week with neither a tabula rasa, nor full mind. Instead, I'm attending to my mind with caution—in two senses that is; with care, and to warn myself: "Beware, self!" I seem to be saying. "Do not get so caught up in all the 'business' and daily struggles!"

Instead, I should pledge perhaps to take solace in my work, no matter the odds of success as they might seem in the present moment. Besides, instead of overworking that mind-machine, should not we both work with it reasonably, but also behold its magic? For yes, that is almost truly what it is—magic. We can live in the past, or in the future, or two or all tenses/times at once.

But in my humble opinion, in concurrence with the virtues of mindfulness—so to speak, the present is unrivaled in its power. Moreover, that power can be multiplied through earnest efforts to behold with suspended judgement. No past, present, or future moments are "good," "bad," "sad," "happy," etc.

Time simply exists, and we exist in it. And we can choose to label its pieces—those moments from which our "lives" arise, as we know them—or simply savor them. And I'll admit, I for one keep grappling with the seemingly paradoxical face of mindfulness. It often feels impossible to savor without labelling. And how can you tell a proud parent of a smiling gurgling baby not to feel happy in that moment, as that baby smiles at them?

In any case, work, by default of its very nature, can surely help us to "simply be." Perhaps we should not strive to enjoy our work (?); perhaps we should simply attend to it carefully, to the best of our abilities, no matter how grueling or fun?

I do not know. I am choosing to simply "be," indeed. Wishing you all a happy work-week! ☺

But alas, therein, in those thoughts,
Lies our foe. That unending and most unyielding of monsters, we brawl.
So behold, and be; simply let be.

Self-Questioning for Creativity-Attempt Outtake 3

State any necessary introductory remarks, and/or answer this question: does it make sense to me now—i.e., in the present—vs. then—i.e., when I wrote it?

Yes, the posts in this C.A.O make sense to me now, overall.

Does it resonate with me now—i.e., in the present—vs. then—i.e., when I wrote it?

Yes, it does. Apparently, I seem to be making an earnest effort vis-à-vis creativity—particularly, via the art of poetry. I also seem to be attempting a practice of mindfulness via reflective writing.

What mood and/or state of mind does the tone suggest?

I seem to be in at least two rather competing states of mind: 1) thinking—or trying to think—hard and/or creatively, and 2) trying to find inner peace and/or balance.

What is/are the direct and indirect relation(/s) to the current chapter via topic(s) and themes?

The posts in this C.A.O seem to be mostly relatable to the sub-topics (/themes) of "SC and Learning," and "Reality And Perception-Processing II: With Metacognition."

Any other evaluations or comments?

N/A

Activities and Discussion Questions

1) Review of journaling activity 2. What can you reveal—i.e. only things that will not embarrass or harm you and/or others—about your life, and how you relate to yourself^ and to others (family, friends, schoolmates, etc.)? (^*Relating to yourself; e.g.: are you a loner, or do you mostly thrive when around people? When alone, what do you do most often?*)
2) From the "Johari Window" (Luft and Ingham, 1977) analytical instrument; between now and next week's meeting, compose a categorized list of traits:
    i.   You do not know about yourself and neither do others,
    ii.   That others might know about you, but you do not (know them about yourself),
    iii. That you know and others know,

iv.   **Only you** know, and others definitely do not know about yourself.

3)   Optional: Continue journaling/creative activity 1 as advised at the end of chapter one above.

4)   Continue activity 6 from chapter one, "Comparison with other types of communication."

REFERENCES

Brown, A. S. (1991). A review of the tip-of-the-tongue experience. *Psychological Bulletin, 109*(2), 204–223. doi:10.1037/0033-2909.109.2.204 PMID:2034750

Gawande, A. (2016). *The checklist manifesto: How to get things right.* Academic Press.

Lindlof, T. R., & Taylor, B. C. (2019). *Qualitative communication research methods.* Academic Press.

Little, W. (2014). *Introduction to Sociology.* B. C. Campus. https://opentextbc.ca/introductiontosociology/

Luft, J., Ingham, J., Hall, J., & Telometrics International. (1977). *Johari window.* Behavioral Science Enterprise.

Obama, B. (2020). *A Promised Land.* Penguin Books.

Piketty, T., & Goldhammer, A. (2017). Capital in the Twenty-First Century (Reprint ed.). Belknap Press.

Thagard, P. (2005). *Mind: Introduction to cognitive science.* MIT Press.

Wrench, J. S., Thomas-Maddox, C., Richmond, V. P., & McCroskey, J. C. (2019). *Quantitative research methods for communication: A hands-on approach.* Academic Press.

KEY TERMS AND DEFINITIONS

Reality and Perception-Processing I; Without Metacognition: Intrapersonal-/self-communication executed or utilized by an individual to process the inputs of their senses and/or their thoughts, without realizing that s/he's in fact undertaking that process.

Reality and Perception-Processing II; With Metacognition: Intrapersonal-/self-communication executed or utilized by an individual to process the inputs of their senses and/or their thoughts, while aware that s/he's undertaking that process.

Self-(/Intrapersonal-)Communication: Communication inside an individual's mind, or outside— e.g., spoken or written—but for that same individual's consumption and/or use.

ENDNOTE

[1]   ***May 14, 2021**: As I wrote my explications of previous voice-notes yesterday, I started experiencing this phenomenon, and eventually—grudgingly—gave up my self-torturous mind-search of the word I was looking for. I knew that it's a word related to **feeling something deeply, in an almost primal way**. I could even remember an instance within the past few months in which I'd used it, to describe the horrendous and callous manner in which George Floyd was killed. And for some reason, I thought the word had a letter "g" in it. Thankfully, as of **a few minutes ago* (approx. 1 AM)**, I was able to successfully sleuth for the word via online dictionaries. The word (?): **<u>visceral(ly)</u>**.

Chapter 4
"What's That Word?! It's on the Tip of My Tongue!":
Intrapersonal Communication/Cognition, Language, and Symbol-Systems

ABSTRACT

This chapter continues the study of the various types of SC instances that human minds produce via the four categories introduced in the previous chapter: 1) "Types of Self-Communication (SC)," 2) "SC and Learning," 3) "Reality and Perception-Processing I: Without Metacognition," and 4) "Reality and Perception-Processing II: With Metacognition." In addition, the chapter examines the role of language and symbol systems vis-à-vis our SC, as well as communication in general. In this regard, the dynamics of reading and writing offer a number of potential insights, and the author examines them via examples involving the first part of former President Barack Obama's presidential memoirs, journaling and blogging, and the use of "to-do" lists for task-completion.

INTRODUCTION

One of the most remarkable traits—possibly *the most remarkable*—of the genius of human beings' socialization is the use of language and symbol-systems. And of course, without language and symbol-systems, communication—i.e., both intra- and inter-personal—would be nearly impossible. Overall, the link between these two concepts—i.e., socialization and communication—is worthy of close(r) study. And without getting into a "chicken vs. egg" origin-story conundrum, I can also observe that socialization plays a big role in the formation of the symbol- and meaning-systems with which we make sense of the world. And yet, each of our minds' complex thought processes define each of our own individual sensemaking. By "complex thought processes," I include the infinite number of thoughts that one mind can produce, per Porpora's explications (Porpora, 2013).

DOI: 10.4018/978-1-7998-7507-9.ch004

The above observations might be helpful as one delves into a typological analysis of his/her/their thought- and SC-processes, and an ontological and epistemological explication of said thoughts' characteristics vis-à-vis the themes below. What are some of the possible insights we might be able to glean from our own individual thoughts—e.g., my own—categorizable under the six themes introduced earlier, that are particularly relevant to the concept of symbol systems in general, and language in particular?

MAIN FOCUS OF THE CHAPTER

Types of Self-Communication (SC)

Nov 30th, 11:40 PM (Voice-Note No.: *Batch 1; No. 2*); Metaphors vs. Analogies versus Similes

This was one of the earliest voice-notes I recorded, and it is one of the briefest among all notes, with a recorded length of three minutes and fourteen seconds. Indeed, the transcript reflects this brevity, with only two succinct points:

- *I looked up the difference between those three concepts online*
- *"One can call an analogy a metaphor, but you cannot call...a metaphor an analogy..."* (^)

The curiosity that led to my lookup of the concepts resulted from my reading of the metaphors chapter in Thagard's text (e.g., refer to section 3.2.1: Nov 30th, 11:25 PM—Voice-Note No.: Batch 1; No. 1). Among other implications (e.g., in relation to reading-comprehension), the content of this note (metaphors vs. analogies vs. similes) is consequential in the current context vis-à-vis its implicit revelation of the complexity of human language and symbol-systems. For instance, one can add the concepts of "idioms" and "proverbs" to the words in question ("metaphors vs. analogies vs. similes") to demonstrate said complexity. And in turn, ultimately, this complexity—among other implications—is one piece of evidence of our brains'/minds' power and sophistication. For in the end, our use of language transcends mere basic description and argumentation, and allows us to verbally demonstrate and illustrate concepts (e.g., using other concepts), regardless of their complexity.

Dec 28, 2:33 AM (Voice-Note No.: *Batch 1; No. 5*); Barack Obama's *"A Promised Land"* and its Relation to the Thaggard (2005) Text

One can extend the above section (4.2.1)'s discussion using the current section's/voice-note's content. By the above title—i.e., " *'A Promised Land'* and its relation to the Thaggard text," I was specifically referring to the excerpt below (from chapter one, page five), which I related to Thagard's "Images" chapter (with the most relevant clause italicized and underlined by me for emphasis):

"With time, my walks down the colonnade would accumulate with memories; there were the big public events of course, announcements made before a phalanx of cameras, press conferences with foreign leaders. But there were also the moments few others saw—Malia and Sasha racing each other to greet me on a surprise afternoon visit, or our dogs, Bo and Sunny, bounding through the snow, their paws sinking so deep that their chins were bearded white. Tossing footballs on a bright fall day, or comforting an aide after a personal hardship.

Such images would often flash through my mind, interrupting whatever calculations were occupying me. They reminded me of time passing, sometimes filling me with longing—a desire to turn back the clock and begin again." (By Barack Obama in *A Promised Land*, Chapter One, Page 5.)

Even though the above excerpt is (apparently) primarily indicative of the power of images and memory, it also demonstrates the aforementioned power and sophistication of the human brain in the context of language and symbol-systems. After all, Obama is using language to paint a set of vivid pictures for the reader. Even more noteworthy for the purposes of this book, he was aided in the task of memorization in the first place by a communication tool or symbol system—language—that enables him to easily "file" away the memories (akin to the "sign and signified" semiotic process, e.g. as expressed by de Saussure, 1966).

Jan 5, 12:05 AM (Voice-Note No.: *Batch 1; No. 7*); Example of Intrapersonal Communication via "To-Do" Lists

The current voice-note then returns us to a simpler (SC-)process, demonstrating its ubiquity and utility. This particular example also makes me realize that it is easy to take SC for granted, despite said ubiquity and utility. Below is the full voice-note transcription entry, with a total of five points (with *emphasis* for the SC-instance):

- *I was writing a two-week to-do list*
- *While writing the dates, my mind wandered—reminiscing about something funny: a news parody show.*
- *Skipped a number of dates because of that distraction*
- *A few moments later, I realized my mistake, retraced my steps to figure out what I had done wrong and why*
- *I realized that "it was because I was not concentrating," and I actually said that to myself—via mumbling*

Jan 6, 3:39 PM (Voice-Note No.: *Batch 2; 2.1.1*); "Tip of the Tongue" Syndrome

Similar to the above ("'to-do' list") type, I also grappled with the current voice-note's phenomenon in the previous chapter, and it is evidently a quintessentially linguistic phenomenon. Looking ahead, I might also have to discuss it under the mental health chapter, considering the mildly-acute temporary neurosis it seems to trigger! For instance, on this occasion, I was dealing with another instance of the problem, and I kept repeating the syndrome's namesake expression. Below is the full transcript, with the episode's pleasant resolution:

- --- Another instance of intrapersonal-comm.: "tip of the tongue" syndrome: "What's that word?! It's on the tip of my tongue!"
- + Editorial note, added during transcription (03/10/21): The word is "zeitgeist"!

Jan 15, 8:51 PM (Voice-Note No.: *Batch 2; 2.1.11*); Suspending and Confirming Understanding—Part I (Suspending)

With a recorded time-length of 24:08, this particular voice-note is among the longer notes, with a total of four topics, one of which—suspending and confirming understanding—is the focus of this section. It should also be noted in earnest that this topic/discussion might be one of the more consequential ones among this book's research findings. While I cannot make the upcoming assertions with absolute certainty, I can provide a fairly strong supporting argument (at least to me, notwithstanding my bias). The assertion: I believe I started utilizing the concept and practice of suspending understanding sometime around the third grade, in my leisure-time reading of novels such as *"The Nancy Drew Files."* Given the stories' USA settings, pop-culture- and other unfamiliar references to me at that time, I—a 9-to-10-year old kid at the time in Uganda (East Africa)—could hardly relate to much of the author's and story-characters' meanings.

And yet despite that hurdle, I was really engrossed by the stories' plots! In hindsight, I realize now that I was only able to enjoy those stories by putting aside my incomprehension of vocabulary words, concepts, references, etc., that did not fit in with my level of knowledge at that time. Rather, I seem to have instead resolved to enjoy "the bigger picture," so to speak. For instance, I allowed my curiosity to be piqued by the big *"Whodunit (?!)"* question, and I allowed my mind and my emotions to be artificially upset while reading the scenes in which the protagonists were in danger.

To this day, I still make good use of that "suspending understanding" technique. For instance, once, a colleague in China asked me if it was really worth it to try reading a book such as *"Goder, Escher, Bach"* (Hofstadter, 2006), given the complexity of the concepts discussed therein. "Yes," I replied. You just have to *suspend* [some] *understanding*, knowing that the puzzle-pieces will fall into place later. Another related question worth considering in this context: can that technique (suspension of understanding) be utilized with academics as well, especially with notoriously difficult STEM and other (e.g., philosophy) subjects? Below is my response to this question, taken directly from the attendant voice-note's transcript:

--- *That method is harder with STEM* [and other] *subjects, whose topics often build on one another, with the need to master lower-level concepts before continuing onto the advanced concepts. E.g.:* [you] *need to understand numbers and counting before learning arithmetic—adding and subtracting, etc., and later, algebra, etc.*

--- *Still, "suspension of understanding" is possible even in STEM subjects: e.g., memorizing formulae and using them to solve problems without really understanding the logic behind those formulae. E.g., memorizing the need to divide both sides of a basic algebra equation, e.g. "2X = 10" using a common divisor on each side, until the number of the right cannot be divided further. In the above example, X = 5. Later, you can learn how the formulae really work; their logic, etc.* (^)

Jan 15, 8:51 PM (Voice-Note No.: *Batch 2; 2.1.11*); Suspending and Confirming Understanding—Part II (Confirming)

Related to the above-discussed technique is another one, namely, *confirming* understanding. At an appropriate time—which only the principal can decide for themselves, one has to also confirm understanding. Again, below is the relevant excerpt from the attendant transcript:

---   *"Confirming understanding": you should confirm that the concepts really mean what you think they mean, e.g. using online or other resources and books, etc. E.g., I often look up various vocabulary meanings while writing, as I find myself using specific words that I think fit the context, but I'm not sure. (^)*

The above assertion bears repeating, with a specific example. Readers of this book should join me in thanking the inventors of the internet in general, as well as sites such as Google and Wikipedia. Why? Because without those resources, this book might either not have been viable, or it would have taken me a much longer time to write it! For instance, in the current "sitting"—my own term for a writing or work session, I have "*Googled*" the following words/concepts' definitions:

- 1—*"Neurosis"*
- 2—*"Word meaning of 'the same name'"*
- 3—*"Another way to say I was so relieved"*
- 4—*"Whodunit"*

SC And Learning

Jan 16, 1:41 AM (Voice-Note No.: *Batch 2; 2.1.12*); Spectrum of Knowledgeability, from "Know Nothing" to Amateur, through Expert

This topic appears severally throughout the current chapter and others, given its versatile applicability. But before I launch its discussion in earnest, it might be helpful to my readers to see the first part of the attendant transcript portion in its raw form:

- *Realization I made about learning processes: knowing a subject **well/thoroughly**—e.g./i.e. as an expert—is different from having a limited amount of knowledge about said subject. Related: my upcoming reading of "30-Second Quantum Mechanics Theory." Reading and completing that book does not make me a quantum theory/mechanics expert! (^)*

Readers should note that a deliberate consideration of such epistemological issues might suggest that a spectrum of knowledgeability is possible with many (or most [?]) topics, regardless of academic discipline or knowledge domain. In other words, knowledge/education is not a simple proposition of whether an individual "IS or IS NOT educated." Rather, it is usually evident that the education level of an elementary-school dropout is different from that of a high-school or college dropout, in large part because of said dropouts' differing knowledge-levels of vocabulary and specialized symbol-systems.

With all this said, so what? How can this meta-knowledge help us? And again, it might be best to return to the raw transcript for the appropriate response:

- *Above point continued: before starting your learning-process of subject/topic X, it might be useful to ask yourself, "where on the 'knowledgeability spectrum'—so to speak, of this topic, am I?" (^)*

Jan 16, 1:41 AM (Voice-Note No.: *Batch 2; 2.1.12*); Use of
Visualization and Concept-Processing using Words while Reading
Fiction and Nonfiction, Academic, and other Genres

Unlike the preceding voice-note/section (4.3.1)'s deliberate examination of the general dynamics of the learning process, the current voice-note focuses on a specific "blow-by-blow account" of learning and cognition processes, depending on the context. The note starts off with an editorial/meta-analytic (sub-)note:

- *Editorial note, explicated at beginning of v-note: For the following months after this recording, from January to March (end-date of recordings), v-note topics will be related to the books I'm reading, as well as other things I'm learning, esp./e.g., software-engineering. (^)*

Thereafter, I proceed with the substantive content of the note, first listing the six general types of human-thought representation and computation, as synthesized by Paul Thagard:

- *Paul Thaggard's breakdown of cognitive-scientists' six general types of human-thought representation and computation, i.e.: logic, rules, concepts, analogies, images, and connections*

I then recap the objective of this book's research to crystalize it further in my mind, and to (at least *try* to) improve the quality of my analysis:

- *Objective of research: application of intrapersonal-communication concepts to the improvement of my own learning processes; essentially trying to figure out, "how do I think?" "How do I think about and/or process the content that I am reading?" Partly, through visualization; I have to visualize what I am reading*
- *E.g. of visualization; John Grisham's protagonist in the "Rogue Lawyer" novel, meeting in the [smoke-filled] diner with someone: how do I process that scene?*

After this focus on visualization and images, I scrutinize the role of abstract and/or non-evocative words in my leisure-reading or academic learning (with added <u>emphasis</u> on the most salient/relevant part, for the current chapter's focus):

- *What about words—what role do they play in my comprehension of the contents that I'm reading? I have to use them to decipher meanings, e.g. in academic reading contexts.*
- *Another way to ask the above question(s)—i.e. How do I think about and/or process the content that I am reading (?): As I'm reading these books, what am I telling myself? Answer: I check my understanding of the content, with fiction and non-fiction, I'll exclaim or laugh, etc. I also use visualization. With more prosaic or academic/noncreative books and learning, I'll use words to make meaning. "When [words] get into my brain, they dissolve into concepts—into abstract concepts." Thus, in my mind, I'm not visualizing words; rather, I'm processing the meanings of the concepts represented by the various words I'm reading.*

An appropriate follow-up question here might be, "*how exactly* do you 'process' the words?" And the best answer might be a neurology-based explanation: my brain's frontal lobe performs that task. Still, somewhat akin to a toddler who indefatigably asks "why," I am tempted to also keep asking "how?" *How* does the frontal lobe perform that task? And to the best of my current knowledge of cognitive science, that latter question is part of the metaphorical heart of the field. Ultimately, I submit that the language and symbol-system-related concepts of meaning, self-reference and meta-cognition, are all insightful tools with which we can abstractly sense and appreciate the metaphorical computing-power of our human brains.

Reality and Perception-Processing I: Without Metacognition

Jan 10, 11:21 PM (Voice-Note No.: *Batch 2; No. 2.1.7*); Journaling and relation between Intra- and Inter-Personal-Comm.; Emotions as Intrapersonal-Comm. Instances, including Guilt

This voice-note and its attendant transcript is labelled with the above date and time—in other words, the voice-note itself was recorded in that timeframe, but as of June (the 5th) of 2021, it resonates deeply with my current depressing situation. In this section (4.4.1), I will attempt both an explication of the voice-note in question, and concurrently, an application of some or most of its core analytical-tools on said situation. To achieve this dual goal, I discuss both the transcript, as well as two public-journal entries (taken from the same collection of the posts displayed in the chapter-end "Creativity-Attempt Outtake" sections). First, I will verbatimly re-transcribe both the voice-note transcript and the journal entries. I will then reflect on the above-listed topics/themes in relation to those two text-sets, and the overarching theme of "Reality And Perception-Processing (I) Without Metacognition."

A. Transcripts

Below are the two transcript-sets introduced above.

Voice-Note Transcript:
--- *"To-do" lists are a form of us talking to our future selves; "Seif, do not forget to do X, Y, Z* [later today, or tomorrow, etc.].*" Journaling is somewhat similar; you're also talking to yourself* [,recalling what your self did in the past]. *However, journaling is also a way for us to preserve our life events for posterity, i.e./e.g. after we die. You're both talking to yourself in the journal, and potentially talking to others in the future.*
--- Many times, when we're talking to other people, we're also talking to ourselves; e.g. the common line in TV/movie dramas: "who are you trying to convince here—me, or yourself?"
    + Intrapersonal-comm. can lead to interpersonal-comm., and vice-versa. In fact, perhaps we're always talking to ourselves, vis-à-vis arranging and screening our thoughts before divulging them?
--- *Emotions; particular case of guilt: "I am a bad person" for having done that bad thing. You do not necessarily have to use those or other words; the mere emotion is* a de facto *intrapersonal-comm* [-instance].

Public-Journal Entries:
Post 20; 05-25-21:
Of Change, Chaos, And Crucible Moments

Starting over the past weekend through the next few months, I am experiencing an event that is ranked (by numerous relevant professionals) as the 2ⁿᵈ among life's most stressful events. For various reasons, I am choosing to be oblique here—not annoyingly coy, at least not on purpose—about the details of said events. Still, given this current extremely tumultuous life-transition in which I am because of said event, I believe it might be worthwhile for myself and my readers, for me to reflect on my current state of my mind, especially in relation to the concept of crucible moments.

The concept of "crucible moments"—which I believe is credited to Warren Bennis as the idea-originator, can be explained using the definition below, courtesy of Will Krieger (link: https://medium.com/@ willkrieger/your-crucible-moments-35994634a312):

A crucible moment is, by definition, a transformative experience through which an individual comes to a new or an altered sense of identity.

These are times when our character is tested. These are times of adversity where great strength is shown.

Those who go on to be great are those who take time to pause and reflect on these moments. These are the moments that make us the leader we're going to be, the parent we're going to be, the person we're going to be.

Incidentally, the culmination of "stressful event X" unfolded alongside other considerable stresses in my life. But at some point over the past few months, I realized something that might be a key trick for my productivity, which (I believe) I have hinted at or mentioned before in another post. The realization (?): I believe our worst days might be our biggest opportunity to succeed in life.

How? If on your very worst day (e.g., because of family issues, or mild health issues, or conflict at work) you still manage to cross at least one item off that "to-do" list, IMHO, you've conquered the vagaries of life. I am also realizing the priceless value of consistency vis-à-vis routines, habits, and even rituals. And yes, many of us have numerous idiosyncratic behaviors, work-processes, etc., that we do not realize are in fact rituals.

Ultimately, Heraclitus was right indeed (paraphrased): change is life's only constant. I would also add—in an existentialist mood, that chaos is the other guaranteed constant of life. Somehow, you have to get to know yourself well, figure out your productivity modus operandi, and just do your best each day. And whenever you find yourself in a discombobulated situation—and you will surely, from time-to-time, stay calm; breathe; think; and again, stay calm. Then do your best to solve whichever problems might have arisen, or to otherwise get yourself back to your state of equilibrium.

"You got this!" 😊

Post 21; 06-02-21:
Easier Said Than Done

The other title I was debating on using for this post is something along the lines of "Practicing What We Preach." When I wrote last week's post, I fully intended—in other words, I had a good faith intention—to indeed do what I was saying in that post.

And to a limited extend, I did. But I guess the issue I want to grapple with in this post is, how well can one accomplish that goal—i.e., of chugging along, regardless of the messiness of life at any given timepoint—in general? How easy or hard is it, and why and how do some of us manage to do it, while others struggle?

All of the above questions have been triggered by a worsening of aforementioned "stressful event X." Interestingly, that worsening unfolded in an eerily similar fashion to the way another stressful event befell me, years ago.

Unfortunately, I do not have any good answers for the questions I pose in the second paragraph farther above. I just know that it is sometimes/often excruciatingly painful and hard to acclimate to our lives' situations. I also still believe in what I said last week. It is vital for us to "pick ourselves up" and "dust ourselves off" and "move on," regardless of the impediments in our paths.

And regardless of all the great advice any expert can give you, only *you* can know how to do what is right for *you.*

Again, I submit: "You got this!" 😌

B. Reflection

The above texts demonstrate the way we often communicate with ourselves without realizing that we are in fact doing it (i.e., communicating with ourselves). We can also utilize the texts to demonstrate the basic mechanics and role of language in intra- and inter-personal communication.

To achieve the latter goal, we need to take a closer look at one of the themes/topics discussed, i.e. emotions. Often, we experience intense emotions without knowing or being sure of those emotions' identities. For instance, it is possible to feel happy or sad—or a combination of the two emotions—without realizing that those two emotions (alone or together) are what you experiencing. In such instances, yes, we might be communicating with ourselves, but we are doing it in an abstract or indirect way. On the other hand, via the use of "to-do" lists and journaling, we are clearly and directly communicating with ourselves using language.

Unfortunately, the main argument I make about the attendant increase of intrapersonal-communication during moments of intense (negative and positive) emotions is being proven to me at the moment via my own reaction to the aforementioned depressing situation I'm enduring. That situation—to which I also refer obliquely in the public-journal entries—is a rather "slow and painful death" of my marriage. Beyond the context of journaling, I often catch myself in clear self-communication: for instance, owning up to my share of the mistakes that have led us—my wife and I—to this juncture, decrying her role and her inability—in my biased opinion—to admit fault, etc.

Ultimately, if utilized correctly (e.g., by trying to avoid obsessively thinking about the situation, and forgiving oneself, etc.), intrapersonal-communication during hardships—as reflected in the public-journal entries above—can markedly help us in coping well with such crises. We can perspectivize our human condition, and we can counsel and motivate ourselves in an arguably much better fashion than many—or *most*—therapists and motivational speakers.

Reality and Perception-Processing II: with Metacognition

Dec 29, 3:03 AM (Voice-Note No.: *Batch 1; No. 6*); NY Times
Article about J.H. Conway's Game called "Life"

As I mention earlier, some of the roles played by language in our lives include its enablement of our socialization, and its role as a semiotic tool in description and memorization. However, one very powerful role we also have to consider is that of general ideation, i.e., the formation and sharing of ideas. With this background, we should also consider the concept of "insight(fulness)." What do we mean when we describe an idea (or observation, argument, etc.) as being insightful?

As of mid-May (05/17/2021 to be precise), I am grappling with this word/concept, and not for mere curiosity or fun. Rather, I have to make an effort for my analysis of this book's research to be insightful. In the end, the use of dictionary-definitions or of other reference resources in this context can only go so far in informing or guiding me. Instead, in addition to those references, I have to also assume the role of an ideational adjudicator. For instance, in addition to one of Google's dictionary-definitions ("a deep understanding of a person or thing"), I can ask myself, "what ideas or things do *I* find insightful, and why?" I can then answer the question with an exemplar—i.e./e.g., the attendant transcript entry for the current section's voice-note, expressed in three points (with special *emphasis* on the second and third points, for the current context):

- *What are the random topics that catch our attention? "Insightfulness." What does it mean for a topic to be insightful?*
- *I found the game of "Life" insightful; [reason:] out of [extreme] simplicity, we can produce [infinite] complexity*
- *Also interesting: recursion and self-reference: related to D. Hofstadter's "Goder, Escher, Bach." (^)*

In fact, the first above-emphasized point understates the insightfulness of the game (i.e., J.H. Conway's game called "Life"). As I started writing the current section (4.5.1), I looked up the game's article on Wikipedia. And as I read the article, I kept mumbling in wonderment to myself, "*Good Lord* (!)," "*oh my goodness* (!)," etc. Part of my wonderment comes from the mathematical beauty of the game, as well as its various connections to numerous mathematical and general STEM (and even art) topics, including computation, inventions by John von Neumann and Alan Turing, systems theory, algorithms, music, biology, chemistry, music, and philosophy, among others.

Regardless of this reflection, unfortunately, I might still find it hard to decide what is insightful and what is not, from my data-analysis (for this book). But I am glad to report that I believe the task will be a little easier.

CHAPTER CONCLUSION

Apparently, symbol-systems such as language "super-charge" both our intra- and inter-personal communication abilities, enabling sophisticated processes of making/understanding and transferring intended meanings. Via this lens, this chapter grapples with concepts such as metaphors and analogies, and tools such as "to-do" lists and journals. Ultimately, such concepts and tools enable us to go beyond merely making/understanding and transferring meanings. Rather, symbol-systems enable us to gain insights into our essences, and to harness our humanity vis-à-vis self- and societal-improvement.

Creativity-Attempt Outtake 4

Post 7; 02-22-2021:
A Meta-Cognitive/Philosophical View Of Writing, From One Writer's Deck

TL;DR Version: Focusing attentively on the way we think, work, and write (AKA metacognition), reveals several essences of our humanity. In many ways, we would not be the awesome humans we are today, if it were not for the concept and activity of literacy.

Over the course of history, humans in the East, Middle-East, and West—among other places—discovered or invented a clever technique of working. This technique allows humans to efficiently use their minds and other tools (e.g. quills with ink, and parchment, or computers today) to virtually travel through time; preserve for the future, or retrieve from the past, or even create and (re-)visit new/hypothetical eras that do not quite belong to the past, present, or future.

In addition to that **magical** (for lack of a better word, really) use of this working technique, humans became creative in many other ways. The technique helped us create more forms of literary and other art-works. In fact, even "science" as we know it would probably not have evolved to the sophistication it has, without the technique in question.

Fine, I'll drop the coyness here, and clearly state the technique in question: writing! Apparently, this art—the art of writing that is, tends to unite numerous concepts and practices.

As I've already mentioned, writing necessarily involves the concept and practice of work. But I for one am also fascinated by the way writing complicates "work." Once, during a grant-writing workshop (at Fort Hays University in Kansas), a colleague decried the fact that grant-donors often demand productivity-reports from scholars (and/or other grant beneficiaries), yet the value towards art, and/or science, and/or other fields from writing and other cognitive works is often hard to quantify.

If I spend a good amount of time and effort on an experiment which eventually does not yield good results, can I be judged to have worked hard, regardless? We can also segue to the concepts of arbitrariness and standards here. For instance, I believe (based on education and experience) the concept of "good" grammar is decided by convention, more-or-less. And by what or whose standards should we judge a poem, or a fiction or nonfiction story?

Relatedly, let us consider the concept of coherence: if you do not understand a poem or rap song*, does that mean that poem or song is meaningless? (*Side-note: "track" or even "piece" might be the appropriate word, if you think rap does not qualify to be called "music." But I wonder if in this day and age, some folks still believe that?) This question also reminds me of the concept of internet "virality" nowadays. For instance, as Daniel Abrahams mentions on his LinkedIn profile, his goal is to simply write—I guess for his own pleasure, as well as that of those of us with whom his writings resonate (?), not for virality. And yet, his works often do go viral!

Considered in relation to the above questions of literary aesthetics and meaning as well as arbitrariness, one can suppose/imagine that in an alternate universe, the works of Shakespeare are simply another set of obscure literary art-works by some obscure/unknown Elizabethan-English playwright! And of course, some of the real and rhetorical questions posed above (i.e./e.g., about literary meaning and hard work or creativity) can be answered by this response: "it depends on the purpose for which the writing is done." If a playwright composes a comedy play that audiences find humorous, then we can perhaps judge that he worked hard, and/or is creative.

And regardless of our evaluations of the aesthetical value of the works of individual authors and other artistic/scientific/other creators, the art of writing (and the fields of art and science in general) also make use of typologies to help humans make sense of things. Hence the existence of "genres," "taxonomies," etc.

In the end, focusing attentively on the way we think, work, and write (AKA metacognition), reveals several essences of our humanity. In many ways, we would not be the awesome humans we are, if it were not for the concept and activity of literacy.

Post 37–I; September 21, 2021:
The Observer (A Ugandan National Weekly Newspaper) 09-19-21 Article* Outline
("When I grow up, I want to be..."*)

To state the obvious, writing is hard. What's even harder—or very complex, is the development of ideas (and that concept and action is of course a parent-concept/action of writing). For instance, here's a question for myself: how did I come up with my latest op-ed column piece?

Answer: I knew the topic I wanted to write about—including the most important points about said topic (career-choice dynamics, especially in comparison between various generations), but I was not sure how to develop it further. Resultant solution: the list below, in which I somewhat broke down my idea into various sub-ideas, then later, ordered them—i.e., via the numbers on the right—in a way that would flow naturally and/or make sense to a reader.

After it's published at some point next week or two weeks from now, I will share the article itself. But for now, I decided to share the outline, as well as the story behind it. I hope other writers can benefit from the technique!

—The statement represents way more than a career-aspiration; ideals of good life in general--1
—Myself: journalist, physician, journalist, writer, professor, writer, thinker———————2
—How/why do evolve to want to be, and to actually be, who we are?————————3
—Possible answers: nature/nurture, formative experiences, etc.————————4
—"Happiness": destination versus a temporary state————————9
—Our gen-Y/Z generation and the complexity we face, re: careers————————8
—Paying bills vs. happiness vis-à-vis vocation————————5

Post 37–II; September 22, 2021:
The Observer Article In Question

Apparently, the editor chose to run my piece immediately, contrary to my supposition (I thought he'd be running it one or two weeks from the current one). So voila, ladies and gentlemen; presenting the final result of the above outline!

What do you want to be when you grow up?

Several years ago, LinkedIn promoted a campaign in which they encouraged users to share their childhood career ambitions. When you were 15, how did you complete the sentence "when I grow up, I want to be…"?

And so, I dutifully shared my own childhood ambition. I had wanted to be a TV journalist, earnestly inspired by Christian Amanpour. But alas, it was not to be! I will return to the issue of what I eventually became. But first, allow me to dissect that statement—"when I grow up, I want to be…"—a little deeper.

I believe this statement represents more than a career wish. To me, it is indicative of an innocent conception of a holistically idealistic life.

Between my early childhood — from around six years — through my teenage years (around 16), I meandered through various career ambitions: TV news anchor, physician, chemical engineer (to specialize in making perfumes), and finally (coming rather full circle), journalism. In college, I first selected journalism as my major, then eventually graduated with a B.A. in English writing (before eventually earning a master's and PhD. In communication).

But if you try to ask me why I wanted to do all the above things, I doubt that I would have an insightful answer. I suppose as a child, TV anchors (with Ugandan English anchor Francis Bbaale as the archetype) represented a neat and dignified persona. Later, I suppose being a physician represented the virtues of extreme intelligence and benevolence.

As for the perfume-making/chemical-engineering ambition? No clue; perhaps I was at the time obsessed with the sciences in high school, not out of sincere fascination or deep interest, but as a status symbol of intelligence. Again, no clue.

The above discussion might provoke a worthy question for us all: why did you want to be what you wanted to be as a child, and why do you want to be—or why do you enjoy being—what you want to be or are, today (as an adult)? Role models perhaps — your parents, or the likes of Francis Bbaale or Christian Amanpour for some of us? Or formative experiences that convinced us of the worthiness of the profession, say a physician or nurse that took good care of someone you loved or a teacher that impressed you so much?

Regardless, for me, the childhood ideal of a good or enjoyable or worthy profession has crashed messily with the cold harsh reality of the need to pay one's bills. Apparently, we tend to seek both career-happiness and a holistically good quality of life—which comes in large part from our careers, side-by-side with financial success.

How should we then balance those demands? At this point in my (early) career and adulthood stage, I am not sure. But I have learned the hard way that we need to look at the big picture in order to realize that regardless of how challenging our careers seem to us in the moment, there are solutions out there.

For instance, whereas I did not thrive there and thus had to leave after a brief duration, I have several colleagues who make a decent living as professors in China. And whereas the gig-economy is not ideal (and is probably often exploitative and "not worth it"), one can take advantage of part-time jobs such as driving Uber.

What is also true, and rather unfair, is that our cohort's generation(s)—i.e., the millennials and gen.-Zs—seem to have missed out on the myriad of easy career-pathways that previous generations enjoyed. I have come across an abundance of anecdotal and scholarly evidence over the years of the fact that the other generations (the Baby Boomers especially, and the members of generation X and others in between) seem to have had far better chances for success compared to our generations.

With nothing more than primary or high school qualifications, many of our parents were able to get well-paying blue-collar jobs e.g. as factory workers. And yet for our millennial generations, the dynamics are a lot more complex.

Apparently, there are plenty more opportunities for them — especially via roles related to the digital and knowledge industries. But these roles also require longer and more specialized education and training. At the end of the day, I have to console myself with the conclusion that in relation to careers and life in general, happiness is a temporary state of mind, not a final destination.

Perhaps you are or will be happy in your chosen or assigned profession; perhaps you are not or will not be. But you can be happy, regardless. Besides, one can argue that ultimately, we are all versions of Peter Pan: we will never truly grow up. Thus, we might as well keep our career-related and general idealism alive. What do you want to be, when you grow up?

Post 38; September 26, 2021:
Article Outlines (& One Resultant Article) 09-25-21
1-   A–Outline: *"Basiima Ogenze"* (translation: unfortunately, our virtues are often realized too late, posthumously); Perhaps That's Okay
  —Per Obama, millions of citizens of ordinary citizens toil in obscurity, day in and day out, looking for no "pat(s) on the back" or extreme wealth, just mere "kawogo"/"daily bread"—3
  —These ordinary hardworking citizens are ultimately the real "army" or pillars that actually grow the wealth/prosperity of nations and enlightenment of nations; after all, Europe came from far, viz the stages of development, as Ken Follet's historical fiction trilogy shows——4
  —BMK———————————————————————————1
  —In actuality, there are probably hundreds like him in towns all over Uganda; successful, but humble and fair—————————————2
  —Let us all take a pledge to step up to that plate each and every day; to endure the grind, do the right thing by our God or spiritual beliefs in general, ourselves, and our families and countrymates——————5

>>>+<<<

1-B–Resultant Article:
"Basiima Ogenze," Indeed; A Tribute to the Unsung Heroes of Our World

The outpourings of mourning and grief were numerous and earnest. Indeed, the country had lost an awesome person; an entrepreneur, a self-made man, a kind man. As I read all these tributes, my first reflex was to reflect on the expression popularized by musician Jose Chameleone several years ago, "basiima ogenze" (people only appreciate your good actions after you've passed). But upon further reflection, I realized that on the contrary, the late BMK and numerous other Ugandans in various sectors (in business, as well as other fields such as academia, healthcare, politics, sports, etc.) are often justly given their due praise. And that is only fair.

But what is also true but rarely mentioned—or perhaps, not mentioned enough—is the fact that there are millions more men and women in towns and villages all across Uganda that have the same commendable traits. Perhaps not in the same breathtaking magnitude of the late BMK and a few other prominent personalities. Nevertheless, these men and women strive to do the right thing day in, day out; work hard, take care of their families, pay their taxes.

In fact, that story—of the unsung ordinary citizen—can be found in all countries around the world. For instance, in his latest memoir, former US President Barack Obama severally gave voice to their toils in the United States. One particularly moving tribute/eulogy was of his late grandmother, who sadly passed away one day before he was elected to the US presidency. Toot (short of "Tutu," the Hawaiian term word meaning grandmother), as she was known affectionately by the family:

"…was one of those quiet heroes that we have all across America, … They're not famous. Their names are not in the newspaper. But each and every day they work hard. They look after their families. They sacrifice for their children and their grandchildren. They are not seeking the limelight—all they try to do is just do the right thing.

And in [the massive crowd of his supporters that day], there are a lot of quiet heroes like that—mothers and fathers, grandparents, who have worked hard and sacrificed all their lives. And the satisfaction they get is seeing that their children and maybe their grandchildren or their great-grandchildren live a better life than they did…"

✳✳✳ ✳✳✳ ✳✳✳

From this perspective, perhaps social-scientists focus way too much on the macro-level structures and dynamics of society—such as "the invisible hand of the market," and GDP and political-governance structures. Perhaps the "magic"—so to speak—of the positive transformation of nations lies in the day-to-day actions of ordinary men and women across time and space.

To be sure, of course, the actions of national (and other—i.e./e.g. traditional, religious, etc.) greatly influence the destinies of our nations/societies. But arguably, kings and bishops without citizens willing to engage in commerce in general—or to pay taxes, or give tithings in church, would be rendered powerless.

One good demonstration of the above argument lies in the historical-fiction-genre novels of Ken Follet. It bears emphasis that these novels are based on real historical events, and they paint vivid snapshots of life in England and Europe at large from the medieval through Elizabethan eras.

What Follet clearly demonstrates in these novels—again, based on factual historical record—is that ordinary citizens helped build Europe into the powerful continent that it is today. For yes, believe it or not, England and Europe were once as poor as the developing countries of "the global south" today. In fact, the global south is "leapfrogging," foregoing various outdated technologies in the march toward socio-economic development.

Regardless, the men and women portrayed by Follet engaged in agriculture and commerce, paid their taxes to the state and their required tithings to the churches. And slowly-but-surely, these men and women gained enlightenment from one generation to the other through accidental or (necessity-)forced and deliberate discovery and innovation, as well as critical thinking—for instance, in regard to questioning received (theological and other) knowledge.

At this juncture, my rather simple thesis—i.e., that ordinary citizens around the world do not get enough credit for their day-to-day toils—gets a little complicated. Perhaps "doing the right thing" sometimes necessarily involves thinking or doing things differently, which would anger political, religious, and other leaders?

But after all is said and done, that argument or rhetorical question might be moot. I believe if/when one strives to make a good faith effort to do the right thing the way the late BMK and billions of other citizens did/do, it is only natural that sooner or later, prosperity and justice will inevitably reign (e.g., as argued by Steven Pinker in his book, "Enlightenment Now").

In Memoriam:
Bulaimu Muwanga Kibirige,
October 2, 1953 – September 10, 2021.

And in honor of all the unsung heroes—the ordinary men and women—in Uganda and across the globe who strive to simply be good citizens each and every day.

Self-Questioning for Creativity-Attempt Outtake 4

State any Necessary Introductory Remarks, and/or Answer this Question: Does it Make Sense to Me Now—i.e., In The Present—versus then—i.e., When I Wrote It?

The posts in this outtake often demonstrate the tension between raw thinking and professional output, especially vis-à-vis writing deliverables. By "raw thinking," I include basic and complex emotions and thoughts in my meaning. For instance, sadness and feeling moved—or stated differently, experiencing catharsis—in reaction to President Obama's description of his then-recently deceased grandmother. And by "professional output" and "writing deliverables," I refer to the end-products of my general/vague and organized thinking, e.g. as demonstrated by the articles. However, the first post (no.-7, published 02-22-2021) is mostly a detached analytical reflection of such thinking and writing processes as recapped above. And for the most part, all the writings still make sense to me in the present, just as they did when I first produced them.

Does it Resonate with Me Now—i.e., In The Present— versus then—i.e., When I Wrote It?

Yes, the posts resonate with me as well as they did then. I believe they appropriately demonstrate and try to analyze the markedly challenging processes of organized thinking and writing.

What Mood and/or State of Mind Does the Tone Suggest?

The tone(s) demonstrated in the posts are by turns detached and analytical, and—as aforementioned, "raw" vis-à-vis emoting and thinking.

What is/are the Direct and Indirect Relation(/s) to the Current Chapter via Topic(s) and Themes?

Overall, the posts in this outtake appropriately demonstrate the concepts discussed in the chapter, with the most important one—the thesis-concept, so to speak, being the power of (spoken and written) language.

Any Other Evaluations or Comments?

N/A

Activities and Discussion Questions

1) How do you utilize language on a daily basis (i.e., both written and spoken) for small tasks/goals, and for big goals—e.g., getting a promotion at work? How can you improve your use of language for both short-term and longterm goals?
2) Brainstorm a set of lists:
   i. List A: Of communication activities that are executed best via only nonverbal, spoken, and written modes, respectively.
   ii. List B: Combine some of the above activities—which you originally wrote on their own list(s)—together, depending on whether or not they can be accomplished using any other modes—e.g., I can express anger or pleasantness both via facial/nonverbal means, and/or via writing.
3) Regardless of whether or not you speak other languages in addition to English, discuss this question: what are some of the features of the language that you find interesting or curious, and why? If you speak other languages, discuss the same question for those languages, then compare the two sets of answers.
4) Optional: Continue journaling/creative activity 1 as advised at the end of chapter one above.
5) Share your Johari-Window list with the group, but do not share items you feel uncomfortable disclosing.
6) Continue activity 6 from chapter one, "Comparison with other types of communication."

REFERENCES

Hofstadter, D. R. (2006). *Gödel, Escher, Bach: An eternal golden braid*. Basic Books.

Obama, B. (2020). *A Promised Land*. Penguin Books.

Porpora, D. V. (2011). How many thoughts are there? Or why we likely have no Tegmark duplicates $$ 10^{\{\{10^{\{115\}}\}\}}$$ m away. *Philosophical Studies*, *163*(1), 133–149. doi:10.100711098-011-9790-6

Saussure, F., Bally, C., & Sechehaye, A. (1966). Course in general linguistics. New York: McGraw-Hill.

Thagard, P. (2005). *Mind: Introduction to cognitive science*. MIT Press.

KEY TERMS AND DEFINITIONS

Reality and Perception-Processing I; Without Metacognition: Intrapersonal-/self-communication executed or utilized by an individual to process the inputs of their senses and/or their thoughts, without realizing that s/he's in fact undertaking that process.

Reality and Perception-Processing II; With Metacognition: Intrapersonal-/self-communication executed or utilized by an individual to process the inputs of their senses and/or their thoughts, while aware that s/he's undertaking that process.

Self-(/Intrapersonal-)Communication: Communication inside an individual's mind, or outside—e.g., spoken or written—but for that same individual's consumption and/or use.

Chapter 5

"Good Lord, I Do Not Belong Here; I'm Not a Professor (/Doctor/Lawyer)...":
Intrapersonal Communication/Cognition, Reflexivity, and Society – Part II: Details Of Tentative Reflections On Self-Concept

ABSTRACT

The author continues the previous chapter's examination of his own SC instances as derived from the voice-note/transcript data via the themes of 1) "Types of Self-Communication (SC)," 2) "SC and Learning," 3) "Reality and Perception-Processing I: Without Metacognition," and 4) "Reality and Perception-Processing II: With Metacognition." Because of its position in the book, the chapter also offers an ideal opportunity for reviewing the categorical appropriateness of the SC instances analyzed thus far. As a result, the author confirms that 1) we can indeed pinpoint clear instances of SC, 2) we can utilize said SC-instances to improve our personal learning methods, 3) we often participate in SC without realizing it, and 4) we can also use SC to deliberately or systematically process our reality and perceptions.

INTRODUCTION

Recap of Voice-Note Timeline/Transcription and Chapter-Writing

Between January 3 and March 27 (2021), I recorded a total of 64 voice-notes. Later—between March and late June (2021), I transcribed the first three batches of that total number of voice-notes, and utilized that transcript with the writing of chapters one ("Introduction"), three ("Cognition, Reflexivity, And Society—Part I; Early Potential Insights"), five ("Language and Symbol-Systems"), and six ("Mental Health") of this book. Thus as of mid-July (2021), I have written a substantive amount of the contents of a total of four chapters, i.e., the above-listed chapters, as well as chapters one and two.

DOI: 10.4018/978-1-7998-7507-9.ch005

Overview and Goals

In light of the above background, the main goals of this chapter (based on its attendant transcript) are twofold: 1) to continue chapter three's examination of how my SC instances demonstrate how our thinking and SC processes are necessarily affected by, and connected to, the societies within which we reside, and 2) to confirm, refute, or otherwise help us better understand the analyses divulged so far. In fact, perhaps the word "refute" should not be used in this context at all. For as instructors often say in a bid to encourage participation in Socratic class-discussions, *there are no wrong answers here.*" In other words, instead of seeking to "confirm or refute" findings, this chapter might be better utilized to shed more light on what we've uncovered so far.

In parallel pursuit of—and/or in line with—the above goal, I intend to connect the content to the overarching theme of effectively building and continuously (re-)fortifying that most mercurial of human social phenomena, the *self*. In other words, how can we relate the topics discussed in this chapter to the those of previous chapters, especially in the context of effectively and continuously defining, crystalizing, and strengthening our own individual self-understandings? Better yet, how do we execute the latter task, while simultaneously pursuing—and hopefully achieving—individual self-improvement vis-à-vis learning and general conscientiousness?

Two other related points are worthy of mention in this context, namely: 1) the search for categorical appropriateness, and 2) particularly-salient potential insights (e.g., time, "to-do" lists, and journals). With regard to the first point, readers should critically examine whether the data findings of this chapter are appropriately discussed under the correct categories, especially the first four below, namely "Types of Self-Communication (SC)," "Self-Communication (SC) And Learning," "Reality And Perception-Processing I: Without Metacognition," and "Reality And Perception-Processing II: With Metacognition."

Readers should also look out for instances that can firm up some of the consequential topics or themes that have featured in previous chapters. For instance, does this chapter add some depth or nuance to the theme of time in relation to recording past events via journaling, and/or reminding ourselves to do things in the future via "to-do" lists? If yes, how?

MAIN FOCUS OF THE CHAPTER

Types of Self-Communication (SC)

This first thematic category enables our pursuit of the goals stated above—i.e., adding nuance to previous chapters' findings, and effectively building our (respective) self-knowledge—in an expansively general introductory manner. Indeed, the instances identified under the same thematic category in chapter three—"Cognition, Reflexivity, And Society Part I" and chapter four—"Language And Symbol-Systems," exemplify this characteristic. These include (among several others) "effects of life's general events/features on learning," "tip of the tongue and meta-analysis," "metaphors vs. analogies vs. similes," and "suspending and confirming understanding (parts I & II)." In light of this characteristic, what more can we glean from the current transcript in relation to the current theme—i.e., Types of SC, and how can we relate it to the above goals for this chapter?

Mon, Jan 18, 2021; 7:01 AM (Voice-Note No.: Batch 2; No. 2.1); Recap: Self-Doubt as Discussed by Barack Obama in his "A Promised Land" Autobiography

This topic demonstrates the potency of recurring themes in our day-to-day SC, particularly in regard to our self-worth-evaluations. However, in this and other contexts of SC, it is important to differentiate between instances of SC in which we can clearly "hear" ourselves communicating, versus instances in which we simply have a vague idea/notion of the ongoing communication. In other words for instance, whereas it might be true that many of us often get "imposter syndrome," can you recall a particular instance in which you "heard" yourself saying, *"good Lord, I do not belong here; I'm not a _____ (professor/doctor/lawyer or judge/etc.)*?

Fri, Jan 22, 2021; 1:11 AM (Voice-Note No.: *Batch 2; No. 2.6*); 1)—Quantification of News Consumed for Knowledgeability Appraisal, 2)—Awareness of One's Lack or Possession of Knowledge, 3)—Self-Debating and Censoring during Learning Discussions (e.g., At School, Church, etc.)

This transcript entry—among others in this particular sub-batch—is characterized by a substantively longer length. And so as to effectively discuss its arguments, I am pasting some of its sections below.

---Point 1: Quantification of all the knowledge we consume: e.g., via news, on a daily basis:
- *Context: I personally consume a lot of news on a daily basis. Not on soc.-media, but* [I'm nonetheless a] *big consumer of news—i.e., in a written format, online.*
- *For someone like me who loves learning as much as I do: how can I harness that news-consumption? I'm guessing that "actionable news" out of the total percentage that I consume is in the single-digit percentage range. No more than 9 or 10%.*
  - *One example of news utility is the current COVID pandemic. Infection rates, vaccines, etc.*
- *But apart from such an example, news-consumption = "knowledge is power," but in the abstract sense.*
- *In the simplistic way we think about intelligence, most of the news we consume is useless... Or is it? "This is an evolving issue in my mind."*
- *It does not have to be a black or white issue.*
  - *Over time, you can harness the news to your advantage. But "there is a deliberate process you have to put yourself through, I feel like." E.g.: with this book's research, I can perhaps utilize the news I consume. But it has to be a systematic periodic review. "What have I been learning via the news? What are the implications with various knowledge-domain subjects?" E.g., in the context of political-science or in general, social-science.* (^)

Arguably, one consequential implication from the above transcript-section is the need for each of us to make an effort vis-à-vis learning our particular interests, and how we can harness them to our advantage. For instance, compared to my love of news and current-affairs, another individual might discover—e.g., early in high-school—an earnest interest in the day-to-day fluctuations (or via general time-periods e.g. calendar quarters, and/or budget- and calendar-years, etc.) of the financial markets. And in turn, this realization might lead to an academic specialization in math, accounting, and finance courses. In fact, related to the above-argued potential implication, in the second point of this transcript entry, I discuss the issue of individual knowledge self-assessment:

---Point 2: Vague or rough knowledge of enormity of general or specific knowledge you do not know.

- ◦ *In general overall, or by subject- or topic-matter. E.g.; random example: fish. What are fish? How do you classify them in the biological taxonomy [As in, as] vertebrae/invertebrate, genus, family (?), etc.*

---For you to become smarter, you first have to grapple with how much you do not know. E.g., in my case, apart from the little I learned in primary school about fish—e.g., the parts of a fish (fins, gills, scales, tails, etc.); they're vertebrae and are aquatic—I do not know much else!

- ◦ *Explication/detail: go to a physicist and ask them about human-comm. theories; they'll probably look at you blankly or puzzled—or annoyed—LOL.*

---*Conclusion of point 2: first, know the limits of your knowledge, thus know what you need to learn more. (^)*

Sun, Jan 31, 2021; 9:49 PM (Voice-Note No.: *Batch 2; No. 2.14*); 1)--Ideas and/ or Knowledge and their Confluence with People in our Minds, 2)--Creativity through Making and Bending, Breaking, and Remaking etc., Your own Rules

In addition to the current thematic category, the first point in this entry is related to two others, namely 1) "Reality, Perception-Processing 1—Minus Metacognition," and 2) "SC And Learning," explicated thus:

- *1: Ideas and/or knowledge and their confluence with people in our minds and later, in our writings. E.g., the Wright brothers or my one of my mentors, Doug Porpora [vis-à-vis critical-realism metatheory].*
- *Explication: it's hard to separate ideas from their discoverers or inventors. I might think about ideas and knowledge, but my mind will inevitably also think about the inventors/discoverers of said ideas and knowledge. Regardless of their ideas, they are fellow human beings, and I relate to them as such. (^)*

Thereafter, the discussion turns to the mercurial concept of creativity:

- *Explication: one technique to help myself learn better is through bending rules, and making my own rules. First of all, what **is** creativity? Regardless, what does this topic have to do with SC? My assertion: a lot. In context of SC, metacognition, and improving our learning methods, creativity has a big part to play. Via SC, "hm...how does this work (?),"/"how can I make this work (?)," we're engaging in a creative process.*
- *Further explication: one can also argue that journaling is creative activity or output. Poetry is also a form of creativity, regardless of whether you're writing the poem to yourself or others.*
- *Most of us that write journal entries or poetry are not famous. There's always a chance that your writings will become famous later. But primarily, that journal-entry or poem is for your own consumption. (^)*

Mon, Feb 1, 2021; 11:23 PM (Voice-Note No.: *Batch 2; No. 2.15*); 1)--"Good job, Seif." 2)--Similar to Comp-Sci/Soft-Dev Concept, "Sanity Check"

Finally, this voice-note and its attendant transcript-entry rounds out our introductory categorization of SC. I discuss the concept and practice of self-praise, as well as a computer-science or software-development practice that might be useful in other spheres of life. Below is the full entry:

- *1: "Good job Seif," after finishing a task. A psychological trick I use to reward myself after successfully completing a task (quantitatively, not necessarily qualitatively—i.e., as long as it's **done,** regardless of quality).*
- *That psych-trick is especially important when you're doing favors or work for others. Metaphor: when you present the cake to them, all they see is the finished product, not all the hard work that went into it.*
- *2: Similar to comp-sci/soft-dev concept, "sanity check." The original concept in soft-dev: executing a print command to check if the code is working correctly, or using said command to debug your code.*
- *Explication: It can be helpful in our work in general for us to borrow that (sanity-check) concept and tailor/apply it to the different tasks we're executing. For me, TDLs help in a way similar to sanity-checks, to help me stay well-oriented and level-headed, knowing well what is done and what remains undone. (^)*

Overall, I submit that the above four transcript-entry discussions are unassailably exemplary vis-à-vis the various types of SC that in which we commonly participate on a day-to-day basis. They demonstrate the nuances of SC's mechanics—especially whether we're explicit via the use of language in our thoughts, or we're noticing our SC taking place vaguely, e.g. via negative or positive emotions about our *selves*. Below, we move on to the other thematic categories, and the voice-notes or transcript entries that can be classified under each.

Self-Communication (SC) And Learning

This sub-genre of SC is the most salient vis-à-vis self-improvement. In addition to their mere (re-)occurrence, the instances below answer the following question: how can we harness our SC instances vis-à-vis the improvement of our personal learning methods and styles?

Mon, Jan 18, 2021; 7:05 AM (Voice-Note No.: *Batch 2; No. 2.2*); Knowledge and Self-Assessment; (Self-)Motivation

Among all the other utilities uncovered in this study, this particular function is worthy of critical examination: how do we self-assess our knowledge levels about particular topics on an ongoing basis vis-à-vis? We can and often perform this self-assessment at the beginning of a particular learning "journey," i.e., of a particular topic or discipline, and continuously thereafter as we gain some mastery of a topic/discipline.

Similar to the discussion above in relation to self-doubt, it is important to note that this self-assessment can be both conscious and passive. My own experience suggests that research and general learning are two contexts in which we perform such self-assessment. But are there other instances of this evaluative SC in other day-to-day life spheres?

Motivation is another area related or similar to self-assessment of knowledge. How do we measure and improve our self-motivation for learning and/or working? Two possible solutions come to mind: 1) the "carrot and stick" concept, and 2) "tough love" towards self. Pasted verbatim below are the details of these possible solutions from the transcript:

- *Low self-motivation as a big impediment to learning: how can self-communication help with the alleviation of this impediment?*
- *"Carrot and stick" system in life; we're forced to work to earn a living.*
- *...*
- *How does self-comm. help with self-motivation? Partial answer: it helps to first figure out the fact that you're indeed unmotivated, that you do not feel motivated.*
- *One way to deal with that motivational deficit is: "Ok, so I do not feel motivated. Well, suck it up, self (/Seif)! I need to learn this, for my own good." (^)*

Sun, Jan 24, 2021; 3:14 AM (Voice-Note No.: *Batch 2; No. 2.7*); 1)—Curiosity and Acting on it or Not, 2)—Self-Coaching, 3)—Porpora's (2011) "How Many Thoughts" Paper

More than others, this instance's transcript section is rather self-explanatory. Admittedly, the point in regard to self-coaching is somewhat of a misfit, but I believe the explication connects it adequately to the thesis of curiosity.

- *1: Curiosity, and choosing to or not act on it. Regardless of the benefit of learning. Explication: part of how we learn is via curiosity. E.g., random example: "I wonder what the farthest distance is that migratory birds can fly?" You can then either look it up if you really want to [know], or not. Or: Having to look up or study something because of an assignment at work or school, etc.*
- *2: Self-coaching, e.g., through time-management. E.g., my 40-min resolution [i.e., for watching at least 40 mins of MOOC videos per day]. Another e.g.: my curiosity about number names between trillion and infinity. Explication: after you realize that you're very inquisitive by nature, it's up to you to self-coach.*
- *Explication of number-names between a trillion and infinity. This curiosity-instance had happened a few days prior, triggered by something I was reading. Thus, I Googled the question, and found the answer.*
- *Conclusion of point 2: if you love learning, it's up to you to nurture the love and follow it up with discipline and self-coaching, time-management, etc.*
- *3: Porpora's "How Many Thoughts" papers. Explication: for the most part, this point is not connected to the ones above; then again, it kind of is. The way Doug and his opponents wrote their papers, is emblematic of how knowledge is discovered and expanded, and it is connected to the above discussion of curiosity. Curiosity leads to our searching, theorizing, formal research, etc. (^)*

In other words, particular aspects of SC—e.g. in this case, sparked by curiosity, can result in a pursuit of knowledge that can improve our personal lives, as well as the lives of the general public at large.

Mon, Jan 25, 2021; 1:04 AM (Voice-Note No.: *Batch 2; No. 2.9*); 1)—Free Will versus Determinism, 2)—Self-Coaching through the Long and Lonely Road of Learning

As I note at the beginning of the entry, the first part of this voice-note/transcript—i.e., the free will vs fate debate issue—was inspired by a thematically-/topically-corresponding public-journal entry on my personal-professional website. Thereafter, I move on to discussing self-coaching in autodidactic learning.

- *1: Free-will vs. determinism: inspired by the relevant journal entry on my website.*
- *Explication of point 1: in undergrad, one of my profs liked asking us about our opinion of that debate. What's the biggest influence on our lives—free will, or fate?*
- *Further explication: if/when we're in classes or other venues discussing such topics/debates, what's our cognitive-SC process?* "I can think out loud, or…before stating [my opinion about the topic at hand,] I have to first…rehearse it in my mind [or] think about it, [then] say, [e.g.,] this is what I think: …e.g., I do not think free will is absolute…[neither is fate.]" *Regardless, I will first have to say that to myself before saying it out loud.*
- *2: Self-coaching through long, hard, and lonely road of learning. Indeed, the road of learning is long and arduous.*
- *Explication: part of the way the analysis [for this research] bears out, the way I come to the insights for the research, is via repetition. [Thus my revisiting of the concept of self-coaching.] In my learning of soft-dev, I often reflect on the fact that, that stuff is hard. Especially algorithms and data structures.*
- *Given the above fact/consideration—i.e., that CS concepts are hard, you have to do a lot of self-coaching as you're learning the topics. You have to remind yourself, "alright Seif, remember, this is for your own good. [It] is gonna pay off in the long run. Come on, you can do this, Seif. You can do this." Regardless of the fact that it's a long arduous road.* (^)

Beyond such positive self-directed messages, how can you actually achieve your learning goals with particularly hard subject-matter? In the following entry below, I offer an answer.

Mon, Jan 25, 2021; 7:00 PM (Voice-Note No.: *Batch 2; No. 2.10*); 1)—"Brute-Force Prolonged Learning" 2)—Research-Design Tailoring through SC 3)—Messiness of that Tailoring through SC Process

Among other methods, this entry suggests a technique I refer to as "brute-force prolonged learning," i.e., repeated attempts using a variety of resources (including modern online platforms such as YouTube), teachers, etc. Somewhat similar to the eventual outcome in one of the methods of computer-password hacking, my belief—explicated in detail below—is that given enough time and resources, anyone (of average intelligence, not afflicted by intellectual disabilities) can learn anything. Caveat: it is not easy, and it can get very messy!

- *1: "Brute-force prolonged learning"; stats and content-analysis: You can learn anything, regardless of how hard it is, if you keep at it for a long time… and [using] a triangulation of methods; teachers, study-partners, etc.*

- *Explication of above point: it's a firm belief of mine, and it's bearing out with my software-development learning. Like most folks, stats—anything to do with math—was for a very long time challenging for me. I had and to some extent, still have, math anxiety.*
- *The starting point for that "phobia"—for lack of a better word—was in the 2nd grade when the teacher at the catholic boarding school would inflict corporal punishment on us for not learning the multiplication table. Ever since, math for me was something to be feared, something hard, not fun, etc.*
- *Turning point: grad school; passing the stats course was the spark for the fire of math love, so to speak. "Ok...I can do this."*
- *Still, stats was until recently a methodology that I try to avoid. Lately though, I've fallen in love with stats. The breakthrough: the text by Wrench et al., "Quantitative Research Methods For Communication," especially the table from chapter 6, describing appropriate measurements for different variables (nominal, ordinal, interval/ratio), research goals (differences or relationships), and variable numbers (two or more variables or groups of variables).*
- *And in recent project-management study, voila, I'm enjoying myself while using statistical content-analysis!*
- *Relation to above point(s) to SC: you have to keep telling yourself, "just keep going, you'll get this sooner or later."[You have to often use] suspending of understanding, ...stick with it, [and use] self-coaching. Much of this voice-note's content might be repetitive, but part of the way you get to insights is through repetition. E.g., Owen's (1984) thematic analysis: repetition, recurrence, and forcefulness.*
- *Conclusion of above point (1): brute-force learning involves a lot of SC.*
- *2: Research-design: e.g., with Te's project. Using methodologists' systems for your own studies involves a lot of SC, to tailoring the methods to your own research-purposes.*
- *3: Messiness of the that process of research-design tailoring through SC. Between all the triangulation, hammering out of details, using some principles and ignoring others, the process is very messy! (^)*

Sat, Jan 30, 2021; 4:24 PM (Voice-Note No.: *Batch 2; No. 2.13*); 1) What is the Average Attention-Span Ability of the Average Joe/Jane? 2) One Possible Answer: via Learning Activities that Hold our Attention. 3) Explaining Things to my Mum; Sub-Point: Arranging Thoughts for Explanation and/or during Arguments etc., Outside of Formal Contexts. 4) Sense of Self, 5) Related to Self-Doubt and other Concepts I've discussed Before. Context: As a Professor. 6) Sense of Time and Place as Well

This entry is the second-longest of all in this transcript. Based on a holistic examination, we can divide its meaning/significance chunks into two parts, i.e., points one to three, and four to six. Below, I re-transcribe the first part:

- *1: What is the average attention-span ability of the average Joe/Jane? Explication: using a number of metrics including the controversial IQ and intelligence in general. Do some people have superior focus? Can they sit through a 3-hour lecture and be alert the whole time [without] their thoughts starting to wander? Or, do most of us in a three-hour lecture inevitably wander off men-*

tally? Further explication: regardless of the answer and negative influence on learning ability, what are some of the ways we can overcome attention-span shortcomings?

- *Further explication of above point: I will try to provide an answer to the above question in my book's analyses.*

- *2: One possible answer: via learning activities that hold our attention. One example for me: reading French news out loud. Explication: I have been trying to learn French for a very long time, and I am currently at an advanced beginner or intermediate level. One of the ways I improve my French is via Radio France Internationale (RFI) and France 24 television. However, I often find myself trailing off via attention span. Not only does that happen with listening and watching broadcasts, but also with **reading** French news stories. In this context—[i.e.,] unreliable attention span via reading, I've found over time that it helps when I read the stories out loud.*

- *3: Explaining things to my mum; sub-point: arranging thoughts for explanation and/or during arguments etc., outside of formal contexts. Explication: a lot of us have trouble connecting with our parents. Being/getting "on the same page" with them, regardless of their education levels. One of my challenges with my mum: our starkly different levels of education. She's very smart, but unfortunately did not go far in her formal education.*

- *Further explication: despite her limited education, she can understand things clearly if you patiently explain them to her, if you **connect** with her. But **that** is the challenge; **how** do you "connect" effectively with her? (^)*

On one hand, I ask a question in the last point above that I have somewhat just answered—i.e., via stating the need to patiently explain ideas to my mother. And yet, the closing question is also valid, and the answer I can hazard here is thus: by using frames of reference she (my mum) can relate to. Thereafter, I extend the discussion to the sub-context of (academic) debates versus inquiries, sense of self and self-doubt, as well as sense(s) of time and place, all in relation to learning and SC.

- *Sub-point: arranging thoughts for explanation and/or during arguments etc., outside of formal contexts. Explication: despite my advanced education, I hate the concept and activity of debate. Why? "I am not quick with words" (or 'quick-witted,' I guess). I am also not big on confrontation. I can do it if I have to, but I tend to be a conflict-avoider.*

- *Further explication: you often know you're right in the argument, ethics dilemma, or vis-à-vis the conflict you have with someone. The other party often has the wrong argument, clearly. And yet, it takes a great effort for you to arrange the words [in your mind], to be able to respond or rebut arguments, etc. Even in formal contexts—e.g., Doug responding to that critic, it takes a great deal of effort and time.*

- *Final point in above point's context. Nowadays, we're lucky to have Google. We often do not realize that it's not necessary to argue; you just have to Google the topic/question, etc.*

- *4: Sense of self; this concept is probably related to symbolic interactionism and pragmatism.*

- *5: Related to self-doubt and other concepts I've discussed before. Specific context here: as a professor, I have no choice than to really immerse myself in knowledge to be competent. Explication: as a professor, one of my worst nightmares is being in that state/situation where you have not prepared well enough, and you find yourself in that "deer in the headlights moment" when a student asks you a question and you cannot answer it.*

- *Further explication: It's ok if it's a philosophical [/"grey area"-related] or debate-related question, or if you in fact prepared well, but the question is about something not in the relevant text chapter(s). But overall, for me—as an instructor in the context of teaching, I believe I have no choice but to be steeped in knowledge, especially the knowledge I'm supposed to teach. Conclusion: in general, self-doubt is related to the general concept of a sense of self.*

- *6: Sense of time and place as well. Learning does not happen in a vacuum. You learn things in particular times and places: at school, work, from your childhood, through your late adulthood years. Explication/implication (partial-relevance) for me: as I am trying to retrieve a piece of knowledge, e.g., "I come across a factoid or question I do not know and I'm trying to answer it in my head, ...a big part of the clues will come from" my recollection of the time-period in which I might have learned the content, e.g., grad-school or undergrad, etc., and try to "connect the dots," construct the answer.*

- *When that happens, your memory is transporting you "across the plane of time." And inevitably your memory is also connected to a location, where you learned that knowledge. However, as I have already mentioned, longterm comprehension of knowledge, and making connections between various knowledge-domains in your mind, is vague and abstract.*

- *Further explication and preview of point-to-come: knowledge is also tied to people, i.e., the people who taught you or helped you gain your knowledge. Despite the fact that some knowledge is tied to time, place, and people, much/most of it is an unclear collection of various facts, topics, subjects, etc. (^)*

Thr, Feb 4, 2021; 12:59 AM (Voice-Note No.: *Batch 2; No. 2.16*); Suspension of Understanding; Break[ing]-up into Bite-Size Pieces, and the Nurturing of Hope/Optimism while Learning

Finally, in conclusion of the current sub-genre ("SC and Learning"), the notes below highlight the above two concepts/practices, which I find very useful in my own learning (i.e., "suspension of understanding," and break[ing]-up into bite-size pieces, [and the nurturing of] hope/optimism [while learning]).

- *Suspension of understanding: E.g. of David Malan* [the popular and gregarious Harvard/Ed-X instructor of CS50, Introduction to Computer Science, a "MOOC"], *"just trust me for now," as he introduces concepts to students that have never done comp-sci. This phrase is similar to my own "suspension of understanding" concept. In this context, related to active listening. Unlike Malan's students—i.e., who are not familiar with comp-sci basic concepts, you understand your interlocutor's points. However, you're also suspending many of your own thoughts, ideologies/ beliefs, knowledge, questions, etc. You're setting them aside to first listen well to the current incoming message.*

- *Primary reasons for your suspension of understanding and/or active listening. Seeking of "meeting of the minds," and looking forward to new pleasant or useful facts, analyses, etc. When you listen actively, you learn things you did not know. You encounter different angles/perspectives to old or already-known knowledge, etc.*

- *Relation of above point to SC: If I am reading a book and I do not understand the concepts therein, and I say to myself, "Ok, I'm going to suspend understanding for now, I do not have to understand each and everything that I'm reading in this book. First, let me read, and eventually it's going to make sense." That's you talking to yourself.*

- *In the case of David Malan, he is not talking to himself; rather, he's lecturing to students. But again, the point stands. Another example: me reading Goder, Escher, Bach. You [just have to] say [to yourself]: "Oh well, I do not understand it all, so I'm going to try understand at least a chunk of it."*

- *3: Break-up into bite-size pieces, hope/optimism, and the knowledge that as a lifelong-learner, time is your best friend. Similar expression with potential utility and/or similar proverbs etc., [to] "time is a life-long learner's best friend."*

- *Explication; application of above concepts to autodidactic-learning of software-development and other topics. How to use these techniques: first, you have to (/can) tell yourself, "Ok, let me break this [soft-dev, French, etc.] into bite-size pieces for me to learn this better." Second, realize and admit that yes, learning is hard, but you have to nurture hope and optimism, telling yourself "you know what, I'll get this; I'm gonna learn this."*

- *Further explication: sometimes, you're studying for an exam or otherwise have a deadline. But if you're life-long-learner, you have to realize that you have the whole of your life ahead of you to do what you love doing, i.e., learning. You do not have to learn everything today or right now.*

Reality and Perception-Processing I: Without Metacognition

In comparison to the others used throughout this book, the current thematic category might be the hardest to discuss vis-à-vis gleaning syntheses to satisfy the goals articulated in the chapter-introduction farther above. To recap, those goals are: 1) to help us better understand the analyses provided throughout the chapters so far, 2) to help us clarify how we achieve the goal of building and continuously (re-)fortifying our *selves*—by default, and for self-improvement via learning and general conscientiousness, 3) to help us crystalize and consolidate the categorical appropriateness we utilize in discussing the book's findings, and 4) to continue pursuing confirmation of particularly-salient potential insights (e.g., time, "to-do" lists, and journals).

The reason for the above assertion (i.e., that the current thematic-category is more challenging than others for achieving the above goals) is the following paradox: unlike the ways in which we implement our analyses with the other categories (in particular, "Types of SC," "SC and Learning," and "Reality And Perception-Processing II: With Metacognition"), this category is particularly defined by its lack of metacognition. In other words, while we can deliberately observe our *selves*—and minds—as we delineate the various ways in which we execute SC, i.e./e.g., use SC with our learning processes, and process our reality and perceptions, it might be particularly challenging to realize our cognitive- and SC-processes, *without a deliberate meta-cognitive/communicative process*!

Regardless, can we in fact try to delineate some of the ways we can—and sometimes or often do—process reality and perception, without realizing that we are in fact doing so? In the current transcript, I identify four examples that (I believe) demonstrate that process (process reality and perception, without metacognition).

Wed, Jan 20, 2021; 11:51 PM (Voice-Note No.: *Batch 2; No. 2.5*); 1)--Need for Self-Communication before Interpersonal-Comm. 2)--Reading Comprehension

This is the first example of the dynamics of the current thematic category. In the recording and transcript, I explicated on the necessity of internal SC before communicating with other individuals, in turn using the actions/examples of free-style rapping or "stream of consciousness" verbal outflows, which was one of President Donald Trump's infamous traits.

- *---Point 1:*
  1) >>"Before communicating with others, you have to communicate with yourself first."
  2) >>"To me, reading-comprehension demonstrates part of or a similar process."
     - >> *"I feel like that note is self-explanatory...I do not wanna um...I do not know, say much more. I feel like I wrote it clearly and the way I just read it out makes sense. So...I'm not gonna say more about it."*
     - >> *Quick revelation about some psychologists' work, re: the formation of sentences in our minds before speaking. Before you (/"one") divulge(s) a sentence, which is arguably the basic unit of communication, it has to make sense to you first.*

- *Comparison of ordinary people's communication process to rappers in performance of "free-style rapping," as well as "stream of consciousness," a la President Trump: thinking out loud. When we think out loud, not all thoughts make sense. Our thoughts [might be] jumbled. "Thoughts in process are not the same as refined thoughts, [or]... 'thoughts ready to consume.'"*

- *Point 2:*
  - *Reading comprehension as an example of the above concept:*
- *Acquiring literacy skills in childhood or early-ed.,*
  - *Deciphering basic meanings via alphabet and words, sentences, etc.*
  - *Deciphering wider-scale meanings, e.g. an emerging novel's plotline, etc. (^)*

In other words, SC helps us clarify and either simplify or explicate concepts before we speak them out loud to others, or before we write them on paper or machines such as personal computers. Moreover, reading-comprehension involves cognitive processes that go beyond mere *"[d]eciphering basic meanings via alphabet and words, sentences, etc."* Rather, the process necessarily requires SC, for a reader to untangle the various layers of meaning, *"e.g. an emerging novel's plotline, etc."* But unless we're in a process of self-study—such as that which I had to embark on for this book's research—or are otherwise mindful of our cognitive and communicative processes in real time, all the above activity is usually done without metacognition, indeed.

Sun, Jan 24, 2021; 3:31 AM (Voice-Note No.: *Batch 2; No. 2.8*);
1)—"Weirdness" of Human Beings, 2)—Learning and/or Working
despite Ongoing Problems in your Life or Distractions in General

At some point within the past 10 to 17 years, I started saying something to close family and friends, a pop-social-scientific perception: "human beings are (so) weird!" In the current transcript section, I reflect on that rumination; what are its origins, or how/why did I start saying it? Thereafter, I shift my focus back to the learning process, postulating on the nature and/or need for resilience.

- *1: Quote by me: "Human beings are [SO] weird!" Explication: Our behavior, apparent thought-processes, etc. Relation to self-communication and learning: it's notable, given the fact and way in which I say that to myself.*
- *Further explication of point 1: E.g., at Walmart in line and someone cuts ahead or otherwise does something bizarre, causing you to say that. "We all (many/most of us) might say a variation of something like this in such moments."*
- *This (above) instance of self-comm. (SC) is not quite related to learning, but this research is not limited to the learning-methods application.*
- *2: Learning and/or working despite ongoing problems in your life (connected to point no. 3 below):*
- *3: Working and/or learning against distractions overall, … in your mind or around you [or] your surroundings.*
- *Explication of both points 2 and 3: regardless of our class, society, etc., we all have problems in our lives. "The question is, how do we plod on, chug on, keep going…in our work and in our learning process?"*
- *Further explication: example of students in college or professional school. College, med-school, etc. You're there, and you have problems in your life going on. E.g., you lose a loved one. How do you keep learning effectively while quieting the mind? For you to suspend yourself from being preoccupied by those other thoughts?*
- *The above questions are open and unresolved questions. Part of the answer vis-à-vis getting over problems in your life is self-coaching, per my notes of the 22nd.*
- + *"You have to tell yourself:* 'alright Seif/John/Mary/etc., look ok, I have stuff [i.e.,
- meaning problems] going on right now, [e.g.,] my parents are going through a divorce, [or] …I just lost my cat who I loved…very much…I have those problems, but I need to just…concentrate…like, push those out of my mind for now and just concentrate…I do not have a choice; I'm trying to do this for my success…so I have to…push those thoughts out of my head for now.' *It's ok to grieve, but that should not then stop you from doing your work.*
- *"It's important that we teach ourselves to do that* [i.e., learn resilience], *I believe" if we're going to learn or work effectively. "You do not want to be the person at work with problems at home, and then you yell at your colleagues." Neither would you want others to do the same—i.e., "coming to work with their problems." (^)*

It should be noted that the irony of the first point above is not lost on me. After all, behold, a man that is earnestly making the case for the normalcy and utility of self-communication—including talking out-loud to yourself—calling other people weird! My response: by "people," I am certainly including myself, first and foremost.

Tue, Jan 26, 2021 (Voice-Note No.: *Batch 2; No. 2.11*); 1)—Obama's Mum about to Die; 2)—The Vague or Abstract and Mysterious Digestion…Of Knowledge

A close examination of these two points from the attendant voice-note transcript section can demonstrate their typicality vis-à-vis reality and perception-processing without metacognition or basic mindfulness.

In other words, they demonstrate a couple of the ways in which we can and often perform SC processes without realizing it.

One of those processes of SC is the use of our emotions while consuming literary and theatrical/cinematic entertainment. In this particular case, I recapped the intense feeling of sadness—followed by weeping—which I experienced while reading Barack Obama's narration of his mum's impending death from cancer.

- *...Obama's mum about to die,* [in] *chapter two* [of his "A Promised Land" memoir]. *Explication: as aforementioned, when we cry while reading a sad book* [or express another emotion], *that is a form of SC. Not quite you speaking quietly in your mind, or speaking out loud, but that emotion is a form of SC.*
- *Further explication: often/sometimes, the emotion is vague; you might be experiencing a mixture of two or more emotions. You might be able to able to divulge the emotion, but* [often,] *it's just that vague feeling in your mind. In the case of Obama's mum about to die, I wept.*
- *Part of the cause of my sadness was the fact that his mum was dying of cancer, and I've been through a similar experience—i.e., having a mum afflicted with cancer. Thus, this is an example of my aforementioned description of SC via emotion.* (^)

I then switch gears—staying within the same theme/topic, but discussing it (our experience of SC unconsciously) in the context of learning. To introduce the sub-topic/theme, I recall a question that a classmate once asked during a class-discussion in our Master's program at Kean University: *"will I retain it"* (i.e., all knowledge that we were consuming in the Master's program)? I thus discuss the rather mercurial or complex processes of knowledge-retention over the course of our lifetimes as children and adults, in relation to the concept and practice of self- and inter-personal communication.

- *1: The vague or abstract and mysterious digestion, processing, connection, interconnection, etc., of knowledge. Explication: H's concern about retaining all the knowledge she was soaking in, in our Master's program: "Will I retain it?"*
- *Explication: my common experience of "tip-of-the-tongue" syndrome. In those moments, you clearly have the knowledge in your mind, and you know what the concept is. But there is a mental block between your knowledge of the concept and the specific [name of the concept].*
- *Further explication 1: while we often have clear access to the knowledge in our minds, most of our education is turned into a vague abstract knowledgeability or collection of knowledge. We're educated in general, but we often perform poorly in contexts like that "Are You Smarter Than a Fifth Grader?" show.*
- *Further explication 2: the physicists who—according to Quora—cannot answer the lower-level physics-topics questions. They are so steeped in their higher-level knowledge, their brains/minds can only hold a certain finite amount of knowledge in the short-term-accessible memory. "You have this mushy soup of all the things you've learned throughout your life." Sometimes, some things are at the top of that "soup," easily skimmable, [but]...you will not be able to retrieve most of the knowledge."* (^)

Thr, Feb 4, 202 (Voice-Note No.: *2.2.16*); Imagining what
Others are Thinking and/or Saying about You

Finally, this voice-note transcript entry demonstrates the breadth of our SC processes on a day-to-day basis, which can and often do include internal (mind-based) musings such as the evaluations of our reputations from other individuals' points of view.

- *1: Imagining what others are thinking and/or saying about you now or before, or in the future, in your presence and absence. Explication/recap: the focus of the book is SC, in the context of improving our own learning methods. [But one] cannot just stick to the application of learning-method-improvement, and the concept is tied up with that of metacognition and cognition in general, and selves. Thus, I will also focus on other areas of life in addition to learning-method improvement.*
- *At any given time, most of wonder what others think of, or say about, us. Random example: "Boy meets girl or in LGBT context, boy meets boy;" a love story. The guy falls in love, then starts to wonder, "I wonder if s/he thinks about me" / "I wonder what s/he thinks about me" / "...what s/he tells her friends about me."*
- *Unless you're self-employed, do not all of us want to know what our bosses think of us? And if you're self-employed, you have employees, suppliers, etc. Regardless most of us—in the context of business or work—also wonder what others think of us. Relation to SC: your self has a primary and secondary self (my terms). The primary self is me—my name is Seif, 36 years old, etc. But it also helps to "get out of your own body/mind" to imagine, via the two selves (primary and secondary): "hm, I wonder what she thinks about me?" That is a clear instance of SC. (^)*

Reality and Perception-Processing II: With Metacognition

Unlike the preceding thematic category (*Reality And Perception...Without Metacognition*), the current thematic category is easier to conceptualize: what are some of the different ways in which we often—or sometimes—*deliberately* use SC to process reality and perceptions (i.e., our own as well as others')? Below are three of most prominent instances I identified in this regard. However, it might be useful to note in earnest that the exemplars below are not straightforwardly emblematic.

In fact, a critical examination suggests that their typology is only slightly more meta-analytic in comparison to the previous thematic category (*Reality And Perception...Without Metacognition*). One other way to critically question this category's instance-typology is thus: in situations such as the self-checking of our own emotional and/or knowledgeability state (/status), do most of us indeed realize that we are performing SC? Regardless, let us explore some of the examples that my analysis highlighted in this context.

Tue, Jan 19, 2021; 6:53 AM (Voice-Note No.: *Batch 2; 2.4*); Checking
Emotional State and Self-Motivation; Critical Self-Questioning of Ideation;
Reading-Comprehension as Example of Checking Understanding

In this first set of instances, I examine situations in which we pause to critically examine our own emotional, intellectual, and comprehension states. In other words, we often do (or should) ask ourselves questions such as: 1)—"should I be driving, given the emotional state I'm in (e.g. extremely sad or angry, etc.)? Or, in relation to ideas:

- *Critically questioning ideas; "I [this author] am a man/person of ideas [,but I often ask myself]":*
  "Hm...is that idea possible/viable?"/Is that a good idea?
  - *This in essence is a discussion between your primary and secondary selves: "I (Self 2) is wondering [aloud to Seif 1] if this idea...is a good idea!" (^)*

However, perhaps reading comprehension provides one of the best exemplars of critical self-examination. In this particular instance, I utilize the example of myself, as I was reading Ken Follet's "Columns of Fire" novel:

- *Another example of checking of understanding: while reading the Ken Follet "Columns of Fire" novel: pausing to make sure I understand who is saying what during the dialogue turns of the book's characters. Thus, reading-comprehension is one learning activity during which we regularly pause to check our understandings. "Am I understanding this correctly?" (^)*

Thu, Jan 28; 5:07 PM (Voice-Note No.: *2.2.12*); Bending, Breaking, or Making your own Rules [During Learning Processes], 2: Journaling; Similar to "To-Do"/ Notes-To-Self, as a Form of SC, 3: Wonderment and Exclamation in Luganda, in reaction to Ken Follet Novel, 4: Porpora's Thoughts and Barry Paradox Papers, 5: "For now, just take it on Faith," 6: Breaking the Rules; Poetry

With this set of notes, I stay within the last theme discussed above—i.e., reading, and I expand it to its general *parent-theme*—so to speak—of learning and self-improvement. First, I focus on the process of learning the two STEM subjects of software-development/"coding" and math:

- *1: Bending, breaking, or making your own rules: e.g., about how amateur coders/programmers should learn by building apps. Explication: everyone learns differently. Learning is inevitably intertwined with the concept of stylistics. Note: that is a layperson's, not a professional educator's opinion or evaluation. E.g., the opinion or myth that some people are visual learners*
- *Further explication/example: my preference of real books over PDFs. Or math. For me, it was very helpful to visualize fractions, so as to learn that concept well.*
- *Many software-developers assert that building apps is better for you learn, versus endless learning of concepts. But my argument is: how am I supposed to build apps, etc., before really learning the concepts first? Thus, in that particular case, that's my personal opinion.*
- *Relation to SC. Each of us has to play out that personal debate in our minds. You know yourself best vis-à-vis personal preferences. (^)*

Next, I focus on the activity of journaling—which can be considered as a general self-improvement tool, as well as some of the internal mind-based/SC-style reactions I often experience while reading novels:

- *2: Journaling; similar to "to-do"/notes-to-self, is a form of SC. It is actually you talking to your future self. E.g., random example of a to-do-list (TDL) note to self. "1) Read XYZ book, 2) Go do XYZ task." I'm telling Seif of the future to go do those things. Journaling is similar.*
- *Further explication of 2; journaling, etc. This point is not related to learning—i.e., the application I'm focusing on, re: SC. Mostly related to general [and/or self-improvement-related] SC. (^)*
- *3: Wonderment and exclamation in Luganda, in reaction to Ken Follet novel. Explication: by "wonderment," I'm referring to the good "pictures" painted by Follet of life in the dark/medieval, pre-Elizabethan, and Elizabethan, eras in England and Europe.*
- *"Columns of Fire" is about the transition from the older members of the House of Tudor to Elizabeth I, and from Catholicism to Protestantism. Most of the wonderment is in English. However, every once in a while, I express said wonderment in Luganda, using a common Ganda wonderment expression, which directly translates as "[my] mother, [dear] lady!" (^)*

Returning to the core theme of academic and other utility-purpose learning, I focus on the consumption (or specifically, reading) of complex knowledge-material. In other words, I implicitly contrast such instances (of learning complex knowledge-material) to the one above, in which we're reading for pleasure or are reading materials that are fairly easy to comprehend (despite the occasional need to pause and check our understanding). Subsequently, the concept/practice of "suspending understanding" reappears (in point five below).

- *4: Porpora's Thoughts and Barry Paradox papers; context: in ref. to my notes of the 26th. Abstract processing of ideas, knowledge, etc.*
- *Those papers discuss very complex philosophy of physics topics. While reading them, one feels as if, s/he's not really understanding said topics. My plan: to revisit the papers...[at some point, to (re-)test my understanding]*
- *Further explication of point 4: Deleuze and Guattari's book did not make sense either, and I highly doubt whether I will eventually understand it, at least without other online resources' expositions. In any case, the instructor—Brent—advised us to also revisit that book in the future, to evaluate our understandings over time.*
- *Relation to SC: my understandings of whatever I'm reading or learning are usually informed/ constructed via SC. "Do I even understand what s/he's//they're trying to say?" There are two parties within you, namely the primary and secondary self. The primary self is asking the secondary self, "do you even understand this (?)," [or] "what's going on here?" Usually, this process is happening by default or reflexively. But as I am doing this study, I realize that that process is playing out. It's very clear.*
- *5: "For now, just take it on faith..." Explication: David Malan often says that; he is the instructor of Harvard's CS50 course, a MOOC on Coursera. That course—along with other resources online—is how I'm learning software-development. He says that phrase as an implicit promise that he'll explain the meanings/details later. It stood out to me, because it's very similar to my own "suspend understanding" maxim/concept.*
- *Further explication of 5: E.g., with Doug's above-mentioned papers, and Douglas Hofstadter's "Goder, Escher, Bach." With the learning involved in those two works, the meanings will not be clear from the outset. One of the tools that help me learn is to say to myself, "hold on, do not*

panic; suspend understanding, the puzzle pieces are gonna make sense later. In the meantime, just try to keep concentrating, reading, learning, and later on you'll make the connections."

- *Clearly thus, I am not the only one with that thought. Further explication: Malan is talking to us and himself that way because he performing a formal lecture. But regardless, that "suspension of understanding" concept is clearly not just my own. It seems to be a useful technique vis-à-vis learning. (^)*

Thr, Feb 4, 2021; 12:59 AM (Voice-Note No.: *Batch 2; No. 2.16*); Me Composing a Question to my Mum in Luganda and English

In this final notes-excerpt for the current sub-genre ("Reality And Perception-Processing II: With Meta-cognition"), the pendulum swings back to general (non-learning-related) contexts. Specifically, I recount an instance in which I caught myself composing a specific question-sentence, which I was planning on asking my mum later that day. Thereafter, I discuss the potential implications of my self-observation in that instance, as well as whether or not the instance is indeed one of SC:

- *4: Me composing a question to my mum in Luganda and English.* "Wawulidde (*did you hear*) *about Kabushenga retiring from the New Vision Group (NVG)?" Clarification: NVG is a Ugandan print-news and general media-company. Explication: I caught myself composing the above question to my mum because of the current research.* [In other words, because of the current research,] *I try to execute metacognition on my own thinking processes.*
- *Further explication. The above words ("Wawulidde about Kabushenga retiring from the New Vision Group?") were the exact words that were forming in my mind, and I was planning to text my mum that question. I was not thinking about the topic in the abstract. With this research, I am trying my best to better understand human cognition and memory.*
- *Further explication; related question: what's the percentage break-down of our thoughts—i.e., by type (memories, future plans, general abstract thoughts, images, etc.). You can divide time into two basic parts: moments before now, and moments after now. And "now" keeps changing. Thus, in your mind, you can think about the time that you've already lived, or you can imagine the future.* [Editorial Note 1: It is rather interesting how the "present/now" is almost none-existent, or is very fleeting, in this breakdown.]
- *Further explication: I do not think the above concept/practice (/concept-practice) is indeed an instance of SC, as I was not composing that message for my own consumption.* [Editorial Note 2: The current self/Seif disagrees strongly with the self/Seif that made the preceding judgement! That instance is both one of SC, and interpersonal-comm.] *For instance, a typical pure-SC instance is one saying "I wonder how that works." But in the above case, I was composing a message for my mum's consumption.*

CHAPTER CONCLUSION

Given its arguably strategic location—i.e., the midway point of the book, I utilize this chapter in an attempt at firming up the topical demarcations that I use to categorize SC. Consequently, we are able to closely examine more instances from the voice-note/transcript data which indeed confirm that 1) we can isolate various instances that can clearly be defined as SC-instances, 2) we can utilize said SC-instances vis-à-vis the improvement of our personal learning methods and styles, 3) we often partake in SC without realizing it, and 4) we also often *deliberately* use SC to process reality and perceptions—i.e., our own as well as others'. It might also be worth mentioning that I utilize a "do not tell, show" method in this chapter, via a more robust use of extensive transcript excerpts. Regardless, the chapter's data re-presents many of the types of SC we have encountered in previous chapters (e.g. "to-do" lists and suspension of understanding), as well as instances that can only result from a heightened awareness of our SC-ability, such as my self-realization of the sentence I intended to divulge to my mother.

Creativity-Attempt Outtake 5

Post 8; 03-01-2021:
Of Connections And Chocolate

And thus it came to pass, that February 2021 ended, and March began. New news stories will be published/broadcast, and new government- and other organization-collected statistics will be released.

What role do I—and you—play in all of this? Are you employed or unemployed? Have you caught and luckily healed from the Corona virus, or did you (G_d-forbid) lose a relative to that dreadful disease?

As a social-scientist, I have always wondered about those interesting moments and interactions in which the micro—that's you and I, meet the meso—that's our groups of neighbors, as well as the macro—our countries, and the world.

Some random examples stand out in my mind: getting pulled over by a cop, visiting the DMV or passport office, voting, etc. But many—perhaps most (?)—such moments and interactions are easy to miss. You being friends with someone who will later become famous, you writing that paper that will later be very useful to that student somewhere, somehow, someplace.

The possibilities and links are endless. I guess Forrest Gump was right about life being similar to "a box of chocolate," indeed.

Here's to a great March, 2021!

Post 10; 03-15-2021:
^^Speaking of Time, What Is It?
[Picture of Wikipedia Article Deleted.]
Link to the above article: https://en.wikipedia.org/wiki/Time

For one reason or the other, my curiosity was piqued at some point yesterday, about the concept of time. What is it?

And so there I was, reading about its definition on Wikipedia: "Time is the indefinite continued progress of existence and events that occur in an apparently irreversible succession from the past, through the present, into the future," the article begins. And in a right margin, Wikipedia provides us with some related concepts, under the topic-subheadings of "major concepts ("Present, Present, Future")," "fields of study"—including history, archaeology, and horology among others, "religion and mythology," and "measurement/standards," among others. Inevitably, the article also discusses complex theories/frameworks such as spacetime.

Unfortunately, the paradoxical nature of some topics—including time—renders them somewhat "un-study-able," or un-understandable, so to speak. However much we study the topics, we cannot seem to definitively understand them. Our understandings might incrementally increase or improve across time and generations, but we can never claim to fully grasp such topics and their implications.

Of course, that does not mean that we should not study the topics. No, far from it, in fact. It means we need to keep studying them, but with a humble stance.

And if that argument does not convince you—as in, you do not think we need to keep studying complex topics such as time, perhaps this will: we have no choice. Otherwise, I can guesstimate that around—or, **at least**—half to two thirds of today's modern technological and other conveniences would not exist.

Besides, in addition to practicality/pragmatism, what would the world—and/or humanity—look like without wisdom? And without learning—including the study of complex topics such as time, wisdom would not exist.

In the mean*time* (😊), I hope you and I both use our time wisely this week. Let's not waste it; apparently, it is so precious and priceless, we cannot even truly measure it! And yet, as every second ticks by, our bones and muscles—as adults, that is—grow weaker. And sometimes unfortunately, our minds are also lost. Thus, let's make it count, this precious time we've been offered by the universe!

Post 11; 03-22-2021:
Of Time, Existence, And Questions

In a continuation of last week's topic or theme, I find myself pondering a few things about time. For instance, if time has no beginning and end, yet all living and non-living things are constantly changed by time—with many eventually ceasing to exist, what does that imply for what we should do with the finite time we have in the universe?

I suppose a related or similar question to the one above is that of meaning. What does or should it mean to be alive, or to exist? Before proceeding with this line of inquiry, I feel compelled to point out to anyone reading this, that no, I do not know why I am asking these questions. And in the end, they might be useless questions. But "let's just say" that "something tells me," that I am not the first person—nor will I be the last—to ask the questions.

A temporary answer to the above existentialism-related questions: unless one chooses to foolishly believe that they have the right answers to the questions, I believe some of the ways to correctly try to deal with the questions and their implications might be: 1) To continuously ask, answer, re-ask, and re-answer the questions—in other words, to keep searching for the truth about our existence, 2) Strive to rigorously live in the most ethical of ways, 3) To balance number one above with an effort at finding contentment with what we have here and now, versus what we seek in the future; and to enjoy and learn from the past, instead of yearning for it or regretting it.

It's funny/interesting how this entry evolved from a consideration of the concept of renewal from mindful repetition. I was thinking about how every time we mindfully close our eyes and savor our breath before opening our eyes again, or start a new "to-do" list at the beginning of a new week, etc., we are somehow renewing our existence. How in the world did my mind go from that concept, to one of existentialism as explicated above?

Self-Questioning for Creativity-Attempt Outtake 5

State any Necessary Introductory Remarks, and/or Answer this Question: Does it Make Sense to Me Now—i.e., In The Present—versus then—i.e., When I Wrote It?

Yes, the posts still make sense to me to date. But for the most part, I find myself struggling to make clear connections between them and the content of the chapter—more on this topic under number four below.

Does it Resonate with Me Now—i.e., In The Present— versus then—i.e., When I Wrote It?

Yes, it still resonates with me in the following sense: I am struck and somewhat befuddled at my own continuous wonderment about various concepts both big and small, concrete and vague, etc., including "connections [between societal structures and our micro-level roles in them]," "time," and "existence."

What Mood and/or State of Mind Does the Tone Suggest?

Intense curiosity.

What is/are the Direct and Indirect Relation(/s) to the Current Chapter via Topic(s) and Themes?

Under the bullet points below, I attempt to discuss the explicit connections I can draw:

- Self-doubt (as discussed in 5.2.1, and post 11):
  After alluding to it in 5.2.1, in the third paragraph of post 11, I concede that self-doubt will always be part of our experience, but we have to keep striving to achieve, regardless of that particular hindrance.
- News (as discussed in 5.2.2 and post 8):
  This connection is looser and rather indirect, but is a connection nonetheless. Whereas I ponder the value of my news-consumption in 5.2.2, in post 8, I wonder reflect on the role that we each play at any given time in the making of history; after all, as Philip L. Graham stated, news is the "first rough draft of history."
- "Ideas" (5.2.3) and "free will vs. determinism" (5.3.3):
  Overall, one can argue that all the posts at the end of the chapter demonstrate my constant grappling with the above concepts.

Any Other Evaluations or Comments?

N/A

Activities and Discussion Questions

1) Make a list of at least 5 things you've learned or thought about while reading the previous five chapters.
2) Share your list with your friends/family, classmates, etc. Get their reactions: can they relate? How/why, or how/why not?
3) Brainstorming mind/consciousness-states, part I; "what do I think/say to myself while...(?)":
   ◦ Sleeping (?): In your journal—or using another document specifically tailored for this activity, try to keep track of your dreaming patterns/behavior, or otherwise give some serious thought your dreaming patterns/behavior; e.g.: Do you dream often? Can you recall some good and/or bad dreams you usually have? Can you detect some patterns vis-à-vis the contents of your dreams, versus your stress-levels in your day-to-day life?
   ◦ Awake (?): i.e., general intrapersonal-/self-communication and thinking patterns.
   ◦ Planning your communications with other people (e.g. your family or friends, fellow students and teachers, boss or colleagues, etc.).
   ◦ Communicating with yourself, and thinking about your thinking and communication processes.
4) ***Optional***: Continue journaling/creative activity 1 as advised at the end of chapter one above.
5) Continue activity 6 from chapter one, "Comparison with other types of communication."

REFERENCES

Obama, B. (2020). *A Promised Land*. Penguin Books.

Porpora, D. V. (2011). How many thoughts are there? Or why we likely have no Tegmark duplicates $$ $10^{\{10^{\{115\}}\}}$ $$ m away. *Philosophical Studies*, *163*(1), 133–149. doi:10.100711098-011-9790-6

Thagard, P. (2005). *Mind: Introduction to cognitive science*. MIT Press.

KEY TERMS AND DEFINITIONS

Reality and Perception-Processing I; Without Metacognition: Intrapersonal-/self-communication executed or utilized by an individual to process the inputs of their senses and/or their thoughts, without realizing that s/he's in fact undertaking that process.

Reality and Perception-Processing II; With Metacognition: Intrapersonal-/self-communication executed or utilized by an individual to process the inputs of their senses and/or their thoughts, while aware that s/he's undertaking that process.

Self-(/Intrapersonal-)Communication: Communication inside an individual's mind, or outside—e.g., spoken or written—but for that same individual's consumption and/or use.

Chapter 6

Mr. Turtle (From "The Tortoise and the Hare" Fairy Tale):
Intrapersonal Communication/Cognition (and Sensemaking) and Learning

ABSTRACT

This chapter explores the potential application of SC by individuals vis-à-vis the improvement of their own learning methods. The author recaps the main characteristics of his own learning processes, and he (re-)introduces the second data-set of the study in earnest, i.e., a corpus of data- and cognitive-science articles. Altogether, the results indicate that individuals can utilize SC and cognitive-science principles in the improvement of their learning methods. The author identifies five main steps that individuals can use via SC, before and during learning processes. And for cognitive science, the author identifies three main aspects that summarize how that interdisciplinary field can be used to improve learning.

INTRODUCTION

In a similar fashion to chapter one's introduction, we can utilize two supporting stories to illustrate the important potential and/or actual role of SC in learning. One of the stories in question is a personal anecdote, and the other is a formal/academic "story" of sorts, a research paper by Warren et.-al (2001) about the role—or "logic" to use the authors' own term—of "everyday sensemaking" in science learning.

Mr. Turtle

First, I will share the personal anecdote. From around the beginning of the fall of 2020, I vigorously re-embarked on my quest to autodidactically learn software-development. As part of that initiative, I am consistently a utilizing a number of free or very affordable* online or practice-based resources and activities, including (but not limited to):

DOI: 10.4018/978-1-7998-7507-9.ch006

- *"Coding Dojo's* (a software-development learning-"bootcamp") "Basic-13" algorithms & Loiane Groner's *"Learning JavaScript Data Structures and Algorithms"* book
- Freecodecamp.org and their YouTube channel
- Javascript.info
- Ed-X/Harvard's CS50 Introduction to web development with David Malan
- *The Udemy The Python Mega Course 2021/2022 by Ardit Sulce, and
- Planning and building actual projects, i.e.:
    1) "Higher-Ed-X" —a website/app for helping prospective US-college-bound international students;
    2) "Pop-Scholar"—a soc.-media and mini-blogging site for lovers of general knowledge; and;
    3) A site for public-users to perform data-analysis and visualizations using Python and Flask.

For all the above three projects, I utilize the three main development computing software-tools (so to speak) for websites: i.e., HTML, CSS, and JavaScript. I also use the "MERN stack," a set of development tools based on the JavaScript programming language.

I also joined an online software-learning platform called "The Odin Project." This platform enables autodidactic-learners of software-development to utilize free quality resources curated from various online sources—including the ones in my list above, and it facilitates collaboration among aspiring software-developers/engineers.

Consequently, I linked up with a number of diverse learners from across the globe—USA, Asia, Europe, Africa, and we formed a study-group via the Discord collaboration platform. And over the course of the past one and a half years, I realized that as a result of my learning style/abilities and personal-economic/financial situation, I have to utilize a "slow-cook" learning method.

Eventually, I gave myself a nickname—"Mr. Turtle," based on the fairy-tale of "The Tortoise and the Hare." The rationale behind that nickname, which I often repeat to my *"Odinson"** study-group buddies to date (*the name we gave our study-group), is thus: they—i.e., my buddies—might rapidly overtake me with their impressive learning speeds, akin to cheetahs or Usain Bolt. But I on the other hand, will eventually win the race through a slow-but-steady run.

And I am glad to report that the approach has yielded substantively-quantifiable results to date. The doofus who could barely start a collaboration session via the VS-Code editor or write a standard "for-loop," can now correctly type out a number of JavaScript algorithms—e.g., printing the maximum, minimum, and average values from a given array, or using functions to somehow process array elements, and can comfortably start building and maintaining a basic web application using. In fact, some of my *"Odinson"* buddies and I are now building the above-listed "Higher-Ed-X," a real e-commerce app using the aforementioned JavaScript-based MERN stack. The app is the culmination of a business-idea I've been nurturing and have field-tested since 2016, with international students who intend to pursue higher education in the USA.

Warren et-Al's (2001) "Logic of Everyday Sensemaking"

As mentioned above, the second "story" is a research article whose thesis I'll utilize in this chapter. The article in question is by Warren et-Al's (2001), titled "Logic of Everyday Sensemaking." In sum, the authors of this paper demonstrate the wisdom of utilizing diverse learners' "everyday sensemaking" in the teaching of science. In other words, students'/(adult-)learners' multilingual and multicultural worldviews can be harnessed by teachers to help learners' understandings of various scientific concepts.

Based on this rationale, in the following sections of this chapter below, I will demonstrate the various unique/personal SC-related sensemaking strategies and tactics I utilize in my own learning in general. In fact, it should also be noted that I have utilized these same sensemaking strategies and tactics in my synthesis of the potential insights or learning strategies I present throughout the chapter.

MAIN FOCUS OF THE CHAPTER

Recap of Book's Research Questions and Relevant Voice-Note Findings

Recap of Research Questions

Before digesting the relevant research-findings, readers might benefit from a recap of the book's overarching research questions, and I will thus re-present them below.

RQ 1: *How can I analyze my own thoughts and/or intrapersonal-communication, to help me improve my learning methods and self-motivation?*

RQ 2: *How can I use the techniques of cognitive-science and other relevant academic and applied fields to help improve my learning processes?*

And as mentioned in chapter one, for each of the above questions, the respective methods I use throughout the book are: 1)—a careful self-study via recording, transcribing, and analyzing of voicenotes; and basic discourse-analysis of private and public journal-entries, and 2)—a basic qualitative content-analysis of academic-journal-published and other online-derived studies, as well as instructional articles on the study of data-science, and about learning and creativity in general.

Voice-Note Findings

Next, before examining the analysis results of the second-listed method above, we should holistically review the relevant findings from the voice-notes/transcripts data for some observations in relation to my learning style. And for that purpose, below is a synthesis of at least 15 potential insights (among others) that we can glean from the three summary-transcripts of the recorded voice-note data (the transcript unit numbers are indicated in parentheses, or at the beginning of the potential-insight; please refer to the appendices A through C for all relevant transcripts):

1) Our lives' varied routines and vagaries (e.g., over the past three years—i.e. 2019 to 2021, the obstacles from the COVID pandemic) necessarily influence our learning motivation and progress (voice-note-transcript no.: batch 1, no.-3; also related to 3.1.6).

2) There are key-points in our learning journeys, from which we gain insights into the various topics or skills we're trying to learn. E.g., my figuring out of the fact that inferential statistics are mostly

about testing for differences and relations between one or more variables or groups of variables of nominal, ordinal, or interval-ratio natures (voice-note-transcript no.: 2.1.3).

3) You can apply theoretical/abstract and empirical reasoning to your past learning experiences to figure out your learning style and the best ways for you to learn, regardless of your supposed IQ- or other limits (voice-note-transcript no.: 2.1.9).

4) Before embarking on the learning of a topic, it might be helpful to ask yourself: *"where on the 'knowledgeability spectrum' [of said topic]…am I?"* (voice-note-transcript no.: 2.1.11).

5) Keep monitoring your emotional and/or mental state, and know that adverse emotional or mental states can impede your learning (voice-note-transcript no.: 2.1.11).

6) Some/often-times, the strategy of suspending understanding can help us; try to set aside that element that you do not understand, and revisit it later, then integrate it with the rest of the contents that you understood. However, that strategy might be harder to implement with STEM subjects, whose learning often relies on prior understanding of lower-level topics. Still, it's arguably possible even with them, e.g., memorizing how to solve for "X" in "$2X = 10$," even though you're not quite fully sure of the rationale or utility of the equation's style or format (voice-note-transcript no.: 2.1.11).

7) From voice-note-transcript no. 2.1.12: It might be helpful to figure out your cognition styles or habits. E.g. that while reading creative fiction and nonfiction, you visualize things. And/or:

With more prosaic or academic/noncreative books and learning, I'll use words to make meaning. "When [words] get into my brain, they dissolve into concepts—into abstract concepts." Thus, in my mind, I'm not visualizing words; rather, I'm processing the meanings of the concepts represented by the various words I'm reading.

8) Despite our self-doubts, we can utilize "brute-force learning" to persistently keep learning various topics or skills (voice-note-transcript no.: 2.1.14).

9) In my own experience, I believe learning is inevitably intertwined with the concept(s) of style or personality; i.e., an individual learns best via their own preferred style (voice-note-transcript no.s: 2.2.12, and 3.1.1).

10) Regardless of how difficult it is to learn any given topic, one can utilize the following two strategies: 1)—Break-up whatever you're learning into "bite-size pieces,"; and
2)—Maintain hope/optimism: realize that as a lifelong-learner, time is your best friend (voice-note-transcript no.: 2.2.16).

11) One has to continuously self-assess in terms of his/her own learning abilities and methods; i.e., so as to ensure ongoing progress and/or for learning-strategy reviews (voice-note-transcript no.: 3.1.3).

12) Food for thought 1: Can one cultivate an intense curiosity in learning contexts, similar to the one we have while reading or watching riveting stories or movies for leisure? For instance, while reading an academic book of a given subject (e.g., chemistry/computer-science/communication, etc.), somehow compelling yourself to look forward to the end of the book to know how that academic book's "story" ends (voice-note-transcript no.: 3.1.6)?

13) Food for thought 2: Considering—for instance, the paywalls of various good-quality journalism-publications, *"how easy is it for a poor person, … or* [an otherwise] *disadvantaged person to implement their goal of becoming a polymath? Ans.: it's not easy for a poor person to cultivate that trait. Apparently, knowledge is* [an economic] *commodity. Sure, there are ways for one to learn without*

paying. But, it's definitely harder [for poor individuals], *compared to more privileged members of society* (voice-note-transcript no.: 3.1.6)."

14) Unfortunately, an intense love of learning—including the desire and eventual pursuit and acquisition of advanced degrees, and/or the desire of being a polymath (as in my case) causes various social problems for individuals e.g. (among many others): social-awkwardness and loneliness, social-judgments of pretentiousness and/or arrogance, or of being "pedantic." Among other remedies, there are at least two that lovers of knowledge can utilize:

    i.   *We should strive to develop our self-confidence; "this is who I am; I love learning, and* [/but] *society tends to shun people that love learning. But I am who I am..." Otherwise, you'll feel like you're doing something wrong, i.e., loving to learn.*

    ii.   *Sometimes, you* [need to] *"shut up"...it's ok to share, e.g. in class, but it has to be relevant, not because you love hearing the sound of your own voice, or other vain or inappropriate reasons. Sharing is ok, but do not overdo it; know when it's appropriate, etc.*

15) I believe software-developers'/computer-scientists' use of the "rubber-duck debugger method" vindicates SC: ultimately, you're talking to yourself, NOT the rubber-duck! And by going through that process, you very often realize and correct your programming error (voice-note-transcript no.: 3.2.6).

Presentation of Potential Insights From Set II Data: Data-Science (and General Self-Improvement Knowledge), and Cognitive-Science Instructional and Research Articles

Introduction to Data-Set Contents

Unlike chapter one's much more succinct counterpart data-set sample, the contents of the data-set utilized in the current chapter—as well as in chapters eight (digital technology and AI) and nine (final potential insights)—are much more expansive, and are composed of three sub-data-sets. Those three sub-data-sets are:

1) i) Two general-knowledge articles/resources (listed under "*Part 1 > I: Listing > A—General/ Other*" in appendix D), which are representative of the rather myriad number of articles/resources I constantly email myself from various online sources, or otherwise consume from within various web applications (especially LinkedIn, Google News, and other sources), namely: "12 Ways to Get Smarter in One Infographic" by Jeff Desjardins, and the creativity quiz from MindTools.com. ii) 40 instructional articles about data science, which I emailed myself over time (see timeframe under 6.3.2 below)—sourced mostly from LinkedIn, where they had in turn been reposted mostly from a portal called "Data Science Central."

2) Cognitive science articles from the "*Cognitive Science*" and "*TopiCS in Cognitive Science*" journals, and

3) Cognitive science articles from Google Scholar, sub-grouped using Thagard's (2005) six cognitive-science mechanisms, namely i) logic, ii) rules, iii) concepts, iv) analogies, v) images, and vi) connections.

Other Details, Categorization, and (Pre-Significance-Analysis) Orientation

The publication-date ranges for all the above data is as follows: 1) 2020-2021 for the general knowledge and data-science articles/resources, 2) January to June of 2021 for "Cognitive Science" journal articles, and 3) from 1993 (earliest) to 2020 (latest) for the Google Scholar articles. Overall, compared to the interdisciplinary but relatively straightforward academic/research field of cognitive science—i.e., wherein the articles are easier to categorize by (sub-)discipline, the topics and themes of sub-data-set 1 are harder to categorize.

Brief Overview of General Knowledge and Data-Science Articles/Resources

Despite the challenge I share at the end of the above paragraph, I identified the main sub-disciplines as mainly two, namely: 1) "Math, Statistics & Data-Science," and 2) "Software-Development," with an additional one added for any articles with markedly ambiguous foci, namely, "Learning and Professional Development." Within those main disciplines, I identified the topics and themes covered as: 1) "learning," 2) "tools and techniques," 3) "careers," 4) "how-to," and 5) "other." It should be noted that the delineation is not clear-cut, with significant overlap among the sub-topics and themes of the different articles in each of these five discretionary groups. Moreover, it can be argued that all articles are in fact about "learning" by default, notwithstanding their main explicit topics and themes. I also noted the following partial implications:

- The articles reflect the breadth and depth of the tools and techniques available for the interdisciplinary field of data science—i.e., mostly involving statistics, computer-science, information-science, and graphic-design (e.g., refer to the relevant *"Data Science"* Wikipedia article).
- They also demonstrate that there is a focus by authors—albeit not as intensive as one would wish—vis-à-vis guiding potential career-changers or early-career professionals.
- Finally, the articles also reflect the dearth of articles that provide step-by-step "how-to" instructions. However, there are plenty of online sources for such instructions, including MOOCs, YouTube, etc., and those other platforms are arguably better-suited for the purpose.

Significance Analysis of Cognitive Science Articles: Brief Re-Introduction and Presentation of Results

On the other hand, there are at least two main characteristics that can be identified among the cognitive science articles. The first is the specific typology of their interdisciplinarity. Among other academic disciplines, I can list at least five major evident ones, namely: psychology, linguistics, education, computer-science, and philosophy.

The second defining characteristic is the authors' apparent propensity for implying the validity of the *computational-representational understanding of mind* (or "CRUM") hypothesis—i.e./e.g. as discussed by Thagard (2005) and Lieto (2021), or for somehow utilizing computational-analysis methods, regardless of their precise disciplines, or their papers' precise topics. To help me gauge the helpfulness of this chapter's data-set—i.e., data- and cognitive-science articles, I had planned to utilize a set of five questions, which I introduced earlier in chapter one. However, given its inefficacy in chapter one's pilot study, I eliminated question five. I also eliminated question three from the current analysis, but I still perform a thematic analysis in chapter nine.

After those two adjustments, below is a recap of the three questions I used in this chapter's significance analysis (please refer to table 6.3.3-I and appendix D for details):

1. Regardless of new vocabulary, formulae and notations, methodologies, etc., do I understand the studies' main findings? (Yes/No/Maybe)
2. Can I restate the findings, or interpret their implications, in my own words? (Yes/No/Maybe)
3. On a scale of 1 to 10, how can I rate the helpfulness, relevance, value, etc., of the studies to me personally—i.e., for my own professional and personal or general growth or improvement?

To enable a more robust discussion of this data-set, I have to utilize a basic mixed-methods format, with the results presented using a limited use of basic descriptive statistics, as well as a qualitative discussion about the scores presented. To execute this strategy, I had to quantify the "Yes/No/Maybe" in questions one and two; "Yes" is represented using the number "1," and "No" and "Maybe" are represented by a "0." On the pages below, I present the quantitative summary, and the (brief) qualitative discussion.

I: Significance Analysis: Brief Discussion, Part I

Similar to the results in chapter one, the data-set reveals that Google Scholar articles tend to be a lot more efficacious to my own learning-method improvement, in comparison to the *Cognitive Science (CS)*" journal articles. *Prima facie*, this makes obvious sense: the metaphor I can use to differentiate

Table 1. "Cognitive Science" journal significance-analysis results (Please refer to Appendix D for details of articles)

Article No.	Qn. 1	Qn. 2	Qn. 3
1.1	1	1	9
1.2	1	0	5
1.3	1	0	6
1.4	1	1	8
1.5	0	0	4
1.6	1	1	9
1.7	1	1	6
1.8	0	0	4
1.9	0	0	7
B.-1 Sub-Total	*6*	*4*	*58*
2.1	1	0	5
2.2	1	1	5
2.3	1	1	6.5
2.4	1	0	5
2.5	0	0	4

continues on following page

Table 1. Continued

2.6	1	1	9
2.7	1	1	5
2.8	1	1	9
2.9	1	0	7.5
B.-2 Sub-Total	*8*	*5*	*56*
3.1	1	0	5
3.2	0	0	3
3.3	0	0	3
3.4	1	0	4
3.5	1	1	7.5
3.6	0	0	4
3.7	0	0	3
3.8	1	1	5
B.-3 Sub-Total	*4*	*2*	*34.5*
Total	18	11	148.5
% (Total/N)	69%	42%	57%

the two sets of articles is: a menu that is forced on you (the CS journal), versus a menu that you choose for yourself (Google Scholar).

In other words, whereas I had no authority vis-à-vis the articles that would be published—as well as which of those published ones would be open-source, I had the freedom to "pick and choose" which articles I wanted to add to the data-set from Google Scholar. However, despite Google Scholar's clear advantage, I am somewhat struck by the degree to which it is not that much more helpful compared to CS—i.e., a mere additional 10 percentage-points.

II: Significance Analysis: Brief Discussion, Part II

What more potential insights can we glean from these articles vis-à-vis the individual significance-analysis questions posed farther above (under 6.3.4)? First, let's examine the scores of the first two questions, i.e., "do I understand (?)," and "can I restate in my own words (?)," respectively. And in this regard, the CS articles are arguably more helpful, given the variability within them as far as the scores for each of these questions.

And within that sub-set, we can use a total of six articles for this further exploration, a side-by-side comparison of two articles from each batch (1 to 3); one, an article with positive scores ("1") for both those two questions, the other with negative scores ("0") for both of those two questions.

Overall, one possible explanation that can emerge from this sample—i.e., vis-à-vis one's ability to understand and restate—is in relation to the methods used by the authors. Apparently, the more the authors make use of (sub-)discipline-specific esoteric terminology and/or methods (in this case, usually computational-analysis), the harder it becomes for me to understand and restate their work.

However, two important caveats have to be stated here: 1) Regardless of the above-described challenges, it was often possible for me to understand the overarching message and/or thesis, thanks to the

Table 2. Google Scholar significance-Analysis results (Refer to Appendix D details)

Article No.	Qn. 1	Qn. 2	Qn. 3
L1	1	1	10
L2	1	1	6
L3	1	1	7.5
L4	1	1	8.5
L5	1	0	5
L-Sub-Total	*5*	*4*	*37*
R1	0	0	4
R2	1	1	10
R3	1	1	9
R4	1	1	8
R5	1	1	8
R-Sub-Total	*4*	*4*	*39*
C1-1	1	0	5
C1-2	0	0	3
C1-3	1	1	7.5
C1-4	1	1	7.5
C1-5	0	0	4
C1-Sub-Total	*3*	*2*	*27*
A1	1	1	8
A2	1	1	7.5
A3	1	0	7
A4	1	1	7.5
A5	1	1	8
A-Sub-Total	*5*	*4*	*38*
I1	1	1	8
I2	0	0	4
I3	1	1	5
I4	1	1	7
I5	0	0	6.5
I-Sub-Total	*3*	*3*	*30.5*
C2-1	1	1	8.5
C2-2	1	0	7
C2-3	1	1	6
C2-4	1	0	0
C2-5	1	1	8
C2-Sub-Total	*5*	*3*	*29.5*
Total	**25**	**20**	**201**
% (Total/N)	**83%**	**66%**	**67%**

"*Abstract* + "*IMRAD*" format of academic writing. 2) There were several instances in which I could go beyond that basic thesis-understanding and was able to intuitively understand (arguably, at least,) several of their main points, even though I could not understand the minutiae of their calculations and explanations.

III: Significance Analysis: Brief Discussion, Part III

Finally, we can scrutinize the data-set to try to deduce what makes some articles more helpful than others. And in that regard, the Google Scholar subset is potentially more efficacious, given its higher performance for that metric. In the table below, I list some of the most helpful articles I identified. Farther below, I try to explain why I gave each of these articles their high scores.

III-a: MacWhinney (2013)

This article offers an accessible overview and explanation of the Unified Competition Model of second-language learning. Without delving too deep into technical minutiae or relying on any one specific method, the author clearly explains what the model is, and how it can be used to help L2 learners. As someone who is passionate about consolidating my French knowledge, and learning Swahili, Arabic, Mandarin,

Table 3. "Cognitive Science" journal Q1 & Q2 high-vs.-low score article-sample

High-Score Article	*Low-Score Article*
(Batch 1)	**(Batch 1)**
The Relation Between Cognitive Abilities and the Distribution of Semantic Features Across Speech and Gesture in 4-year-olds; authors: Abramov et Al (2021).	Do Humans and Deep Convolutional Neural Networks Use Visual Information Similarly for the Categorization of Natural Scenes? Authors: Cesarei, Cavicchi, Cristadoro, and Lippic (2021)
(Batch 2)	**(Batch 2)**
Concept Appraisal; author: Sapphira R. Thorne, Jake Quilty-Dunn, Joulia Smortchkova, Nicholas Shea, James A. Hampton (2021).	Modeling Misretrieval and Feature Substitution in Agreement Attraction: A Computational Evaluation; authors: Paape, Avetisyan, Lago, and Vasishth (2021).
(Batch 3)	**(Batch 3)**
Are There Cross-Cultural Legal Principles? Modal Reasoning Uncovers Procedural Constraints on Law; authors: Hannikainen Et Al. (2021).	Monotone Quantifiers Emerge via Iterated Learning; authors: Fausto Carcassi, Shane Steinert-Threlkeld, Jakub Szymanika (2021).

German, and other languages, I found this high-level explanation very helpful.

III-b: Koedinger, Corbett, & Perfettic (2012)

The main reason for this article's high score is its deft explanation of various new concepts, alongside its reinforcement—i.e., for me personally—of concepts with which I've already been familiar or well-versed. For instance, the authors use the concept of "learning events" to discuss memory, integration of new concepts and *sensemaking* (special emphasis on the last concept, given my prior expertise in it).

Table 4. Google Scholar high "helpfulness"-Score article sample

Logic: The Logic of the Unified Model; author: Brian MacWhinney (2013). **Score: 10**
Rules: The Knowledge-Learning-Instruction Framework: Bridging the Science-Practice Chasm to Enhance Robust Student Learning; author: Kenneth R. Koedinger, Albert T. Corbett, Charles Perfettic (2012). **Score: 10**
Concepts: Linking Cognitive Science to Education: Generation and Interleaving Effects; authors: Lindsey E. Richland, Robert A. Bjork, Jason R. Finley, Marcia C. Linn (2005). **Score: 7.5**
Analogies: Analogy, higher order thinking, and education; authors: Lindsey Engle Richland and Nina Simms (2015). **Score: 8**
Images: Capturing human categorization of natural images by combining deep networks and cognitive models; authors: Ruairidh M. Battleday, Joshua C. Peterson, & Thomas L. Griffiths (2020). **Score: 8**
Connections: Making Connections in Math: Activating a Prior Knowledge Analogue Matters for Learning; author: Pooja G. Sidney and Martha W. Alibali (2015). **Score: 8**

III-c: Richland, Bjork, Finley, & Linn (2005)

Similar to the above paper, I also deem this article very useful, thanks in part to its smooth integration of new concepts into my knowledge repertoire. But for the most part, I rate it highly because of its reinforcement—with some helpful evidence-based caveats—of the utility of "learning by doing."

III-d: Engle and Simms (2015).

Unfortunately, the main reason for my high score for the current article seems to reveal a pattern of bias for knowledge that reinforces what I already know. After all, Engle and Simms (2015) demonstrate the utility of analogies. However, they also caution educators to be careful of their (i.e., analogies') use in teaching, especially in regard to whether or not the chosen analogy is appropriate for the subject-matter at hand.

III-e: Battleday, Peterson, & Griffiths (2020)

Unlike the above articles, I have to concede that for the most part, I rate this particular article highly because of mere fascination. In other words, I only have a very basic understanding of the subject matter. But because it is related to machine learning—specifically, to convolutional neural networks—both of which I find fascinating, I rated it highly.

III-f: Sidney & Alibali (2015)

Finally, I am particularly proud of my high rating for this particular article, given its (i.e., my rating's) basis in logic/rationality. The reason (for the article's high "helpfulness" score): *it makes sense*! Not to mention, it reminds me of the concept of "frame of reference" (e.g., Majid et. al, 2004; also refer to the relevant Wikipedia article, "*Frame of Reference > Disambiguation*").

CHAPTER CONCLUSION: RELATION BETWEEN SET-I AND SET-II FINDINGS AND/OR POTENTIAL INSIGHTS

Chapter Recap

In addition to the general/other utilities that SC can help us with, the improvement of our own personal methods of learning—i.e., in addition to, and/or versus other generally-accepted theories and practices of pedagogy and adult-learning—is a particularly important one. Given that importance, this chapter attempts to holistically (re-)examine some of the relevant observations that can be applied to that purpose—i.e., the improvement of individuals' own learning methods via SC. Each of the data-sets' results above suggests a number of strategies and tactics that can be utilized via SC and cognitive science theories or hypotheses to help us improve our own learning methods, and the two data-sets' results complement each other.

Summary and Significance of Set-I Data

One way to conclusively utilize data-set I's findings is thus: we can glean from them a set of guidelines that can be used cyclically throughout the courses of learning various topics/skills over one's lifetime. Out of the 15 listed findings, we can eliminate some, then group many of them together and suggest:

I. Before embarking on a learning journey:
   1) Perform a baseline analysis of how well you already know (about) a given topic/skill, and your purpose and level of motivation for learning the topic/skill. However, I should add here that learners should also try to avoid becoming victims of the Krueger-Dunning effect (1999); i.e., over-rating their knowledge. In other words, strive to be intellectually honest and humble with yourself.
   2) Review your relevant past experiences and best practices or hindrances vis-à-vis learning, thus improving your learning methods in this new learning round.
II. During the journey:
   3) Monitor your ongoing understanding of topic/skill "X," as well as your general emotions, as well as your emotional reactions to your progress vis-à-vis learning topic/skill "X." E.g., are you pleased (maybe even getting complacent), or you are sad about your sub-par progress (?).
   4) Regardless, keep focusing on the longterm goal, but break up that goal into easily-achievable sub-goals.
   5) If you do not fully understand some of the constituent sub-topics of topic/skill "X," do not lose heart. Try to first understand the big picture, and/or the basic building blocks of the topics/skills in question, then try to figure out how all constituent parts are related to each other, and to other topics/skills and subjects.

These steps can be used repeatedly/cyclically to (re-)learn or improve/consolidate our knowledge and technical know-how for the same or different individual topics, disciplines, knowledge-domains, etc.

Summary and Significance of Set-II Data

Beyond that cyclical process, data-set II extends our understanding of how we can harness the mechanics of our cognition processes to improve our learning methods. Overall, three specific aspects can be reviewed in this regard, namely: 1) the efficacy of utilizing our power of agency or choice, vis-à-vis learning materials and methods, 2) the utilization of social/formal conventions as cognitive map-tools during learning, and 3) the use of various specific evidence-based strategies and tactics, e.g., via Thagard's (2005) mechanisms-grouping.

One can argue that the key lesson we can take from Google Scholar's higher helpfulness score is in relation to the earlier-stated metaphor of its articles being a self-chosen vs. imposed menu. In fact, we can expand the lesson learned thus: as various adult-learning theories caution, adult learners (and perhaps, younger ones as well) can be assisted in their learning by sharing control/ownership of their learning processes with their instructors (e.g.: theories and best-practices such as andragogy, transformative learning, and experiential learning, as reviewed by New England Institute of Technology 2021, and Western Governor's University, 2020).

But even with that agency, learners have to constructively engage with the conventions/ rules/frameworks of learning that most authors use to package knowledge. As such, learners will more-than-likely find it beneficial to acclimate themselves to those features of their chosen knowledge domains and their publications or learning resources. For instance, in my case—as I point out farther above, the *"Abstract + IMRAD"* format was helpful vis-à-vis my attaining of an overall or basic understanding of the cognitive science articles, despite my inability to fully grasp many/most authors' methods.

Finally, section 6.3.5-III above demonstrates the usefulness of the various strategies and tactics that we can glean from cognitive science studies vis-à-vis the improvement our learning methods. Most of those six articles (besides the one by Battleday, Peterson, & Griffiths, 2020) help me improve my understanding of how minds work with respect to the learning process, and I can in turn harness that understanding in my learning-method strategy-reviews going forward.

Comparison to other Relevant Studies/Findings

So as to try and ensure a fairly rigorous—albeit basic—review, I solicited advice from some collaborators in regard to any studies they can easily cite which discuss best-practices vis-à-vis learning methods. In this regard, Dr. Michael Wolfe[1]'s advice was especially invaluable (personal communication, April 2021). His advice helped me identify seven impactful studies from the interdisciplinary field of instructional psychology/cognition, which can also help us farther our understanding of the mechanics of learning and cognition.

All-in-all, the results discussed farther above do not contradict these studies' results. Rather, similar to the complementarity of the above-discussed data-sets' results with each other, this chapter's overall findings complement those of the seven studies in question.

Altogether, the findings of these studies suggest that:

1) Simplifying tasks/topics for novices can help them learn well in the beginning, but making the topics/tasks gradually harder helps learners retain those tasks details better over the longterm (Mannes & Kintsch 1987, and Kintsch 1994);

2) Repeated testing—i.e., of the same material—over time improves retention (Roediger and Karpicke 2006);

3) Distributed practice is better than overlearning, and spacing helps with smoother induction of new (sub-)topics/skills (Rohrer & Taylor 2006, and Kornell & Bjork 2008);

4) Interleaving of content helps learners solve problems better via improving their judgements of the appropriate procedures or problem-solving solutions they should use (Taylor and Rohrer 2010); and

5) More deliberate self-monitoring of our understanding of the topics/skills we're studying is indeed efficacious for the improvement of our own learning methods (Thiede, Anderson, Therriault 2003).

Final Reflection on the Concept and Practice of Metacognition in General for Learning-Improvement and other Contexts

As I briefly mention in the introduction chapter, as well in as chapter two, this book's focus—i.e., intrapersonal-/self-communication—necessarily involves the study of our cognition processes. In other words, in addition to focusing on SC in a pure sense—i.e., as sets of messages to ourselves composed of clearly formed words (both spoken and/or written, or clearly articulated within our minds) or even abstract and clear feelings, the book's research is also focused on the dynamics of the cognitive processes from which those SC messages arise.

Given that fact, I would be remiss not to focus directly—albeit in a basic manner—on the body of knowledge that has been collected so far on the concept and practice of metacognition (e.g., Frazier, Schwartz, and Metcalfe 2021). Simply put, metacognition refers to the idea and practice of *thinking about the way we think, and how we can improve those processes.*

Throughout this book—and in the current chapter especially, I have more-or-less harnessed that idea by default—often indirectly, but sometimes directly, in my investigations of SC. And as this chapter demonstrates, we can indeed critically examine our cognition processes, patterns, flaws, etc., to solve a specific problem—e.g. in this case, the improvement of our learning methods. However, this promising application highlights other potential problems that might also benefit from the use of metacognition in their amelioration. For instance, can we utilize it—i.e., metacognition—to tackle our mental health challenges, and to improve our designs and interactions of/with digital systems, including AI? In the next two chapters, we explore this problematic in earnest.

Creativity-Attempt Outtake 6

Post 36; September 14, 2021 (In Lieu of Timely Written Post):

My First National-Newspaper Article; Passing An Exam; A Luta Continua

And so it came to pass that a couple of weeks ago, my first op-ed article was published in a national (Ugandan) newspaper, "The Observer," (https://observer.ug/viewpoint/71047-in-defence-of-children-and-adults-of-kyeejo) and today—September 14, 2021, I did and passed my CAPM certification exam. In between those two events, I was hospitalized due to stress-related illnesses, and of course, I continuously deal with life's vagaries day in, day out.

One of the subliminal-yet-significant points we can highlight from the above anecdote(s) is the fact that the outcomes we call "achievements" are often muddied by life's other less glamorous outcomes. And yet all of it together—the "good" and "bad"—is the holistic or even wholesome package that is defined as life, for better or worse. One has to—even if in a resigned manner—embrace that complexity and messiness.

There I go, gloriously winning with those two achievements, and there I will go, failing spectacularly with other pursuits. *A luta continua, victoria e certa.* Just keep chugging along!

Post 27; July 12, 2021:
"Dear Journal"

As I continue writing my intrapersonal/self-communication book, I have to constantly review my strategy, in a bid to ensure that I finish the book on time, and to also ensure—perhaps most importantly—that the book makes sense. And sometime within the past couple of weeks, I realized that I might not have enough "raw data" to help me achieve the contract-mandated minimum, i.e., 85,000 words. I was alarmed; "how am I going to pull this off (?)," I wondered.

Then, serendipitously, I realized that I that I can document the growth and evolution of my intrapersonal/self-communication habits ever since my late teenage years, aided by the journals and blogs I've written over the years. It's an amazing feeling, reading the entries one wrote while in high school. You can hear echoes of your current evolved self, but you're also witnessing—from a temporal distance—the mental and general struggles of another person, in a sense.

Well, thank goodness for this medium of communication, writing. Let's see where and how the book effort ends, and I guess I can also look forward to reading these words 20 or more years from now, G_d willing.

Self-Questioning for Creativity-Attempt Outtake 6

State any Necessary Introductory Remarks, and/or Answer this Question: Does it Make Sense to Me Now—i.e., In The Present—versus then—i.e., When I Wrote It?

Yes, the posts make sense to me now, just as they did when I first wrote them.

Does it Resonate with Me Now—i.e., In The Present—versus then—i.e., When I Wrote It?

Yes, they still resonate with me today (late Nov, 2021), vs. when I first wrote them, i.e. mid-September and July (of 2021), respectively.

What Mood and/or State of Mind Does the Tone Suggest?

>> Post 36: Whereas I vividly recall being very happy—even *ecstatic* perhaps, because of both achievements—especially the CAPM, this post demonstrates an effort vis-à-vis equanimity. In other words, despite my marked happiness, I am striving not to enjoy the moment too much, and I state the reason

for that restraint. I know that the situation is temporary, and I should be expectant and mentally prepared for the upcoming low-tides.

>> Post 27: this post also reflects a practiced restraint; but in this case, the restraint is aimed at blunting a mood or state of mind of relief, given my then-recent serendipitous realization (that I could use my journal/blog-entries as raw data).

What is/are the Direct and Indirect Relation(/s) to the Current Chapter via Topic(s) and Themes?

Both posts are directly related to the chapter in the following ways:

>> Post 36: even though I do not state it explicitly in the post, I was only able to pass that (CAPM) exam because of a rigorous study and review regimen, which is of course a typical method of formal learning.

>> Post 27: The relation in this case is much more subtle, but typically demonstrative of one of the ways in which I realize my learning milestones or achievements: instead of rigorous memorization and utilization via exams or other assessments, I realized that I already had a conceptual/literary tool with which I could solve a relevant conceptual/literary problem.

Any Other Evaluations or Comments?

N/A

Activities and Discussion Questions

1) Brainstorming mind/consciousness-states, part II; "what do I think/say to myself while…(?)." Review and discuss your list with the group/class.
2) What do you enjoy learning about most? And/or: what do you enjoy reading most? And/or: what do you know about your learning style—what kind of learner are you? For instance, do you think you learn better via visual aids, or via trying or doing things, instead of passive learning? Discuss.
3) Introduction to journaling activity 3: Keep a journal for the next week—from today (e.g. Monday/ Tuesday/Wednesday etc.,) to the same day next week. Write as much as you can about the following topic: what are your favorite books, or movies and/or TV shows, and what do they make you think about while/after watching them?
4) *Optional*: Continue journaling/creative activity 1 as advised at the end of chapter one above.
5) Continue activity 6 from chapter one, "Comparison with other types of communication."

REFERENCES

Battleday, R. M., Peterson, J. C., & Griffiths, T. L. (2020). Capturing human categorization of natural images by combining deep networks and cognitive models. *Nature Communications, 11*(1), 1–14. doi:10.103841467-020-18946-z PMID:33110085

Frazier, L. D., Schwartz, B. L., & Metcalfe, J. (2021). The MAPS model of self-regulation: Integrating metacognition, agency, and possible selves. *Metacognition and Learning*, *16*(2), 297–318. doi:10.100711409-020-09255-3 PMID:33424511

Groner, L. (2018). *Learning JavaScript Data Structures and Algorithms: Write complex and powerful JavaScript code using the latest ECMAScript* (3rd ed.). Packt Publishing.

Kintsch, W., Weaver, C. A., Mannes, S., & Fletcher, C. R. (1994). *Discourse comprehension: Essays in honor of Walter Kintsch*. L. Erlbaum.

Koedinger, K. R., Corbett, A. T., & Perfetti, C. (2012). The Knowledge-Learning-Instruction framework: Bridging the science-practice chasm to enhance robust student learning. *Cognitive Science*, *36*(5), 757–798. doi:10.1111/j.1551-6709.2012.01245.x PMID:22486653

Kornell, N., & Bjork, R. A. (2008). Learning concepts and categories: Is spacing the "enemy of induction"? *Psychological Science*, *19*(6), 585–592. doi:10.1111/j.1467-9280.2008.02127.x PMID:18578849

Kruger, J., & Dunning, D. (1999). Unskilled and unaware of it: How difficulties in recognizing one's own incompetence lead to inflated self-assessments. *Journal of Personality and Social Psychology*, *77*(6), 1121–1134. doi:10.1037/0022-3514.77.6.1121 PMID:10626367

Lieto, A. (2021). *Cognitive Design for Artificial Minds* (1st ed.). Routledge. doi:10.4324/9781315460536

MacWhinney, B. (2013). The logic of the unified model. In *The Routledge handbook of second language acquisition* (pp. 229–245). Routledge.

Mannes, S. M., & Kintsch, W. (1987). *Knowledge organization text organization*. Academic Press.

Richland, L. E., Bjork, R. A., Finley, J. R., & Linn, M. C. (2005). Linking cognitive science to education: Generation and interleaving effects. In *Proceedings of the twenty-seventh annual conference of the Cognitive Science Society* (*Vol. 27*, pp. 1850-55). Academic Press.

Richland, L. E., & Simms, N. (2015). Analogy, higher order thinking, and education. *Wiley Interdisciplinary Reviews: Cognitive Science*, *6*(2), 177–192. doi:10.1002/wcs.1336 PMID:26263071

Roediger, H. L. III, & Karpicke, J. D. (2006). Test-enhanced learning: Taking memory tests improves long-term retention. *Psychological Science*, *17*(3), 249–255. doi:10.1111/j.1467-9280.2006.01693.x PMID:16507066

Rohrer, D., & Taylor, K. (2006). The effects of overlearning and distributed practice on the retention of mathematics knowledge. *Applied Cognitive Psychology: The Official Journal of the Society for Applied Research in Memory and Cognition*, *20*(9), 1209–1224. doi:10.1002/acp.1266

Sidney, P. G., & Alibali, M. W. (2015). Making connections in math: Activating a prior knowledge analogue matters for learning. *Journal of Cognition and Development*, *16*(1), 160–185. doi:10.1080/15248372.2013.792091

Stoltz, R. (2021, December 28). *What is Adult Learning Theory?* NEIT. https://www.neit.edu/blog/what-is-adult-learning-theory

Taylor, K., & Rohrer, D. (2010). The effects of interleaved practice. *Applied Cognitive Psychology*, *24*(6), 837–848. doi:10.1002/acp.1598

Thagard, P. (2005). *Mind: Introduction to cognitive science*. MIT Press.

Thiede, K. W., Anderson, M., & Therriault, D. (2003). Accuracy of metacognitive monitoring affects learning of texts. *Journal of Educational Psychology*, *95*(1), 66–73. doi:10.1037/0022-0663.95.1.66

Warren, B., Ballenger, C., Ogonowski, M., Rosebery, A. S., & Hudicourt-Barnes, J. (2001). Rethinking diversity in learning science: The logic of everyday sense-making. *Journal of Research in Science Teaching*, *38*(5), 529–552. doi:10.1002/tea.1017

Western Governors University. (2020, October 20). *Adult Learning Theories and Principles*. https://www.wgu.edu/blog/adult-learning-theories-principles2004.html#close

KEY TERMS AND DEFINITIONS

Metacognition: The process of deliberately analyzing our thinking processes.

Reality and Perception-Processing II; With Metacognition: Intrapersonal-/self-communication executed or utilized by an individual to process the inputs of their senses and/or their thoughts, while aware that s/he's undertaking that process.

Self-(/Intrapersonal-)Communication: Communication inside an individual's mind, or outside—e.g., spoken or written—but for that same individual's consumption and/or use.

Significance Analysis: An individual's evaluation of sources to gauge their level of helpfulness to him/her, for their specific purposes at hand—e.g., for improvement of one's own learning methods.

ENDNOTE

[1]   A professor of psychology at Grand Valley State University, Michigan, USA, and active member of the Society for Text and Discourse, which is actively engaged with research initiatives related to cognitive learning-processes.

Chapter 7

"Come on Seif, You Can Do This!":
Intrapersonal Communication/ Cognition and Mental Health (Including the Concept of Mindfulness)

ABSTRACT

This chapter examines how SC can assist individuals with managing their mental health. The author presents the findings under one overarching theme titled "thinking-triggered/enhanced equilibrium and distress." He identifies several instances that demonstrate how individuals' mental health is affected positively or negatively by SC. The specific triggers or catalysts in this context include self-checking, temporary memory dysfunction-triggered neurosis, meditation, positivity and distress, and resilience, e.g., via a method the author refers to as "brute-force learning." The author also discusses the concept/ practice of self-doubt and the role of SC in extremely distressful situations—both as a trigger/catalyst for worsening our states of mind or improving them.

INTRODUCTION

As of June 2021, news in the USA and the UK seemed to reflect a growing acceptance of the open discussion of mental-health issues without prejudice. One particular example in this regard is the self-disclosure (via the interview with Oprah Winfrey and in other forums) by Prince Harry and his wife Meghan Markle, about their own acute respective struggles with depression.

And yet, self-disclosure of mental-health struggles is still a minefield for most ordinary citizens. Declaring that one struggles from conditions such as clinical depression and anxiety, or phobias, etc., is akin to one declaring that somehow, s/he is "crazy," "soft/spoilt," etc. Despite this societal stigma against transparency vis-à-vis mental-illness, I have already shared my struggles in chapter two in regard to clinical anxiety and depression. Regardless, mental-health is not the/a primary focus of this book.

DOI: 10.4018/978-1-7998-7507-9.ch007

But can/should one attempt to accomplish a fairly thorough analysis of intra/self-communication and cognition without at least briefly considering the related dynamics and implications of mental-health, vis-à-vis the main topic (intra/self-communication and cognition)? In my personal/professional opinion, the normative answer to that question should be a "no." In other words, we have to at least briefly consider the *hows* and *whys* of the health and illness of our minds, in relation to intra/self-communication and cognition.

Among other reasons, such a focus is apt because as I discuss in a previous chapter under section 4.4.1 and elsewhere in this book, so far, my autoethnographic self-study seems to suggest that SC instances are more frequent and earnest during moments of intense (positive or negative) emotions. For instance, as I point out in section 4.4.1, over the past and current months of 2021, I have been, and I am still experiencing acute negative emotions because of my failing marriage.

And as a result, I am noticing a marked frequency and intensity of SC. I also distinctly recall similar upticks of SC during joyous episodes in my past—e.g., the day on which I simultaneously found out about the birth of my niece Amber, as well as my acceptance into two Ph.D. programs, and the delivery of my green card. Moreover, as demonstrated in chapter two, a careful review of one's past SC-instances via journaling can reveal characteristics (again, I hesitate to call them *symptoms*) of mental illnesses.

Chapter Structure and Overarching Theme

Unlike the preceding chapters—in which findings have been organized under the five categories introduced in chapter three, the current chapter will examine my voice-notes'/transcripts' mental-health-related SC instances under one general grouping. This general grouping is tied together by a theme to which we can refer as *thinking-triggered/enhanced equilibrium and distress*.

The instances suggest that via cognition and intrapersonal-communication, we can trigger and/or maintain mental—and to a significant extent, corporal—equilibrium, and/or distress. Altogether, this grouping can be summarized by the following eight titles: 1) self-checking, 2) temporary memory-dysfunction neurosis, 4) meditation, 5) positivity and distress, 6) guilt and intense emotions, 7) "brute-force" learning, and 8) yearning for coaching and self-doubt.

MAIN FOCUS OF THE CHAPTER

Thinking-Triggered Equilibrium and Distress

1: Self-Checking; e.g., Voice-Note No.: Batch 1; No. 7 and No. 8

Both these voice-notes suggest that often, we realize that we are not thinking about the topic we should be thinking about in a particular situation, or that we are not thinking in the fashion in which we should or want. For instance, in voice-note 1.7, after realizing that I had skipped a few dates on the draft of my "to-do" list, *"I realized my mistake, retraced my steps to figure out what I had done wrong and why."* (^) And in voice-note 1.8, I am forced to figure out the rationale of my e-filing system: *"What did I mean by this? Why did I save this file within the subfolder in which I saved it?"* (^) I also (in voice-note 1.8) give an example of a common situation in which we find ourselves, and are forced to remember our purpose-at-hand: *"Why did I come to the kitchen?"* (^) Both these sets of instances are good examples of self-checking. We find ourselves proverbially or mentally lost, and the remedy is to check ourselves, to re-discover the proper topics or ways of thinking we need for a particular situation.

2: *Temporary Memory-Dysfunction-Triggered Neurosis*; e.g., Voice-Note No.: 2.1.1 and 2.1.6

The prime exemplar of this instance-typology is "tip of the tongue" syndrome. In 2.1.6, I discuss the aforementioned concept of primary- and secondary-self in the context of the "tip of the tongue" phenomenon: *If you're (your primary self is) referencing your secondary self..., then that is* [an instance of] *intrapersonal-comm* (^). And in 2.1.1, I discuss the phenomenon further: *"What's that word?! It's on the tip of my tongue!"* Later, I added an editorial note, relishing my victory over that instance of memory-paralysis: "+ *Editorial note, added during transcription (03/10/21): The word is zeitgeist!* (^)."

Another related temporary-neurosis trigger related to this concept is the acutely distressing situation of temporarily forgetting a(n)—usually small—item's location, e.g. one's keys, especially if there are other stress-factors at play, such as being pressed for time (e.g., while going to work or an appointment, etc.). Yet another example in this context is the phenomenon referred to as "writer's block." We can also include the concept of indecisiveness in this discussion.

3: Meditation; e.g., Voice-Note No.: 2.1.2

In my personal experience, among other techniques, meditation—particularly via meditative breathing—provides a markedly helpful method vis-à-vis nurturing an ongoing mental equilibrium. And yet, the successful or consistent implementation of a meditation regimen is quite challenging, to say the least. And in the above-referenced voice-note/entry, I hone in on this challenge. Below is the full verbatim transcription of the note/entry:

- *I realized something during or shortly after meditation: it's hard to "quiet" the mind, to avoid thinking*
- *Analogy: the way our minds work in that regard—always thinking even when we're trying to meditate, might be similar to a computer opening various random files or performing various other random functions, even though you're taking a break and are not using said computer* (^)

Without getting into a debate or discussion about the appropriateness or effectiveness of the computational-representational hypothesis—which would be triggered by the second point in the above transcript entry, I have to note how remarkable I find that metaphor. If the metaphor is apt indeed, one can surmise that the mind is a (potentially) dangerous tool. For instance, would anyone be comfortable with the proposition of a heart surgeon or passenger-aircraft pilot being distracted during an operation or flight?

4: Positivity and Distress; e.g., Voice-Note No.: 2.1.5, and 2.1.15

Altogether, the above three mind states suggest a complex interplay between cognitive passivity and activity, and the constant search for cognitive equilibrium. And in our daily lives, this interplay is often forced to the forefronts of our minds. Depending on our goals (e.g., while trying to force yourself to go to the gym), we are forced to encourage ourselves, or to generally be optimistic or to feel motivated. And at other junctures, we are forced to accept the negative or less-than-ideal state of affairs in our environments and/or in our bodies and minds.

The preceding discussion is demonstrated arguably well via voice-note entry 2.1.5. Despite the presence of other points, as well as the chronological order of all the points in the attendant transcript entry, there are three points that are most relevant to the current discussion. Below, I present and re-arrange them, despite their actual chronological order in the transcript:

- *Self-questioning, e.g., "I wonder if I can do that?"*
  - *I—[/Seif/[Your Name] the primary self, wonder if I—[/Seif/[Your Name]—secondary self-] can do that (^)*
  [Point no. 2 in the transcript.]
- *Self-encouragement: "Come on, Seif! You can do this!" That is definitely an instance of intrapersonal comm. (^)*
  [Point no. 1 in the transcript.]
- *Corporal states: acknowledging a corporal* [and/or mental] *state to yourself—e.g., "Oh my God, my head hurts!"* [Or, "F&#k! I feel SO angry!"] *That's also an example of intrapersonal-comm. (^)*
  [Point no. 3 in the transcript.]

To recap the current discussion using voice-note 2.1.5's relevant points in their order above, as well as our previous "going to the gym" example: 1) We often find ourselves in moments of self-motivation and confidence deficits, e.g. via questioning our abilities: *"I wonder if I can do that?"* 2) In such moments, we often come to our own rescue, rallying ourselves forward: *"Come on, Seif! You can do this!"* 3) But at various other junctures, we have to simply acknowledge—then proceed to try to treat, or otherwise deal productively with—our corporal-mental states of being: *"Oh my God, my head hurts!"*

Beyond the above-exemplified isolated instances of self-doubt, this phenomenon—i.e., self-doubt—often manifests itself over a long period of time. For me, it is often coupled with what I termed as "yearning for coaching," a feeling that I am not (yet) good enough, so I need to be coached in whatever endeavor I am trying to accomplish. These dynamics are demonstrated partially in voice-note number 2.1.15:

- *Related to learning: "Yearning for coaching." E.g. with me learning algorithms and data-structures: "In your most intense moments of self-doubt, or feeling frustrated, you [at least I, personally, often] yearn for coaching: 'if only I had somebody that could help me figure this out.'"*
- *But it depends on your own learning style. E.g., I for one am not afraid of asking for help if/when I need it, but other people want to figure it out on their own. But regardless, I believe most of us at one point or another—maybe consciously or subconsciously—experience that phenomenon.*
- *In my own case, I was able to satisfy that yearning over several months in late 2020, hiring a software-development coach. But what happens if you experience that need; e.g., in a class where you do not get along with the teacher?*

Apparently, such feelings of self-doubt often grow after we experience them consistently over a long period of time. In his *"A Promised Land"* (part-I presidential-memoir) autobiography, Barack Obama provides one apt example, as he compares his own self-doubts with those of his wife, Michelle (nee Robinson) Obama (also an excerpt from voice-note no. 2.1.15):

Michelle never worried about selling out, because growing up on the south side [of Chicago] meant you were always at some level an outsider. In her mind, the roadblocks to making it were plenty clear. You did not have to go looking for them. The doubts arose from having to prove no matter how well you did that you belonged in the room. Prove it not just to those who doubted you, but __to yourself__ (emphasis by this author). (^)

However, this reference to self-doubt is in the context of "impostor syndrome," unlike my own reference to challenges of learning.

5: Guilt and Intense Emotions, Part A; e.g., Voice-Note No.: 2.1.7, 2.1.8, and 2.1.11—(2.1.7, 2.1.8)

Somewhat similar to self-doubt and other negative feelings about our *selves*, the particular emotion of guilt seems to trigger more marked depths and breadths of SC instances. In voice-note/transcript-section 2.1.7, I explicitly give voice to the core underlying message we grapple with during moments of guilt, i.e., *"I am a bad person" for having done that bad thing. You do not necessarily have to use those or other words; the mere emotion is a de facto intrapersonal-comm.* (^)

In voice-note/transcript-section 2.1.8, I extend the discussion via a particularly comical scene from the popular *Seinfeld* sitcom: *Related: second-guessing yourself, or blaming yourself later after an argument, a la George Costanza: "The jerk store called, they ran out of you!"* (^) In other words, we often regret having said or not having said something, and we blame ourselves for that unwise statement or omission.

The above *Seinfeld*-esque self-blame concept can help us segue to another concept beyond guilt (in transcript-section 2.1.8), namely, intra-cognitive/communicative role-play. Via a concept that my colleague/ friend (name withheld for privacy-protection) termed as *"ethnography of mind," he discovered that he often speaks to "guest speakers" in his mind. For instance, envisioning himself talking to someone else [in the past or future, or both?].* (^) It should be noted that this concept somewhat echoes the theories of dramatism and dramaturgy, as discussed by Burke (1969) and Goffman (1967). In other words, because of our (i.e., socialized humans') sophistication with language, emotion, and interaction rituals (e.g. Goffman, 1967), we seem to have a predisposition towards intrapersonal rhetorical-empathic role-play: *"I cannot wait to tell my friends about my trip to Paris. But I hope they do not think of me as a braggart!"*

6: Guilt and Intense Emotions, Part B; e.g., Voice-Note No.: 2.1.7, 2.1.8, and 2.1.11

From the above context of calm calculative intrapersonal thinking and communication, our pendulum swings back to un-ideal cognitive situations, in particular concepts such as neurosis and distress as discussed farther above in 7.3.2 and 7.3.4. However, it should be noted that in this context, the state of being "un-ideal" is not solely related to negative emotions. Rather, the emotions can also be positive, e.g. being overjoyed, or in love. Beyond this introductory exposition, I believe the verbatim transcript-section can "speak for itself":

- *Realization: a lot of intrapersonal-comm. happens as we experience emotions—happiness, sadness, anger, etc. Implication for learning-process improvement: keep the above fact in mind as you read or study; you're more likely to be distracted when emotional, which interferes with your learning process.*

- *Related qn.: "In an average day [or other time-period, e.g., a few weeks/months, etc.], what percentage of the time do we have intense emotions? Intense sadness, intense happiness, etc." Another related question in same context: how much do we communicate with ourselves during these intense emotions?*
- *Depending on the answer here, we have to keep it—and the above realization—in mind, as we go about our learning processes. E.g., if I am taking a course—e.g. for continuing-ed.-credits or for working training, and I lose a loved one during that same time-period ...* [Incomplete statement completed during transcription: *I need to realize that I'm distracted by that sad event; thus, either pay more attention, or take the course later, etc.*]
- *Example of communication with self during intense emotions. E.g., when someone angers you: "I cannot believe she did that to me!" Or, "I feel so sad." Resultant argument: in these instances, you're definitely communicating with yourself, albeit unintentionally.* (^)

"Brute-Force" Learning

Beyond the above reactive situations and states of mind, one of the most powerful tools we can wield in our struggle for equanimity is resilience. But how exactly can we foster or enact that concept—i.e., resilience? One answer presents itself in one of the current chapter's attendant transcripts, vis-à-vis a concept to which I refer as "brute-force learning." The key argument of this concept is as follows: given enough time and effort, most of us can learn any given topic or skill:

- *Learning: e.g., me trying to solve a number of algorithm challenges on FreeCodeCamp. "One of our worst enemies in the learning process is self-doubt; can I really do/learn this stuff?" My new hypothesis of longterm "brute-force" learning: regardless of the intelligence or IQ of a given person, we are all capable of learning any given concept, as long as we're exposed to it for a long period of time.* [Editorial note: further research might be needed to confirm this hypothesis for individuals with learning disabilities. My own anecdotal evidence suggests that there might be limits to what they can learn.]
- *Meaning of "brute-force" execution; e.g. with hacking passwords: prolonged engagement and attempt at doing something. The key advantage that aspiring polymaths have, is that you can take your time—as long as needed, to learn the various concepts you're trying to learn, regardless of how hard-to-learn those concepts are.*

Yearning for Coaching and Self-Doubt

Be that as it may—i.e., even if it's true that we can learn most topics or skills if we utilize the necessary time and effort, we still have to confront questions of how exactly we can/should do that learning. Chief among these questions is the consideration of whether we should learn on our own (and if so, how exactly), or with the aid of teachers, coaches, study-partners, etc. And regardless of the methods we choose to utilize in our learning journeys, we shall doubtless encounter cognitive road-blocks, which might in turn trigger self-doubt:

- *Related to learning: "Yearning for coaching." E.g. with me learning algorithms and data-structures: "In your most intense moments of self-doubt, or feeling frustrated, you [at least I, personally, often] yearn for coaching: 'if only I had somebody that could help me figure this out.'"*

- *But it depends on your own learning style. E.g., I for one am not afraid of asking for help if/when I need it, but other people want to figure it out on their own. But regardless, I believe most of us at one point or another—maybe consciously or subconsciously—experience that phenomenon.*

- *In my own case, I was able to satisfy that yearning over several months in late 2020, hiring a software-development coach. But what happens if you experience that need; e.g., in a class where you do not get along with the teacher? Or what about outside that class, where your problematic relationship with the teacher is not necessarily an issue?*

- *Self-doubt: relevant reference in Obama's "A Promised Land," specifically in the context of discussing his wife background compared to his. "Michelle never worried about selling out, because growing up on the south side [of Chicago] meant you were always at some level an outsider. In her mind, the roadblocks to making it were plenty clear. You did not have to go looking for them. The doubts arose from having to prove no matter how well you did that you belonged in the room. Prove it not just to those who doubted you, but to yourself" (emphasis by this author). This reference to self-doubt is in the wider context of "impostor syndrome," unlike my own reference to challenges of learning.*

- *Further explication of above discussion of self-doubt: "it [the above discussion] goes to prove that…the enemy here [i.e., in the contexts of learning or career/general success and fighting impostor syndrome] is not other people many times; it's yourself."*

CHAPTER CONCLUSION: WHAT DOES SC LOOK LIKE AT POINT 10 OF OUR AFORE-INTRODUCED SCALE OF MENTAL DISTRESS (FROM 1 TO 10)?

Despite the potential usefulness of this chapter's foregoing research-analyses, one can rightly point out that I have not specifically tackled the relationship between SC and severe mental illness. This absence is especially glaring in light of my earlier self-disclosure vis-à-vis my struggles with clinical anxiety and bipolar-depression (i.e., "type II/cyclothymia/mixed-features"), as well as my current general life-struggles with my career and marriage. And despite the awful nature of these misfortunes and the events I am about to discuss below, it is a rather serendipitous coincidence they have in fact taken place this year, enabling me to discuss them in this chapter.

First, it should be noted that indeed, a deliberate use of SC (by my *self*), mixed with my own interpretations/executions of mindfulness, has been extremely helpful to me this year, given the severe crises I am dealing with. Some of the key techniques I utilize include: deep breathing exercises, constant self-reminders to stay calm and breathe deeply, taking walks outside, reminding myself to stay in the moment—rather than obsessing about the past and future, and other general positive self-encouragement messages.

However, one particularly seminal event this year was especially helpful vis-à-vis improving my understanding of the positive and negative occurrences of SC. First, it should be noted that even though the event involves someone else, I can guarantee that there is absolutely no risk of their identity being divulged. And yet, I was also a key player in this event, and as I mention above, it is instrumental to this book's/chapter's discussion.

On August 31, a very close friend of mine attempted suicide after a very distressing event at her job. Later, she explained to me that after being unfairly reprimanded—via a final warning letter—by her supervisor for defending her colleague (who was often bullied or belittled by clients), she thought about the fact that without that job, she would be destitute, and yet she had no one else to rely on. And having experienced a series of career-travails similar to my own, she was afraid that she would not be able to get another job.

Thus, this event—the threat to her livelihood, was a proverbial last straw (mentally) of sorts. As she explained it to me, my friend wondered: *"what's the point of living if I have no chance to make it in this world* [career-wise]*?"* In other words, this was a clear instance of severely negative SC, which convinced my friend that her life was not worth living anymore. Luckily, we (myself, and one of her close relatives) found her before the drug and alcohol overdose had taken full effect, and the paramedics rushed her to a nearby hospital.

Later, despite the COVID restrictions, her social-worker arranged for me to visit her at the psych ward where she had been confined involuntarily, given the self-risk she posed. And during those visits, I encountered at least two individuals who were partaking in SC that were afflicted with severe mental illness. In particular, they were demonstrating behaviors/symptoms of schizophrenia. For instance: 1) a middle-aged man was yelling in a scared manner at imaginary people or entities attacking him, and he at times paced the hallway of the ward, agitatedly arguing with an invisible person/entity: "but I do not have time (!), I do not have time!" 2) There was also an older woman who pointed at me and incoherently accused me of having come out of a tent in a parking lot. And of course, I could not tell—nor did she explain, why that alleged (hallucination-induced) event had distressed her.

Luckily, my friend was able to return to a calm state while in this psych ward. This result—per her revelation to me later, was mostly precipitated by the realization that despite her unfortunate employment prospects and other misfortunes, she was markedly fortunate compared to many of the individuals with severe or chronic mental illness in that institution. In other words, she experienced a series of self-motivational SC-instances that returned her to an ideal/calm state of mind.

Creativity-Attempt Outtake 7

Post 9; 03-08-2021:
To Seek Or Not to Seek Perfection: That Is A (Not *The*) Question

Would you ever ask a heart-surgeon, who's about to perform a delicate surgery on you, not to seek perfection? I suppose under ordinary circumstances*, anyone in their right mind would have to be suicidal, if they answered that question in the affirmative. (*But if the surgery has to be performed in the middle of an unprotected battle-field, clearly, one has to lower their expectations.) But should you, who's writing a research paper, seek perfection, and without it, fail to start/complete said paper? A variant of the above question occurred to me while jotting down a quick entry in my research notes for my cognitive-science book project.

The topic of said entry: over the years, I adopted the use of a placeholder—"XX," whenever I could not think of the appropriate word in my writings—be it an email, or paper, etc. I will then come back later to fill in the appropriate word, depending on my intended meaning.

At the end of the day, I suppose we all have to decide—depending on the task at hand, whether or not we should indeed aim for perfection, or just "get the job done," so to speak. And I suppose I am conflating two ideas here, namely: 1) getting started with a task—whether or not one seeks perfection, and 2) performing said task with a given degree of rigor.

In other words, the search for perfection is mostly relevant to number two above, but perhaps number one as well, depending on the situation. Regardless, I for one decided a long time ago to mostly follow the Nike slogan, "just do it;" start the task, then keep improving it over time^^. So, are *you* a perfectionist? Why, or why not?

Post 15; 04-18-21:
Thou Shalt Breathe, Or Die

And on and on, the "to-do" list goes on. As does time, as one moves from one work-week into another—or simply, one week; for time knows no purpose, apart from that to which we attach it. For me, that—or those—purpose(s) at the moment consist of mainly the intrapersonal-communication and cognitive-science book-manuscript—that awesome labor of love, as well as a new 90-day software-development and data-analytics challenge.

I do not know if I'll ever enjoy the kind of job-security that many of my family-members, friends, and colleagues seem to enjoy. If you're a believer in G_d or fate, be my guest, and interpret such questions in a way that makes sense to you. Regardless of the presence or absence of such a coveted prize (job-security), all I can or should do—in addition to other necessary and healthy pursuits of course, is to try my best to ensure that I am somehow equipping myself for success as best I can. Hence the data-analytics challenge and the book-manuscript.

*** *** ***

Speaking of necessary and healthy pursuits, I so miss my swimming, which I of course cannot partake in without an institutional membership—as in, for the gym and pool benefits. But apart from the gods of weather, no one can stop me from taking a leisurely evening walk! Thus, I have been indulging in those of late, now that the weather is getting warmer.

In those walks, I enjoy nature, contemplate my life, breathe, and try to relax. I have to admit, I do not know if I'll ever fully understand the philosophy of mindfulness. Being in the moment, avoiding mind-wandering, both savoring and/or ignoring emotions; all these are hard and confusing. During my evening constitutionals for instance: should I try to push out the thoughts from my mind about my current travails and victories; my past, present, and future; and my intellectual ponderings (among other numerous thoughts)—thoughts that seem to flow through the mind so naturally and harmlessly as I simultaneously commune with nature?

Ah, who cares?! The truth is—if you can believe or understand it, I actually do understand that powerful and positive life-philosophy (mindfulness that is) just as much as I do not understand it all the same! One basic interpretation I can confidently state: just do your best, or at least try to live up to those ideals.

I for one believe that that life-philosophy's goal—among others (goals)—is to help alleviate the constant anxiety that afflicts us. It's indeed a powerful tool with which we can try to fight back against the ugly underbelly of our awesome minds' excesses.

*** *** ***

Which then begs the question: is it ironic that I have to add mindfulness-practice to my "to-do" list as well? Does my inclusion of it on that list somehow weaken its power? I do not think so. And yet I know that I have to indeed constantly remind myself to be mindful. Yes, in one sense, I think of the "to-do" list as that "almighty TDL" indeed. But one small victory I've eked out lately, is that I do realize that that "almighty" designation should only be treated as the tongue-in-cheek expression it is—not a creation or reification of a meaningless power over me. An incompletely crossed-out "to-do" list will not kill you; but a failure to breathe will. So, breathe, dear reader (yes, YOU! ☺); breathe. Your life depends on it.

Post 38; September 26, 2021:
A Respectful-But-Earnest Proposition to Current And Future Leaders

"TL;DR" Version: if/when given a leadership role in the future, will **YOU** be able to step up to the plate and meet the … challenge [discussed in the article]? And to reiterate, that challenge can be summed up using the motto: quality over quantity, and heart over mind.

———————————————

True story: Once upon a time—but very recently, a friend of mine tried and almost succeeded in taking her own life, because of the pressures of her job, after experiencing first-hand the acute challenges faced by millions of American workers across the country while simply trying to do the right thing at work. And yet, these same unsung heroes simultaneously have to satisfy myriads of often arbitrary and unrealistic performance metrics imposed by managers, shareholders, and consumers. In other words, the horror stories we often read about Amazon warehouse workers having to pee in bottles because of strict rules regarding time-off of assembly lines, aren't isolated. In fact, Amazon's practices are simply archetypical though exaggerated, in comparison to several professional roles in our private, public, and nonprofit sectors.

My friend made this realization while working in a certain low-level capacity in a colossal, wildly successful, and well-reputed company, a corporation with shareholders' assets of several *T*rillion dollars. I will not disclose any more about the company for fear of retribution, but I can reveal that it is headquartered in the metro area of one of our country's largest cities on the east coast.

Specifically, she made that unfortunate suicide attempt because of: 1)—unfair and inconsiderate treatment by customers, 2)—more unfair and inconsiderate treatment by manager(s), who could have used their discretion to interpret and implement rules in a less-harsher way, and 3)—the aforementioned often arbitrary and unrealistic performance metrics imposed by managers, shareholders, and consumers. These experiences exacerbated my friend's pre-existing but previously well-managed mental illnesses; they pushed her to the edge, mentally.

Dear supervisors and managers: we get it; rules are rules, and we all have to "play by them"—so to speak, for the smooth operation of organizations and society at large. But cannot you honestly use your G_d-given brain/heart, to enforce them with common-sense discretion, and perhaps most importantly—**real** (not feigned-, or "lip-service"-) empathy?

Individuals in leadership roles need to understand something: we're all built differently—i.e., mentally and otherwise, but if/when given the chance and the right tools, most of us can succeed professionally. And sure, I suppose ultimately, you too—i.e., supervisors/managers—are victims of the whims of your own leaders, shareholders and consumers, and other stakeholders. But I get the feeling that a well-organized groundswell of collective advocacy might succeed in educating all these stakeholders, helping us all to understand the ultimate consequences of our unfettered demands/desires in the economy/marketplace.

In the meantime, all I can ask of myself and everyone reading this article is the following question: if/when given a leadership role in the future, will YOU be able to step up to the plate and meet the above challenge? And to reiterate, that challenge can be summed up using the motto: quality over quantity, and heart over mind.

Self-Questioning for Creativity-Attempt Outtake 7

State any Necessary Introductory Remarks, and/or Answer this Question: Does it Make Sense to Me Now—i.e., In The Present—versus then—i.e., When I Wrote It?

Yes, the posts still make sense to me in the present, similar to when I first wrote them.

Does it Resonate with Me Now—i.e., In The Present— versus then—i.e., When I Wrote It?

Yes, the posts still resonate with me now. In two of them, I grapple with the concept of equanimity and mindfulness; living in the moment, not fussing unnecessarily about details or imperfectly-executed tasks, etc. And in the other, I reflect on my friend's mental-health/near-death ordeal. As the expression goes, "given the chance to re-do them," I would write them with similar messages and tones.

What Mood and/or State of Mind Does the Tone Suggest?

Mindfulness-aspiration—i.e., for posts 9 and 15, and aggrieved and reflective, for post 38.

What is/are the Direct and Indirect Relation(/s) to the Current Chapter via Topic(s) and Themes?

Out of all the book-chapters' C.A.Os, this particular C.A.O is probably the most apt. The posts directly mention numerous topics and themes mentioned or implied throughout the chapter especially mental health and illness, and mindfulness.

Any Other Evaluations or Comments?

N/A

Activities and Discussion Questions

1) What do you know about your own thinking and emotion patterns? E.g., what makes you happy, sad, etc.?
2) Discuss: How can understanding your own thinking and emotion patterns—e.g. what makes you happy, sad, etc.—help you better in various spheres of your life, i.e. work/school, home, etc.?
3) *Optional*: Continue journaling/creative activity 1 as advised at the end of chapter one above.

4)   Continue activity 6 from chapter one, "Comparison with other types of communication."

REFERENCES

Burke, K. (1969). *A rhetoric of motives*. University of California Press.

Goffman, E. (1967). *Interaction ritual; essays on face-to-face behavior*. Doubleday.

KEY TERMS AND DEFINITIONS

Thinking-Triggered/Enhanced Equilibrium and Distress: Mental states of mind marked by either tumultuousness or equanimity, triggered or enhanced by our thinking process.

Chapter 8
"The [Expletive] Thing Seems to Have a Mind of Its Own!":
Intrapersonal Communication/Cognition and Improvement of ICT, Including AI

ABSTRACT

In this chapter, the author applies the findings of the previous chapters' research analyses to the topic of the improvement of ICT, including AI systems. He delineates two main significant findings from Data-Set I, i.e., the unintentional and deliberate use of SC, and the vital role of language and symbol systems, and from Data-Set II. He also recaps the three main significant findings, namely 1) our agency's efficacy vis-à-vis knowing how best to help ourselves learn, 2) adapting to the conventions of formal learning to help guide our learning, and 3) utilizing the mechanisms of cognition—e.g., as summarized by Thaggard to help us efficiently comprehend, integrate, and utilize the content we're learning. In addition, the author lists the problems and solutions discussed in eight particularly relevant articles from Data-Set II.

INTRODUCTION

Role of ICT in Human Communication

As I discuss in chapter one and throughout this book, human beings do not simply talk to each other (via int*er*personal communication). For that to happen, we have to talk to our own *selves* first. Following intra*personal-/self-communication and int*er*personal communication, we can/do also form social units—i.e., families and communities—for companionship, security, and convenience. Moreover, our superior intelligence also enables us to manufacture basic and advanced technologies, which help us perform a variety of tasks, including communication. In that last regard (communication), writing was a key breakthrough in helping our SC and interpersonal cognition/communication processes. But by far to date, the computer is the most impressive tool in this context. Remarkably, we now not only have the

DOI: 10.4018/978-1-7998-7507-9.ch008

ability to communicate directly with ourselves and other humans, but we can even communicate to and via those machines—computers (including smart phones), to help ease our personal and professional lives.

We can choose to do so in a bid to deliberately/systematically process our thoughts/feelings (i.e./e.g. via journaling), and/or to store our thoughts and communications for later consumption, and/or to use the technologies instead of communicating directly to other individuals (e.g., because we are shy or afraid to do so, or feel lazy to go to another room, albeit in the same house, instead of simply texting the interlocutor). Indeed thus, we have achieved impressive progress in the sphere of external cognition-processing and communication technology. But in the context of SC and interpersonal communication, are these systems and processes—as well as our usage of them—perfect "as is"? If "no," how can we improve them?

Chapter's Foci

As the chapter-end's "creativity outtake attempt" alludes, my original intention (for this chapter) was to focus on AI alone in the discussion of how we can harness SC to solve various problems, regardless of how complex they are or seem to be. And despite the expansion of chapter's focus—i.e., to discuss ICT at large versus AI alone, the premise is still the same. However, as I note at the end of chapter six, we can garner more and better results by also expanding our "independent variable," so to speak: instead of focusing solely on intrapersonal-/self-*communication*, we should also focus on intrapersonal-/self-(meta-)*cognition* (i.e., how we think and communicate that thinking to our *selves*). We can thus add another defacto research-question for the chapter to state: based on the book's analysis results, how can we overcome the challenges of our SC/cognition to so as to improve our ICT design and utilization strategies and tactics? In the sections below, I reexamine the voice-note/transcript results for particularly relevant instances and discussions. Thereafter, I reexamine the second data-set—both holistically, and with a particular focus at some texts with potentially exceptional insights. Finally, I conclude with an attempt at synthesizing some of the relevant implications from both data-sets' findings.

MAIN FOCUS OF THE CHAPTER

Summary of Voice-Note Findings' Relevance

Unintentional and Deliberate SC

Overall, the findings of this book's voice-note/transcript data can be summarized thus: regardless of whether or not we realize that we have the ability, humans often partake in SC, and there are ways to improve how we execute it, which can in turn help our various personal/social/work/other functions. Among other forms and functions, we partake in SC while making sense of our day-to-day experiences—including our perseverance through, or succumbing to adversities such as the COVID pandemic (e.g., as discussed in chapter three, section 3.2.2). Our minds also communicate recollections of previously memorized or learned facts (chapter three, section 3.2.3); we pause and check the rationales for our actions ("why did I come to the kitchen?"); we feel certain relevant emotions while reading fiction and nonfiction works (chapter three, section 3.2.5); and we make use of "to-do" lists and journaling (chapter three, sections 3.2.6 and 3.2.9).

The Vital Role of Language and Symbol Systems for SC, and Review of Specific Exemplary Instances

Moreover, our minds utilize language and symbol systems in sophisticated ways to process and integrate, store, and recall knowledge. For instance, we often utilize analogies and metaphors (chapter four, section 4.2.1), and we either creatively/randomly or deliberately search our minds for the appropriate words/concepts for particular situations—i.e., as demonstrated by the "tip of the tongue" phenomenon (as discussed in chapter four, section 4.2.4 among others). Furthermore, we often make sense of the world by using what I can describe as cognition- and SC-powered organic versions of Bayesian inference and deep learning, e.g. as discussed under chapter four's section(s) 4.2.5 (i.e., the discussion of "suspending and confirming understanding"), and we often perform inventories of our levels of knowledge of various topics/skills. Finally a review of all the voice-note transcripts highlighted two instances that are particularly worthy of mention in this discussion, namely 2.1.2, and 2.2.12:

2.1.2:

- ○ *I realized something during or shortly after meditation: it's hard to "quiet" the mind, to avoid thinking*
- ○ *Analogy: the way our minds work in that regard—always thinking even when we're trying to meditate, might be similar to a computer opening various random files or performing various other random functions, even though you're taking a break and are not using said computer*

2.2.12:

- ○ *Many software-developers assert that building apps is better for you learn, versus endless learning of concepts. But my argument is: how am I supposed to build apps, etc., before really learning the concepts first? ...*

Based on the above considerations, how can we improve ICT systems' designs and functions? I return to this question in section 8.4, after reexamining the relevance of data-set II's findings.

Summary of Data-Set II's Relevance

The corpus of data- and cognitive-science articles analyzed in chapter six applies SC to a particular human challenge, i.e. learning. It highlights three main aspects to which we can pay attention as we learn a given topic/skill, namely: 1) our agency/power's efficacy vis-à-vis knowing how best to help ourselves learn, 2) adapting to the conventions of formal learning to help guide our understanding of the learning materials we are consuming, and 3) utilizing the mechanisms of cognition—e.g. as summarized by Thaggard (2005) to help us efficiently comprehend, integrate, and utilize the content we're learning. Scholars and engineers trying to improve ICT systems should keep these strategies in mind. However, beyond those strategies, there are some particularly important potential insights that we can glean from a total of eight articles from data-set II. In appendix "D," the articles are listed under Part 1-*I-B* (i.e., Part 1—"...self-emailed articles..." > Sub-part I—"grand-total: 42 articles" > Sub-part B—Data Science). Specifically, we can summarize the problems and solutions discussed therein using the table below.

Table 1. Problems And solutions from eight select articles from data-set II

Articles	Highlighted Problem(s) And Suggested Solutions
Article 1	***The Brain's Most Precious Resource: The role of attention in neuroscience, deep learning, and everyday life.*** <u>Author: Manuel Brenner</u> **Problem:** The brain's overtaxed processing and storage functions, and need for parsimony thereof. **Solution(s):** Careful use of our attention to guide us to the execution of our goals.
Article 2	***Hundreds of AI tools have been built to catch COVID. None of them helped.*** <u>Author: Will Douglas Heaven</u> **Problem:** AI tools failed to predict COVID infections, and some even made the problem worse. **Solution(s):** Better collaboration among AI engineers, and between engineers and clinicians.
Article 3	***Why AI tools Failed to Help With Detecting COVID.*** <u>Author: Ajit Jaokar</u> **Problem (similar to article 2 above):** AI tools failed to predict COVID infections, and some even made the problem worse. **Solution(s):** Better data quality, and use of tools such as Bayesian models.
Article 4	***A Step-By-Step Guide To AI Model Development.*** <u>Author: Jaimin Dave</u> **Problem:** Effective AI model development. **Solution(s):** **1)**—6 steps, namely: i—Problem-identification, ii—Data-collection, iii—Data-processing, iv—Model-building and training, v—Model-testing, vi—Model-deployment. **2)**—Hiring engineers who use well-defined processes, such as RACI charts.
Article 5	***Is Machine Learning an Art, a Science or Something Else?*** <u>Author: Vincent Granville</u> **Problem:** Should machine learning (ML) be classified as art or science? **Solution(s):** It is both, but it's better viewed as a craft at three levels. In the end, ML code and outputs—in particular, visualizations—display remarkable artistry, e.g. the butterfly fractal, which displays "the energy of electrons in an atomic lattice."
Article 6	***Content rewriting techniques using NLP paraphrasers.*** <u>Author: Annie Moore</u> **Problem:** How can you paraphrase other writers' content without committing plagiarism? **Solution(s):** Try using natural language paraphrasing tools such as "paraphraser.io" and "rephrase.info."
Article 7	***Three Steps to Addressing Bias in Machine Learning.*** <u>Author: Vamshi Ambati</u> **Problem:** Addressing bias in machine learning. **Solution(s):** **1)** Acknowledge bias. **2)** Understand bias types: sample, prejudice, measurement, algorithm, and exclusion. **3)** Continuous elimination.
Article 8	***How Artificial Intelligence is Revolutionizing Mental Healthcare.*** <u>Author: Aliha Tanveer</u> **Problem:** General prevalence of mental health around the world, especially in areas underserved by mental health professionals. **Solution(s):** AI can via (among other measures): 1)--helping to categorize mental illness via research analysis, 2)--potential for help via tools like chatbots, 3)--early-warnings, e.g. via language-analysis and monitoring one's activity, 4)--reducing bias and human error, and 5)--integrating mental healthcare with physical healthcare.

CHAPTER CONCLUSION: IMPLICATIONS OF RESEARCH FINDINGS FOR IMPROVEMENT OF ICT SYSTEMS

Overall, my close examination of this chapter's foregoing discussion highlights two main potential overarching solutions that we can apply to the problematic at hand (improved design and use of ICT systems in the context of SC), namely: 1) individual (self-directed) mindfulness and individual and collective soul-searching, and 2) consolidation and more mindful use of ICT.

In the sections below, I apply each of these two solutions severally to the conclusive reviews of the chapter's findings, grouped under the introduction sections of two main research questions, namely: 1) "In the context of SC and interpersonal communication, … how can we improve ICT systems (?)," and

2) "How can we overcome the challenges of our SC/cognition to so as to improve our ICT design and utilization strategies and tactics?"

How can we Improve our Designs and Use of ICT Systems?

A review of both data-sets suggests that this question is best answered using data-set II's articles. Out of the eight articles identified, article one can be interpreted as classifiable under both solutions one and two introduced in the above paragraph (i.e. and individual and collective mindfulness and soul-searching, and consolidation and more mindful use of ICT, respectively).

On the other hand, article five does not seem to fit either solution category. Thereafter, we are left with six articles to categorize. Among these, only one article—i.e., seven—appears classifiable under solution one (individual and collective mindfulness and soul-searching), while the rest of the articles (i.e., three to five, as well as six and eight) fit the category of consolidation and more mindful use of ICT.

I: How can we Overcome the Challenges of our SC/Cognition to so as to Improve our ICT Design and Utilization Strategies and Tactics?

This particular question seems to be more suitably answerable by data-set I's findings. In a nutshell, section 8.2.1 above is classifiable under individual (self-directed) mindfulness and soul-searching, while the first part of 8.2.2—i.e., not inclusive of the specific instances—is more appropriate under consolidation and more mindful use of ICT. This last point is worth elaborating: first, consider how often you've found yourself in an argument with a relative or friend about the meaning of a word or concept, historical fact, etc., yet the two of you could easily just "Google" it on your smartphones! Or, take my example below; as of writing this chapter (late December 2021), I am preparing for a project-management-related presentation and course (as the instructor), and I realized that I need to memorize the 10 most important aspects of project-management, apart from "project life-cycle" and "professionalism (and ethics)." My solution? I "Googled" mnemonic generators, and found one that proved fairly useful indeed:

Embedded Exemplary Activity 8.4.2: Use of a Mnemonic Generator
Time & Date: 12/12/2021, Approx. Between 11:01 to 11:05 PM
Problem: Turning the following words into a mnemonic:

Integration, Scope, Schedule, Cost, Quality, Resource, Communications, Risk, Procurement, Stakeholder.

Solution: Website, i.e. https://www.mnemonicgenerator.com/
Website's Mnemonic Results
Result 1: *Iceman Spanked Shamans Conversely Queasy Roosters Chose Raspy Pistols Suspiciously.*
Result 2: *Insensitive Sabretooth Stabbed Creamy Quilts Regarding Chieftains Regarding Peaches Selfishly.*
Result 3: *Irritated Snowmen Stroked Cuddly Queenie Rashly[;] Consequently[,] R2-d2 Plotted Stealthily.*
Result 4: (My Favorite) *Irishmen Squeaked Since Creepy Queens Rejected Cold Ruthless Padme Sheepishly.*
Result 5: *Irishmen Scribbled Since Clumsy Queens Rotated Confused Rough Piranhas Suspiciously.*

II: How can we Overcome the Challenges of our SC/Cognition to Improve our ICT Design and Utilization Strategies and Tactics (?); Specific Instances

Finally, each of the two highlighted instances from the voice-note transcript—i.e., 2.1.2, and 2.2.12, can respectively be classified under solution one (self-directed mindfulness and soul-searching for 2.1.2), and solution two (consolidation and more mindful use of ICT for 2.2.12).

Final Reflection

As we discuss in the introduction of this chapter (and in chapter one), humans are quintessentially a social species. But like all other animals, humans have both good and bad characteristics. And unfortunately, one of those negative characteristics, i.e., flawed thinking processes—including logical fallacies and cultural prejudices, often negatively affect the designs of our ICT (and other technological) systems. Moreover, regardless of the presence or absence of defects in the ICT systems, we often start misusing those systems—e.g., as pointed by Turkle (2011) via alienating ourselves, or the countless maladaptive effects of social media, as often exposed by media reports over the past several years (e.g., regarding misinformation in general, and during elections).

In fact, many readers can probably relate to the "texting vs. talking (while in the same house/room!)" example highlighted in the second paragraph of section 8.1.1 farther above. In any case, this chapter has grappled with how to solve the numerous problems at the intersections of SC and the optimization of ICT-systems' design and usage. And apparently, scholars and engineers are already actively trouble-shooting this problematic, with results gradually unfolding.

Still, as Thagard (2005) and others point out, it is probably very unlikely that we will create AI with human-like sentience any time soon—i.e., as depicted in movies such as the *"Terminator"* and *"The Matrix"* series. But even though we cannot create machines that understand or feel emotions the way we do, should we at least create machines that can help us regulate our own emotions?

In this regard, various authors in data-set II and elsewhere allude to progress already being made. In fact, smart-watches that regularly remind you to take deep breaths can be counted in this category, and news reports suggest that one of Elon Musk's companies is working on implantable brain chips (e.g. Reuters, April 2021).

But regardless of any progress we continuously achieve in this sphere, we should beware of un-expected consequences. It can also be argued that scholars and engineers should strive to design ICT systems that mirror the variety of humans' thinking styles, versus limiting them to engineers' or other usual stakeholders' ideas/ideals. One other related point can be made in this regard: if CRUM is valid indeed, how can we harness computers' ever-increasing processing and storage capacities to simulate our infinite thoughts in more sophisticated ways?

In the above considerations, our two highlighted solutions farther above—individual (self-directed) mindfulness and individual and collective soul-searching, and 2) consolidation and more mindful use of ICT—can guide us. And they're both closely related to an overarching caution: it's important to keep harnessing the power of metacognition. We need to continue critically examining how we think, and how we can improve our thinking processes in all domains of our lives.

Creativity-Attempt Outtake 8

Post 32; August 14:
On AI and Writer's Block

Definitively completing the first draft of chapter one (of this book: https://www.igi-global.com/book/social-scientific-examination-intra-inter/264268) felt good indeed. But the question I'm now grappling with is, how in the world am I going to write that AI chapter? What the freak do I know about AI? Never mind that I actually earned a certificate of merit for an introductory AI course (from the University of Helsinki). I guess I have to think up something to write about the topic, indeed! One possible remedy: expanding the focus to digital technologies [at] large, versus AI alone.

Self-Questioning for Creativity-Attempt Outtake

State any Necessary Introductory Remarks, and/or Answer this Question: Does it Make Sense to Me Now—i.e., In The Present—versus then—i.e., When I Wrote It?

Yes, the blog post makes sense to me now. I also realize that it was a breakthrough vis-à-vis brainstorming a solution for that quandary, i.e., how to write an entire chapter on something as complex as AI.

Does it Resonate with Me Now—i.e., In The Present—versus then—i.e., When I Wrote It?

Yes it does. In fact, I am still fearful—even after carefully drafting the chapter and repeatedly reviewing/editing it, about whether or not it makes sense!

What Mood and/or State of Mind Does the Tone Suggest?

Slight anxiety/apprehension and eustress, but mostly, a problem-solving state of mind.

What is/are the Direct and Indirect Relation(/s) to the Current Chapter via Topic(s) and Themes?

This CAO might be the most appropriate vis-à-vis the match between its topic and the chapter's topic(s). And to me personally, the "aha (!)" moment—i.e., with the solution of discussing ICT at large versus AI alone—also corroborates with solution two as introduced in section 8.4 above, i.e., consolidation and more mindful use of ICT.

Any Other Evaluations or Comments?

N/A.

Activities and Discussion Questions

1) Think about the various human traits we possess, esp. vis-à-vis our thinking, sensing, feeling/emoting, etc. If you had the ability—or you were making suggestions to a very smart software-engineer, which of the above human traits would you wish for computers to have? Now think about computers' various abilities vis-à-vis memory and processing, efficiency, connectivity, etc. Which of those traits/characteristics would you wish for humans to possess as well?

2) *Optional*: Continue journaling/creative activity 1 as advised at the end of chapter one above.

3) Continue activity 6 from chapter one, "Comparison with other types of communication."

REFERENCES

Reuters. (2021, April 10). *Elon Musk's Neuralink shows monkey with brain-chip playing videogame by thinking*. https://www.reuters.com/technology/elon-musks-neuralink-shows-monkey-with-brain-chip-playing-videogame-by-thinking-2021-04-09/

Thagard, P. (2005). *Mind: Introduction to cognitive science*. MIT Press.

Turkle, S. (2011). *Alone together: Why we expect more from technology and less from each other*. Basic Books.

KEY TERMS AND DEFINITIONS

AI: Systems capable of executing various functions such as complex calculations, language translation, and visual and speech perception, processing, and categorization.

ICT: Software and hardware used for generating, processing, storing, transporting, and exchanging information.

Chapter 9
"Am I Missing Something?":
Final Potential Insights About Intrapersonal Communication/Cognition

ABSTRACT

This chapter presents a holistic review of the book's research findings. First, the author articulates or (re-) emphasizes some important points that arise from the previous chapters' analyses. Next, he summarizes the findings of Data-Set I under five key typological groups, namely 1) unintentional and deliberate SC, 2) intra-/inter-personal communication spectrum, 3) language and symbol systems, 4) applications, and 5) other considerations. He also interrogates the findings via three alternative interpretive frameworks, namely Austin-Lett and Sprague, Vocate, and Linde. The author also summarizes the themes that arise from the second data-set under three main overarching groups, namely "Learning About Our World," 2) "Shared Meaning-Making," and 3) "Metacognition." Finally, he discusses the book's key takeaway in regard to SC, i.e., the concept of neutral, positive, and negative thoughts/SC, as well as the pervasive yet often-disguised appearance of SC-discussions in the media, especially via self-help literature.

INTRODUCTION

Brief Previous-Chapters' Review

The preceding eight chapters, divided into three parts, have introduced the concept of intrapersonal-/self-communication (SC); explored the ways in which we partake in this type of communication; and the ways in which we can harness SC to our advantage in various domains of our lives—i.e., personally, professionally, and other contexts. And at this juncture, there are various points that are worthy of emphasis or re-exploration, regardless of how much (or little) we have encountered them in the preceding chapters, and/or their specific forms and functions—so to speak—in the various chapters:

DOI: 10.4018/978-1-7998-7507-9.ch009

- While socialization plays a pivotal role in the formation of our psyches—not to mention the symbol and meaning systems with which we make sense of the world, each of our minds' complex individual thought- and SC-processes define our own unique sensemaking. By complex thought and SC-processes, I mean the infinite number of thoughts—and thus, SC instances—that one mind can have, e.g. per Porpora (2013).

- Moreover, intersectional power dynamics inevitably affect the outcomes—or "quality" of our socialization, as well as the perceptions highlighted most prominently for each of us. E.g., does a Black man in America look at a cop—of any race or gender—the same as a Caucasian man?

- Regardless of the above facts, it should be noted that the concept and practice of learning can be viewed as a product of both socialization and person-specific meaning-making systems/processes. In other words, I submit that it would be nearly impossible—if viable at all, for an individual to learn effectively without socialization, combined with meaning-making systems/processes tailored by individuals to suit their own personalities/preferences/styles. And this latter set of phenomena—i.e., meaning-making systems/processes—would arguably be impossible without SC. However, other elements also assist us considerably vis-à-vis sensemaking and problem-solving in our world, both generally and in contexts of interpersonal communication/interaction. For instance, how can we define the concepts of creativity and mindfulness, and how exactly can/do they assist us in our sensemaking and problem-solving?

- Or granted, creativity and mindfulness might seem mercurial and impractical. But what about some of the other specific potentially-useful applications uncovered in the preceding chapters, including our ability to realize our distractions and reactions during reading, and our constant checking(s) of our understandings, etc.? Better yet, have not the preceding chapters (re-)vindicated Archer's (2007) theory of reflexivity, which is arguably a system or result of SC?

- Overall, if this book's thesis—as well as (some or all of) the above points—has/have some merit, then more autoethnographic self-studies of SC should be done. Interdisciplinary teams of scholars (and/or practitioners, clinicians, etc.) can design, widely distribute (i.e., to large and diverse groups of individuals), and analyze findings of SC-autoethnographic self-studies such as the one reported in this book.

Chapter Introduction

With the above considerations in mind, in the current chapter, I attempt to holistically review the findings of all the preceding chapters, in an attempt to pinpoint the observations that are potentially the most valuable. This task will be executed via three main sections, namely: 1) a review of data-set I's findings—i.e., from the voice-notes/transcripts, which are utilized throughout most chapters in sections II and III; 2) a review of data-set II's findings—i.e., from the data- and cognitive-science resources/articles, which are mostly discussed in chapters six (learning), and eight (ICT and AI). 3) Finally, in the last section, I attempt to synthesize the biggest takeaway from this book's study of SC, namely, the possibility of pinpointing neutral, positive, and negative types of SC. I also briefly look at the rather pervasive yet disguised presence of direct and indirect discussions of SC in the mass-media, especially via self-help literature and other texts.

MAIN FOCUS OF THE CHAPTER

Conclusive Interpretation of Meaning and Importance— AKA, "*So What* (?)," of Set I Data From Chapters Three to Five and Seven (and Two, to a Lesser Extent)

Final Summary of Findings

On balance, we can delineate the findings from data-set I—i.e., the voice-note/transcripts—into five typological groups, namely (in random order): 1) unintentional and deliberate SC, 2) the intra-/inter-personal communication spectrum, 3) language and symbol systems, 4) applications, and 5) other considerations. Below is a brief discussion of each of the five elements.

The first category—unintentional and deliberate SC—appeared early during the initial in-vivo and axial coding stage, and I was able to define various SC instances throughout section two as constitutive elements of one or both of the groups. The resultant two analytic categories are "Reality And Perception-Processing I: Without Metacognition," and "Reality And Perception-Processing II: With Metacognition."

Next, the intra-/inter-personal communication spectrum category might sound familiar to the reader from chapter one (see section/diagram 1.2.1-II). In a nutshell, this category discusses SC instances as belonging to one or more spectrum-sections on diagram 1.2.1-II. In other words, we have the ability to either think/communicate to ourselves alone, and/or other individuals, and various combinations of those two types of communications in between.

The examples of this element are numerous throughout the voice-note/transcript data. Severally, I think about and/or internally talk to myself—then relay those thoughts and communications via the voice-note recordings—about previous, ongoing, and/or future communications with my mother, my friends/colleagues, etc. Specifically, I often review and plan communications strategies and tactics, and specific messages; e.g., the instance in which I catch myself composing a question to my mother in both English and Luganda.

The next category—i.e., language and symbol systems, focuses on an indispensable framework with which we communicate not only to other individuals, but to ourselves as well (first, before communicating with others). Ultimately, whereas many self-directed communications from our bodies and minds are not transformed into verbalized messages (e.g. corporal sensations, and/or mentally-derived emotions or feelings), it is apparent from this study that many messages are indeed transformed into that format—i.e., verbalized messages. We thus utilize language and symbol systems to partake in SC, e.g. via journaling, or speaking to ourselves within our minds or out-loud (like Mr. "K," from chapter one).

The other clearly identifiable element from the analysis of the voice-note/transcript data is the various functions that can be improved via more systematic studies of SC. In this book, I isolate three specific examples in this regard in part three, with each of these examples belonging to their own chapter, namely: 1) learning, 2) mental health, and 3) digital-technology and AI. I earnestly hope that future studies such as these can continue exploring these and other applications of SC. In fact, whereas my main secondary focus in this book—in addition to studying the general dynamics of SC—was learning, I submit that the topic of mental health might be the most important application of SC.

Finally, as the analysis of section 9.1.1 above demonstrates, there are numerous other considerations that we can/should keep in mind in regard to the voice-note/transcript data. In other words, when closely examined as more than the sum of its parts—e.g. as delineated in the above paragraphs, what can this book's voice-note/transcript data-collection teach us about SC in general?

Introduction to other Interpretive Frameworks; Austin-Lett and Sprague (1976), Vocate (1994), and Linde (1993)

In this section, I attempt to highlight the relevance of the findings of the voice-note/transcripts via the lens of these three frameworks as introduced in chapter one. Given the specific foci of the frameworks' questions/arguments/postulations, it appears that chapter two's data is also relevant to them, and we can also utilize the creativity-attempt outtakes (CAOs) from the endings of each of the book's 10 chapters in this context. Thus, in addition to reexamining the voice-note/transcripts in my search for answers to the questions/arguments/postulations, I will also holistically reexamine those two data-sources (chapter two's journal data, and the CAOs).

I: Questions from Austin-Lett and Sprague (1976)

First, below is a list of the reflection-questions about SC from this study, as introduced in chapter one:

- 1. Individuals should strive to really learn their personalities; how and why do you constantly and predictably think and behave the way you do, and what do others know about your thinking, emotions, and behavior? Relatedly, what are your likes and dislikes, goals, fears?
- 2. Do you realize the effects of your perceptions—i.e., your interpretations of your reality, in your various environments (e.g., at home, work, in public places, etc.)?
- 3. In an expansion of point number 1 above, do you know how your physical, psychological, and other needs, as well as your values (e.g., passed down to you from family and/or society), influence your actions?
- 4. Do you realize the influence of abstract symbol-systems such as language—which are also influenced by culture and other variables—on your self-expression?
- 5. Finally, do you realize and appreciate the never-ending nature of intrapersonal communication and growth, and can you confidently share and appreciate or understand your own and others' intrapersonal experiences as stated above, in your social-interactions?

II: Answers to Austin-Lett and Sprague's (1976) Questions

Based on all the relevant data examined in this book, below are my answers to questions posed above.

1. I think and behave the way I do based on my life's childhood and youth experiences, socialization and acculturation, education and talents/abilities, genetics, and personality, e.g. as described by the O.C.E.A.N psychological framework (e.g., Paunonen and Jackson, 2000). At this juncture of my life, my likes and dislikes, goals, and fears can be summed up as: i)—likes: career and general stability and prosperity, ii)—goals: for a successful academia/other suitable career, iii)—fears: failure of the latter.
2. Yes. Two relevant examples in this regard are the common assumptions I have often had in the past about students of mine that are not "good" students, as well as the assumptions I often have about others' perceptions of me (e.g., "people think I'm weird," etc.).

3. Yes. My basic needs—e.g., the bottom levels of Maslow's pyramid (Maslow, 1943)—unfortunately often deter me from staying focused on my longterm goals.

4. Indeed, especially after the critical self-reflection I've performed over the course of writing this book, I realize the influence of my societies' abstract symbol—and even cultural—systems and paradigms on me. Some of the systems and paradigms in this regard include: i)—the languages of English and Luganda, and lately French, ii)—the Western world's Judeo-Christian-influenced and other general moral/ethical and philosophical frameworks, and iii)—my American-centric way of viewing the global systems of commerce and diplomacy, etc.

5. My interpretation of this question triggers two key concepts from my mind, so as to appropriately answer it: i)—self-love/care, and ii)—empathy. I should strive to improve the way I judge myself and others. Sure, having high standards for yourself and others is beneficial, but remember that there are multiple metaphysical layers with any given reality!

I: Arguments from Vocate (1994)

Next, below is the list of arguments I synthesize from Vocate's (1994) study:

- 1. Despite the stark hindrances in our way as social-scientists to study the neurological, bio-physiological/chemical, and other *brain*-based features, we can certainly explore the cultural aspects that influence our intrapersonal-communication dynamics;

- 2. The study of intrapersonal-communication is more effective if executed via a triangulation of methods;

- 3. The study of intrapersonal-communication does not mean that we have to ignore other individuals' perspectives; rather, we need to recognize the role that others play in our internal dialogues;

- 4. Despite this point's irrelevance to the current study, it is noteworthy nonetheless: the study of intrapersonal-communication should be carried out with human participants from their early years of childhood, through later years of adulthood.

II: Responses to Vocate's (1994) Arguments

Below are my responses to the arguments above, based on all the relevant data examined in this book.

1. I agree with this argument, and I hope that I have managed a fairly thorough attempt to implement the suggestion via this book—at least partly, among the book's overall goals. As I note above, socialization and acculturation are indeed vital influences on the psyche, as well as our SC processes. And throughout the book, I have explored (among other aspects): my personal/family background in Uganda, my education in Uganda in the USA, and the staggered vocational path I'm struggling with. One question that I can pose to future researchers who review the work is thus: are the SC instances shared throughout the book "typical," if compared to similar individuals' SC? Regardless of answer ("yes"/"no"/other), how?

2. Agreed. And I hope my triangulation in the preceding chapters has produced satisfactory results. To recap, in this book, I have utilized the data/methodological analyses listed below, with i) and ii)'s data examined using basic narrative- and qualitative-analysis (transcription, coding, and topical analysis): i)—journal/blog data, ii)—voice-note/transcript data, and iii)—a corpus of data- and cognitive-science articles, examined via discourse-analysis.

3. Again, I agree. And in section 9.2.1 above, I discuss this aspect under the intra-/inter-personal communication spectrum topic.

4. In the end, I believe I also partly satisfy this argument's suggestion. Whereas the data examined in this book does not quite cover my early childhood—apart from minor references to it here-and-there, the journal data stretches back around 20 years from the current date (in 2021), from my high school years. Thus, I believe readers/researchers are offered a decent attempt at a fairly thorough examination of my SC-processes' depth and breadth from my youth years, through today.

Brief Application of Linde's (1993) Life-Story/Coherence-System Method; *The Miseducation and Re-Education of Seif Sekalala, Ph.D.*

From the outset, it should be noted that my study's methodological framework does not quite neatly fit with Linde's. The major reason for this fact is thus: as she makes clear, her study was specifically examining spoken life-stories, garnered in an interview format—i.e., by the researcher speaking to the research participants. However, the contents of my journal-/blog-entries and voice-notes/transcripts provide a fairly thorough self-examination, with a few events and/or elements that can arguably be described as (potentially) reportable—i.e., warranting attention. One major element in this regard for instance is my self-disclosure and examination of mental health since high school, through today. With this caveat in mind, we can review the findings from the relevant data in the preceding chapters via the perspective of Linde's study.

Via such a review, and after thinking critically about my life for the past one year through various intense hardships, I am under no illusion vis-à-vis neatness/obviousness of any given life-story, including mine. In other words, at the moment, I do not have a firm coherence system to which I can give full credit for the outcomes of my personal and professional life.

However, I should note that this was not always the case. At various pre-inflection points throughout my 37 years of life to date (as of mid-2022), I have had held hopes of how my life—career-wise, and in general—would play out. And at this particular juncture of my life, I realize that I have "won some" and either "lost some," or, I am yet to win those other rewards. For instance, I was able to achieve my dream of getting an American college education, and I subsequently earned American citizenship. But at least at this point (i.e., 2021/2022), I have not achieved my dream job, a tenure-track professorship (having graduated six/seven years prior, in 2015).

And in the above context, the role of my own agency in the story is limited. For instance, I did not choose where, when, and to whom I'd be born, but I was nonetheless lucky to be born to upper middle-class parents in Uganda, during a stable era (after the years of Idi Amin and the civil wars of the 1960s-80s). As a result of that luck, I was able to get a good education, and I had the resources to come to the USA. Subsequently, I did in fact work hard in school, enabling myself to earn a Ph.D. And I had discovered my talents for the language arts and writing early on in Uganda, hence my choice of English-Writing and communication-studies as my majors.

From that point however, the story gets complicated. In particular, for the past couple of years, I have grappled with the question of how/why I do not yet have that dream job. Maybe I'm not good enough, after all? Perhaps my race plays a part? (And as I revealed in the introduction chapter, yes it does.) Perhaps my personality also plays a part (after all, my overly chit-chatty sentimental manner probably grates many people's non-dramatic sensibility)? Or perhaps some or all of the above factors, as well as my choice of research and sheer bad luck—among many other factors—have aligned, resulting in that outcome?

Overall, I am learning the hard way that there are no easy answers. In fact, one of the SC messages I constantly say to myself lately is: *"2021 is the year I've learned not to judge."* Of course I always knew better, but I suppose my good fortune had been an impediment vis-à-vis empathy. But when a social-worker and psychiatrist glibly suggest that one should try looking for some other "less-desirable job" (to be fair, I am merely paraphrasing their words), if only to pay the bills in the meantime—without bothering to ask if the individual has already tried that, then that individual attains a special understanding of the frustrations of America's working poor, and other frequently-judged individuals/populations.

Conclusive Interpretation of Set II Data From Chapters Six and Eight

One can argue that the corpus of the data- and cognitive-science articles examined in chapters six and eight tell a story of the interaction between our minds, and the world that those minds—ensconced within our physical bodies—find themselves in. The data-sets can help us answer the following two questions: 1) How do human beings—via their minds—make sense of their world, both in isolation via SC, and together (via interpersonal communication and ICT)? 2) In addition to making sense of their world, how do human beings survive and thrive in that world—i.e./e.g./especially, via functions such as companionship or socialization, as well as commercial/economic activities?

In regard to the first question above: readers can see from the table titled *"Cognitive-Science Article Basic Thematic Analysis"* in appendix "D" that as part of my thematic analysis process, I attempted to highlight themes and connections from and across all the cognitive science articles. As a result, there are six groups of randomly grouped articles, titled together by "Arguable Random Common Theme[s]." However, we can consolidate these six groups of articles into three overarching thematic groups as follows: 1) Group one—titled "Learning About Our World"—consisting of the sub-groups of "knowledge" and "epistemology." 2) Group two—titled "Shared Meaning-Making"—consists of the sub-groups of "Learning and Pedagogy" and "Meaning." 3) And group three—titled "Metacognition"—consists of the sub-groups of "Empirical Models" and "Cognition and Experimentation."

In the first overarching group, we can see how the authors study the concepts of curiosity, self-understanding, and socialization, and how all these concepts contribute to our knowledge of the natural and social world. In the second overarching group, the authors strive to understand how we make and convey meaning, and how we strive to improve those processes via pedagogy. Finally, in the third overarching group, the authors strive to clarify their own and their readers' understanding of how our minds work.

Next, let us focus on the second question posed in the current section's introduction paragraph above. In this regard, the data-science sub-corpus is arguably more salient, even though the above discussion (about the cognitive-science articles) can also be applied to the question. My rationale for that judgement is related to the functions and/or domains of professional life that the sub-corpus can help us with, which I listed in chapter six, namely (excluding "other"): 1) "learning," 2) "tools and techniques," 3) "careers," and 4) "how-to."

Finally, how can I relate the above discussion to my own learning-process improvement? My answer can be summed up via two concepts combined, namely, metacognition and sensemaking. It is apparent to me that I cannot improve my learning processes without a deliberate and systematic consideration of how I habitually think, conceptualize, learn, and utilize knowledge. Rather, I have to constantly re-assess whether or not my ways of thinking and learning are working or not, and/or how to improve them.

CHAPTER CONCLUSION: NEUTRAL, POSITIVE, AND NEGATIVE THOUGHTS/SC, AND SC IN THE MEDIA

Neutral, Positive, and Negative Thoughts and SC

Over the course of writing this book, I believe the three most important realizations I have made about SC are:

1) Apparently, it is a given that at any time while we're conscious—even in our sleep while we're dreaming, arguably—our brains will produce thoughts (also known as the "stream of consciousness"; e.g. James, 1890) or SC instances.

2) However, it is up to us to realize which of the three broad categories these thoughts or SC instances belong to, namely i)—neutral/other, ii)—positive, and iii)—negative.

3) Thereafter, in addition to utilizing the thoughts or SC instances as best we can—if they are indeed "actionable" (e.g., "*Shoot, I need to take out the trash!*") we also need to remember that these thoughts or messages are simply that, nothing more.

In other words, the fact that I often think about that time I got mugged in Newark (NJ), or those times I got bullied in school, or the eventuality of losing a loved one, etc., does not mean I should let myself become sad or worried, or anxious in reaction to those thoughts.

Positivity/Self-Counseling Mantras

In fact, over the past several years, I have come up with several mantras that help me survive adverse situations and/or turbulent or negative states of mind, e.g.:

- "They are just thoughts, nothing more."
- "Peace and discipline shall set you free" (which can be paraphrased as: be consistent and try to achieve your goals, but try to also make peace with the way things, not the way you want them to be).
- "With good planning, almost anything is possible."
- "Defy the odds and expectations; [*Expletive*] it, just do your best, and do not focus (too much or negatively) on how you're perceived."
- "Breathe." Related: "Stay calm; always stay calm" or "chill, Seif; relax!"

SC In the Media

In chapter one, in addition to the story of Mr. "K," I also highlight the advice given by Schwartz & Pines (2019) via their Harvard Business Review (HBR) article. But readers should note that I did not add that reference during my first draft of the chapter. Rather, I discovered and added it recently (sometime around late 2021 to early 2022). Perhaps because of the fact that I have been working on this book for over a year and four months to date, lately, I keep noticing several examples throughout online print media of references to SC.

Most of these references are indirect; the authors do not refer to the concept via that name or even at all. But it is clear to someone attuned to the concept of SC that essentially, SC is the phenomenon being referenced. In addition to that HBR article, I can give two more examples in this regard.

In a New York Times profile of the actor, David Marchese interviews Matthew McConaughey about his various re-inventions over the course of his acting career. Overall, the author/interviewer and actor delve deep into topics of self-reflection in general, and in the context of a famous Hollywood figure such as McConaughey.

In fact, the first question posed by Marchese to the McConaughey arguably cuts to the heart of the matter, so to speak: "Was there a turning point when you realized your future was going to be about something other than acting?" McConaughey responds by asserting that above all, in or outside the context of acting, he prefers being true to himself (with select clauses boldened and underlined by me, for emphasis below):

When I do the best performance in a movie, I tap into what's most myself in that role. So, do you need another name and wardrobe and someone else's script to tap in further? **_What about the monologue with myself? What if that's the dialogue?_** *What if that's the performance in this life? How can I live? Can that be my greatest art? How do I become living art? That's what I'm discovering.*

(Marchese, 2021.)

In other words, McConaughey constantly challenges himself—via SC—to enact his real ethos, his genuine self, regardless of context.

The last example in this context is an article published in "*Success*" magazine, by LaRae Quy, a self-help guru. Quy is described as a "former FBI counterintelligence and undercover agent and founder of the Mental Toughness Center." The article is titled "7 Mental Hacks to Be More Confident in Yourself," and the third tip (/"hack") provided by Quy states:

3.   Talk to yourself.

This might seem crazy, but it works. Talking to yourself can make you smarter, improve your memory, help you focus and even increase athletic performance. The documentary The Human Brain claims we say between 300 to 1,000 words to ourselves per minute. The Navy SEALS and Special Forces use the power of positive self-talk as a way of getting through tough times.

For example by instructing recruits to be mentally tough and speak positively to themselves, they can learn how to override fears resulting from the limbic brain system, a primal part of the brain that helps us deal with anxiety.

How to make it work for you:

Be positive, because the way you talk to yourself influences your neurobiological response to it. When you say, I know what to do here or see things as a challenge rather than a problem, you've turned your response into a positive one.

Regardless of one's stance (i.e. skepticism/favor/other) towards the self-help industry, the above examples seem to suggest that the best life coach you can utilize might be your own *self.*

Creativity-Attempt Outtake 9

Post 28; July 18, 2021:
"Dear Journal," Continued

Writing an autobiographical (/autoethnographical) book—and/or even the mere re-reading of one's past journal-reflections—can make one somehow feel somewhat, that they live in multiple dimensions in this life. One part of me traverses time and space as the ordinary me—wake up each day, learn and/or work, reflect, go back to sleep, rinse-repeat. And the other part of me watches the latter-mentioned me doing all those things.

And from time to time, those two selves converge, resulting in the kinds of deep reflections I am producing in the book manuscript. One positive upshot of the above processes is this: with time, the puzzle pieces seem to oh-so-neatly (more or less) fall in place, and as long as one establishes a realistic paradigm, one can and often does achieve that thing we crave so much in our lives: happiness.

Here's to a week of more living, more reflection, and more (realistic) happiness.

Post 17; 05-06-21:
Is It Okay To Fail?
If "Yes," How–And How Often, When, And Why?

Believe it or not (dear reader), despite the above title's relation to one of the major themes of this post, I did not compose it for that purpose! But it's "interesting"—for lack of a better word—how often such connections arise spontaneously. Neat, quirky, interesting, beautiful, and I guess "weird" connections—"here and there," "all over the place (ha-ha)."

The theme (mentioned above) in question? Deadlines. And the connection in question? The journalistic-lead format's five key questions, of which three are featured in our title above!

Apparently, for two weeks in a row, I've missed a self-imposed deadline, i.e., posting an entry to this page by around 11 PM on Monday of each new week. Do I feel bad about that? Yes, a little.

Only "a little," as I have learned over time, to try to put things in the right/appropriate perspective. Thankfully, I did not mess up something related to a human or other—e.g., a pet's—life; i.e./e.g., surgery, flying a plane, etc. And part of the reason I feel bad is simply related to my ego.

I guess the substantive damage to my psyche emanates from the fact that this slight failure of discipline can lead—if not fixed over time—to bigger and bigger failures. You know, kind of like the 'good ol'' "slippery slope" (https://en.wikipedia.org/wiki/Slippery_slope)! ☹

And I guess—in this case at least, Freud's framework works for me. The high-achiever in me wants to post well before the deadline each week, while the slacker does not want to post at all—seriously though, who would want to miss out on this much fun (ha-ha)?! And right down the middle is the realistic referee, keeping me in check, helping me do my best and re-adjust constantly, aiming for improvement, etc.

Yes, it is ok to fail sometimes. I happen to be a student-pilot—for a private license, that is, and I can attest to the deliberate and methodical way in which one is trained gradually, leading up to a commercial-license. From what I've read/heard, healthcare-workers go through the same routine. How many mistakes do we usually make along the way? "A dime a dozen" of course. The goal isn't to be perfect right from the beginning, and/or with each try.

Random off-key thought/question to close us out. I wonder if one can somehow relate the above sentence ("The goal isn't to be perfect…") to that kid's statement in "The Matrix," about the spoon? 🤪 (https://www.youtube.com/watch?v=uAXtO5dMqEI)

Have a great end-of-week, and a great weekend to boot, everyone!

Self-Questioning for Creativity-Attempt Outtake

State any Necessary Introductory Remarks, and/or Answer this Question: Does it Make Sense to Me Now—i.e., In The Present—versus then—i.e., When I Wrote It?

For the most part, reading all the above blog-posts makes me ponder the mysterious nature of the processes of thinking; knowledge intake, integration/utilization/retrieval, etc. Somehow, in the above posts and at other junctures, I find myself somehow making connections or smoothly retrieving and utilizing concepts I have previously learned. But I am really mystified as to the precise cognitive steps or mechanisms of that retrieval and utilization. I hope this response make sense!

Does it Resonate with Me Now—i.e., In The Present— versus then—i.e., When I Wrote It?

Yes, but please refer to my answer above.

What Mood and/or State of Mind Does the Tone Suggest?

I believe the tones are suggestive of a self-reflective mood or state of mind.

What is/are the Direct and Indirect Relation(/s) to the Current Chapter via Topic(s) and Themes?

The link(s)/relation(s) is/are indirect. In the chapter, one can argue that I am trying to provide (a) clear answer(s) to the question of, "what's the biggest takeaway from this book?" But in the posts above, I am trying to creatively retrieve knowledge from my mind, to turn it into wisdom.

Any Other Evaluations or Comments?

N/A.

Activities and Discussion Questions

1) Recap, part I: Go back and read through some or all—or as many as you can analyze—of your answers to the discussion-questions for chapters 1 to 4. Then, re-consider the questions, and try to revise your answers in a bid to holistically incorporate the things you've learned overall in the book, beyond those four chapters. Discuss your revised answers.

2) ***Optional***: Continue journaling/creative activity 1 as advised at the end of chapter one above.

3) Continue activity 6 from chapter one, "Comparison with other types of communication."

REFERENCES

Archer, M. S. (2007). *Making our Way through the World: Human Reflexivity and Social Mobility.* Cambridge University Press. doi:10.1017/CBO9780511618932

Austin-Lett, G., & Sprague, J. (1976). *Talk to yourself: Experiencing intrapersonal communication.* Houghton, Mifflin.

James, W., & Drummond, R. (1890). *The principles of psychology.* Henry Holt and Company.

Linde, C. (1993). *Life stories: The creation of coherence.* Oxford University Press.

Maslow, A. H. (1943). A Theory of Human Motivation. *Psychological Review*, *50*(4), 430–437. doi:10.1037/h0054346

Paunonen, S. V., & Jackson, D. N. (2000). What Is Beyond the Big Five? Plenty! *Journal of Personality*, *68*(5), 821–835. doi:10.1111/1467-6494.00117 PMID:11001150

Porpora, D. V. (2011). How many thoughts are there? Or why we likely have no Tegmark duplicates $\$\$ 10^{\{10^{\{115\}}\}} \$\$$ m away. *Philosophical Studies*, *163*(1), 133–149. doi:10.100711098-011-9790-6

Schwartz, T., & Pines, E. (2019, April 17). *Harvard Business Review*. Retrieved from https://hbr.org/2019/04/great-leaders-are-thoughtful-and-deliberate-not-impulsive-and-reactive

Vocate, D. R. (1994). *Intrapersonal communication: Different voices, different minds.* Erlbaum.

KEY TERMS AND DEFINITIONS

Coherence System (by Charlotte Linde, 1993): The "system[s] of beliefs and relations between beliefs" (p. 163)—that individuals use to make sense of their life stories.

Deliberate SC: SC that is undertaken by an individual on purpose, alongside metacognition.

Self-(/Intrapersonal-)Communication (SC): Communication inside an individual's mind, or outside—e.g., spoken or written—but for that same individual's consumption and/or use.

Unintentional SC: SC that is undertaken by an individual without their realization.

Chapter 10
"You're on Your Own":
Conclusion

ABSTRACT

In this chapter, the author performs a final review of the goals of the book, its research findings, limitations, and avenues for future research. In a nutshell, the book attempted to answer the following three questions: 1) What is SC? 2) How is it done, and what can it help us achieve? 3) How can we improve it? Most of the book's research data is derived from this author's life-story context, as well as his thoughts and self-directed communications. The other data is from a qualitative basic meta-review of cognitive-science scholarly articles, and articles of instruction for data-science and other related topics—e.g., theorization about learning and creativity—curated over a period of two years. Overall, the research findings suggest that SC is a powerful tool for various functions—including learning, ICT, and AI. However, its most helpful function might be to support our resilience and mental health.

CHAPTER INTRODUCTION

Recap: Focus of the Book

Around mid-to-late 2019, the idea of this book sprung in my mind, having germinated over a period of several years of studying and teaching communication. In summary, my rationale for the importance of this book's research is: the study of intrapersonal-/self-communication (SC) has not been studied sufficiently, yet it is the foundation on which all interpersonal and other communication types are built, including those which enable the development of sophisticated systems such as AI. Consequently, I intended to study: 1) what SC is, 2) how it is done, and 3) what it helps us achieve, and how we can improve it.

Overall, roughly 80% of the book's research data is derived from this author's life-story context, as well as his thoughts and self-directed communications. The other data (roughly 20%) is from a qualitative basic meta-review of: 1) scholarly articles from two cognitive-science journals, and 2) articles of instruction about data-science and other related topics—e.g. theorization about learning and creativity, which I have been emailing myself over roughly two years, sourced from *LinkedIn* and other online sources.

DOI: 10.4018/978-1-7998-7507-9.ch010

MAIN FOCUS OF THE CHAPTER

Final Research Findings Review, Limitations, and Avenues for Future Research

Data-Set I—Voice-Note/Transcript Findings

As I sum up in the previous chapter under section 9.2.1, the contents of this book's first data-set—i.e., voice-note/transcripts—can be succinctly described using five typological groups, namely: 1) unintentional and deliberate SC, 2) intra-/inter-personal communication spectrum, 3) language and symbol systems, 4) applications, and 5) other considerations. These groups became apparent over the course of my analysis and writing of the first seven chapters, and I believe one can categorize most or all SC instances from the voice-notes and transcripts (appendices A through D) appropriately in one or more of those groups.

Data-Set II—Data- and Cognitive-Science Resource-/Article-Corpus Findings

The findings of this data-set can be summarized under three overarching thematic groups, namely: 1) "Learning About Our World," which consists of the sub-groups of "Knowledge" and "Epistemology, 2) "Shared Meaning-Making," which consists of the sub-groups of "Learning and Pedagogy" and "Meaning," and 3) "Metacognition," which consists of the sub-groups of "Empirical Models" and "Cognition and Experimentation."

Limitations

At this juncture, I can think of at least three major impediments that might have negatively affected the quality, and/or depth and breadth of this book. However, I am optimistic that they can be ameliorated via future related research; in this regard, please refer to the next section (10.2.4) for details.

The first impediment is time and resources. Even though the publisher was generous vis-à-vis deadline extensions, I believe a project such as this can always benefit from more time and other resources. And by "other resources," I am not necessarily thinking of financial support per se or alone, but that can also be helpful. After all, if one is working a full-time non-academic job, s/he has squeeze in research and writing work throughout their weekly schedule, alongside their other "gainful employment" work. Generally, an academic-job environment would have helped; for instance vis-à-vis library-database access, as well as contractually-guaranteed time for research and writing.

The other impediment might be debatable, and/or might not technically be an impediment as such, depending on one's leaning vis-à-vis epistemological theory/ideology. The impediment in question can be defined as the issue of perspectives or points of views. Granted, the methods and theoretical frameworks I chose to use for this book's research—i.e., autoethnography and cognitive science—are indeed appropriate for the topic and research questions. However, I would have wished to have (an) other scholar's(/s') perspective(s) throughout the book. For instance, I try to be self-critical in reaction to the chapter-end CAOs. But can one successfully execute such self-criticism without another scholar's/individual's perspective?

Avenues for Future Research

Given the above considerations, I believe there are numerous opportunities for future related research. My biggest hope in this regard is for me to collaborate with other interdisciplinary researchers. Despite my firm belief that SC studies ought to indeed be executed by individuals themselves about their own SC processes, I am intensely curious vis-à-vis the results that can be garnered by combining self-studies— e.g. via this book's methods, with other methods. For instance, how can brain-imaging technology be applied to a study such as this? And if a critical mass of scholars and other individuals pursue such self-studies, what can we learn from meta-reviews of such studies?

Thus, over the next several years, I will try to team up with colleagues in the USA and elsewhere who are willing to further explore the topic(s) discussed herein. But I also hope that other scholars can pursue similar studies independently, given the necessity of SC in our lives, both for communication purposes, as well as learning, mental health, and other functions.

CONCLUSION

As of December 2021, the bleak employment prospects I reveal in chapter one have only improved slightly. And as I have mentioned in other chapters, I am also going through a marriage dissolution. Partly as a result of those stressors, the mental-turbulence severity scale I introduce in chapter two is currently at (roughly) 8 out of 10. And granted, throughout these and other challenges, I have been blessed to receive generous moral and tangible support of various family members and friends.

But oftentimes, even one's most ardent supporters succumb to donor fatigue. Case in point: while concluding chapter seven—i.e., around early July 2021, I experienced an interaction that can be described as a life turning-point or even a crucible-moment. The only aspect I feel comfortable revealing about this interaction is that it involved hearing a hard truth from my mother, or a message that can be described as *unintentional* "tough love."

Moreover, thanks to various factors—some due to my personality traits, others due to lifestyle and my current career/life-trajectory, etc., I currently do not have a substantive number of family and friends that I can confide in. Ultimately, the biggest crutch throughout these stressful events—and overall, over the course of my life to date—has been and remains that aforementioned "secondary self," which constantly takes a step back to make sense of life, and to (re-)interpret life to my (primary) *self.*

As a result, I find myself increasingly utilizing intrapersonal/self-communication to cope. I constantly self-soothe, remind myself to breathe, and remind myself to think about my life from a "big picture" and longterm perspective: "this too shall pass." After all, over time, regardless of various setbacks, my life has gotten—and will thus (hopefully) keep getting—better and better, as long as I keep working consistently and strategically. Somewhat similar to Tom Hank's character Chuck Noland in "*Cast Away,*" I have turned my secondary self into my own version of Chuck's buddy (the volleyball), Wilson.

Overall thus, on balance, perhaps in some extreme cases, some types of SC are indeed symptoms of severe mental illness. But after all is said and done, in addition to the sophisticated technologies such as AI that SC ultimately helps us create, human *selves'* self-communicative sentience might be our most reliable perseverance tool. And instead of ignoring it or undermining its value, we ought to study it and utilize it better.

Creativity-Attempt Outtake 10—Part I

Post 3; 01/24/21:

Life On Autopilot

One of my favorite undergraduate English professors (Dr. Connor, retired Prof. of English at Kean U.) used to enjoy fostering debates among his students—i.e., during his class sessions—about what we believed in more, fate/determinism, vs. free will. At the time, I don't think the early twenty-something version of myself had a strong opinion about the topic.

But I do now, after going through the past 13 years of graduate school, work, and life in general. And while that life experience is the primary source of my evaluations/judgements or basic analyses of the topics, I am also aided in my thinking by a better understanding since those undergrad years, of various philosophical and social-science topics, including (among others, in random order): existentialism, pragmatism and symbolic interactionism, and critical-realism. With all that said, below are some of my thoughts about the roles of fate/determinism, as well as free-will, in each/most of our lives.

Unfortunately, many of us are born with the odds stacked against us. You don't choose to whom, where, and when, you're born. You could be born to an indentured servant early in the previous century, or a fisher-man/woman in some poor part of Africa today, or you could be born into luxury as a royal prince in any time period, or to a billionaire or millionaire today, or in any century.

Unfortunately, those of us who are born in such "short end of the stick" situations—e.g., to poor folk in any era, and/or to anyone (rich or poor), but in an era in which science hasn't advanced to help us avoid unnecessary diseases—will have to "work twice as hard" to overcome our disadvantages. And of course, even those of us born with fortunate odds—e.g., rich, or in advanced economies with good health, education, and social safety-nets—still have to work hard, and/or "work smart" to achieve "desirable" lives.

What is or isn't a "desirable life" is a topic for another journal/blog post. Regardless, here's what I've noticed or deduced about all our lives, in relation to the above concepts:

Regardless of how hard you work, someone, somehow, often/always gives you a helping hand along the way. It could be the government, a kind benefactor, that one teacher who looked out for you, etc.

This is a key point, something I've been considering for a few months/years to date: Regardless of the quantity, or the quality of those fancy plans we make for our lives, there seems to be a natural unstoppable and unflappable rhythm which pushes our lives forward. In the end, yes, we often achieve our goals. But the road to those achievements is most often a meandering one. This is the context in which concepts such as habitus, probability, sensemaking, and existentialism, can help us navigate our way through the world. Based on the society and family in which you were born, do you know why think/believe what you do, and—most importantly—how those thoughts/beliefs positively or negatively influence your life on a regular basis? How well do you "play the odds"? I believe a famous baseball player—Babe Ruth, is it (?)—said something to the effect of, the more pitches you make—sorry baseball fans, I'm not very good with the *linguo* and rules—the better your chances for scoring higher.

The need for humility, and perhaps—perhaps (!)—belief in forces/powers "above/beyond self": All that wealth and power you have can one day just disappear. Think of that archaeological discovery, of mummified rich Italian ([?]/Greek) rich masters dead because of a Volcanic explosion, right next to their poor servants. In the end, we're mere creatures in a vast universe, which is both predictable and also (very) unpredictable!

Creativity-Attempt Outtake 10—Part II:

Post 20; 05-25-21:
Of Change, Chaos, And Crucible Moments

Starting over the past weekend through the next few months, I am experiencing an event that is ranked (by numerous relevant professionals) as the 2nd among life's most stressful events. For various reasons, I am choosing to be oblique here—not annoyingly coy, at least not on purpose—about the details of said events. Still, given this current extremely tumultuous life-transition in which I am because of said event, I believe it might be worthwhile for myself and my readers, for me to reflect on my current state of my mind, especially in relation to the concept of crucible moments.

The concept of "crucible moments"—which I believe is credited to Warren Bennis as the idea-originator, can be explained using the definition below, courtesy of Will Krieger (link: https://medium.com/@willkrieger/your-crucible-moments-35994634a312):

A crucible moment is, by definition, a transformative experience through which an individual comes to a new or an altered sense of identity.

These are times when our character is tested. These are times of adversity where great strength is shown.

Those who go on to be great are those who take time to pause and reflect on these moments. These are the moments that make us the leader we're going to be, the parent we're going to be, the person we're going to be.

Incidentally, the culmination of "stressful event X" unfolded alongside other considerable stresses in my life. But at some point over the past few months, I realized something that might be a key trick for my productivity, which (I believe) I have hinted at or mentioned before in another post. The realization (?): I believe our worst days might be our biggest opportunity to succeed in life.

How? If on your very worst day (e.g., because of family issues, or mild health issues, or conflict at work) you still manage to cross at least one item off that "to-do" list, IMHO, you've conquered the vagaries of life. I am also realizing the priceless value of consistency vis-à-vis routines, habits, and even rituals. And yes, many of us have numerous idiosyncratic behaviors, work-processes, etc., that we don't realize are in fact rituals.

Ultimately, Heraclitus was right indeed (paraphrased): change is life's only constant. I would also add—in an existentialist mood, that chaos is the other guaranteed constant of life. Somehow, you have to get to know yourself well, figure out your productivity modus operandi, and just do your best each day. And whenever you find yourself in a discombobulated situation—and you will surely, from time-to-time, stay calm; breathe; think; and again, stay calm. Then do your best to solve whichever problems might have arisen, or to otherwise get yourself back to your state of equilibrium.

"You got this!" ☺

Post 21; 06-02-21:
Easier Said Than Done

The other title I was debating on using for this post is something along the lines of "Practicing What We Preach." When I wrote last week's post, I fully intended—in other words, I had a good faith intention—to indeed do what I was saying in that post.

And to a limited extend, I did. But I guess the issue I want to grapple with in this post is, how well can one accomplish that goal—i.e., of chugging along, regardless of the messiness of life at any given time-point—in general? How easy or hard is it, and why and how do some of us manage to do it, while others struggle?

All of the above questions have been triggered by a worsening of aforementioned "stressful event X." Interestingly, that worsening unfolded in an eerily similar fashion to the way another stressful event befell me, years ago.

Unfortunately, I don't have any good answers for the questions I pose in the second paragraph farther above. I just know that it is sometimes/often excruciatingly painful and hard to acclimate to our lives' situations. I also still believe in what I said last week. It is vital for us to "pick ourselves up" and "dust ourselves off" and "move on," regardless of the impediments in our paths.

And regardless of all the great advice any expert can give you, only *you* can know how to do what is right for *you.*

Again, I submit: "You got this!" ☺

Post 42; October 26, 2021:
A Beautiful And Sad Poem, In the Wee Hours of the Night

It's 4 AM in the morning, raining outside, and my mood is similarly dour. To match that mood and weather—or perhaps, to try to fight it (and/or for some other reason[s]), I am hearing the refrain of that beautiful and sad Dylan Thomas poem, playing in my mind. You know, that sad and beautiful poem:
Do not go gentle into that good night…
[Link Embedded: https://poets.org/poem/do-not-go-gentle-good-night]

Self-Questioning for Creativity-Attempt Outtake

State any Necessary Introductory Remarks, and/or Answer this Question: Does it Make Sense to Me Now—i.e., In The Present—versus then—i.e., When I Wrote It?

I believe the posts above are an appropriate book-end indeed to this text. I believe they sum up the spirit or ethos that I hope to popularize about SC, namely, that it can help us achieve and maintain balance and self-mastery. And yes, they—i.e., the posts—definitely still make sense in the present.

Does it Resonate with Me Now—i.e., In The Present— versus then—i.e., When I Wrote It?

Yes, the CAO still resonates with me now. The positive message is appropriate for any time-period.

What Mood and/or State of Mind Does the Tone Suggest?

Deep reflection.

What is/are the Direct and Indirect Relation(/s) to the Current Chapter via Topic(s) and Themes?

Unfortunately, I am not sure that the posts directly or even indirectly corroborate the contents of the chapter. But as I mention in answer one (10.6.1) above, they are nonetheless appropriate, as they sum up a—or *the*—key message about SC that I hope to popularize via this book.

Any Other Evaluations or Comments?

I earnestly hope that readers will find this book helpful not only for research/academic/professional purposes, but holistically as well. My firm belief is that we can find and utilize our SC voices to improve ourselves, and thus our societies at large.

Activities and Discussion Questions

1. Recap, part II: Go back and read through some or all—or as many as you can analyze—of your answers to the discussion-questions for chapters 4 to 8. Then, re-consider the questions, and try to revise your answers in a bid to holistically incorporate the things you've learned overall in the book, beyond those four chapters. Discuss your revised answers.
2. *Optional*: Continue journaling/creative activity 1 as advised at the end of chapter one above.

Appendices A – E

APPENDIX A

Voice-Note Recordings Summary-Transcript 1; (*Total: 24 V-Note Summaries*)

Batch 1

1) ***Voice-Note No.:*** Batch 1; No. 1
   Date & Time Recorded: <u>Nov 30th, 11:25 PM</u>
   Audio-Length: <u>4:36</u>
   Main Topic/Theme: <u>Paul Thagard's example of a metaphor to help solve a problem: X-rays to kill cancer cells</u>

Sub-Topic/Theme Summaries:
- --- Paul Thagard's book
- --- Use of X-rays to kill cancer cells
- --- Dictator's impregnable fortress
- --- The key to solving the tumor problem is in the "dictator's fortress" story/metaphor; small groups—but not big groups—can pass over the mines the dictator has planted in the roads to his fortress
- --- Short bursts of rays, as opposed to long uninterrupted bursts, can kill the cancerous cells without destroying the healthy ones

*** *** ***

1. Alright time and date is November 30th, uh…the time is 11:25 [PM?] as of the—beginning of the
2. As of the start of this recording
3. So, I just came across **the most**--like one of—
4. must be one of the most insightful things I've ever read in my life
5. Um…so…in the Paul Thagard uh-text Mind: Introduction to Cognitive Science
6. Introduction to Cognitive Science second edition um…by Paul Thagard um page uh…87 and 88
7. It is so interesting how…[sniffs] the scenario—the problem given, right, the doctor who…has to kill
8. A tumor…inside of a patient…um…but he has to use the full strength of you know those-like X-rays

9. And…uh…which then would then which would then also kill the normal cells
10. On their way to the tumor, okay, so that's the problem and it-just…f-feels unsolvable, ok…
11. I mean I was imagining-I was thinking of like maybe…some kind of like-targeted uh…
12. Some way to use the rays in a way you know like…a method of using the rays
13. In such a way that like they maybe somehow bypass the cells-which-now that I'm thinking out loud
14. About it…I mean its-that's rather impossible but…I don't know I'm not a physician-I'm not a like
15. A physicist or whatever…

*** *** ***

2)   ***Voice-Note No.:*** <u>Batch 1; No. 2</u>
        Date & Time Recorded: <u>Nov 30th, 11:40 PM</u>
        Audio-Length: <u>3:14</u>
        Main Topic/Theme: <u>Metaphors Vs. Analogies Vs. Similes</u>
Sub-Topic/Theme Summaries:
      ---   I looked up the difference between those three concepts online
      ---   "One can call an analogy a metaphor, but you cannot call…a metaphor an analogy…"

*** *** ***

3)   ***Voice-Note No.:*** <u>Batch 1; No. 3</u>
        Date & Time Recorded: <u>Dec 11, 4:10 AM</u>
        Audio-Length: <u>7:19</u>
        Main Topic/Theme: <u>Effects Of Life's General Events/Features On Learning</u>
Sub-Topic/Theme Summaries:
      ---   I had the Corona virus
      ---   I realized that life's general routines and vagaries affect our learning processes //
      ---   Of equal impact: emotions, our other (non-learning-related) thoughts
      ---   Learning processes are just one part of general cognition
      ---   Distractions during reading //
      ---   In addition to disciplines such as software-development and math, language-arts, etc., I decided to also analyze the fiction and nonfiction books I'm reading

*** *** ***

4)   ***Voice-Note No.:*** <u>Batch 1; No. 4</u>
        Date & Time Recorded: <u>Dec 15, 6:45 PM</u>
        Audio-Length: <u>1:05</u>
        Main Topic/Theme: <u>Later Automatic-Recollection Of Learned Facts</u>
Sub-Topic/Theme Summaries:
      ---   After learning/reading something, the concepts can and very often do come back to you spontaneously later—depending on the context, as you write, discuss, give a lecture, etc. //

*** *** ***

5) ***Voice-Note No.:*** <u>Batch 1; No. 5</u>
   Date & Time Recorded: <u>Dec 28, 2:33 AM</u>
   Audio-Length: <u>3:42</u>
   Main Topic/Theme: <u>Barack Obama's "A Promised Land" and its relation to the Thaggard text</u>

Sub-Topic/Theme Summaries:

--- Obama was talking about memories he'd formed over time...

--- *"With time, my walks down the colonnade would accumulate with memories; there were the big public events of course, announcements made before a phalanx of cameras, press conferences with foreign leaders. But there were also the moments few others saw—Malia and Sasha racing each other to greet me on a surprise afternoon visit, or our dogs, Bo and Sunny, bounding through the snow, their paws sinking so deep that their chins were bearded white. Tossing footballs on a bright fall day, or comforting an aide after a personal hardship. **Such images would often flash through my mind** interrupting whatever calculations were occupying me. They reminded me of time passing, sometimes filling me with longing—a desire to turn back the clock and begin again..."*

--- Images as mechanisms of cognition

*** *** ***

6) ***Voice-Note No.:*** <u>Batch 1; No. 6</u>
   Date & Time Recorded: <u>Dec 29, 3:03 AM</u>
   Audio-Length: <u>5:33</u>
   Main Topic/Theme: <u>NY Times Article About Game Called "Life"</u>

Sub-Topic/Theme Summaries:

--- What are the random topics that catch our attention? "Insightfulness." What does it mean for a topic to be insightful? //

--- I found the game of "Life" insightful; out of simplicity, we can produce complexity

--- Also interesting: recursion and self-reference: related to D. Hofstadter's "Goder, Escher, Bach."

*** *** ***

7) ***Voice-Note No.:*** <u>Batch 1; No. 7</u>
   Date & Time Recorded: <u>Jan 5, 12:05 AM</u>
   Audio-Length: <u>3:13</u>
   Main Topic/Theme: <u>Example of intrapersonal communication via "to-do" lists</u>

Sub-Topic/Theme Summaries:

--- I was writing a two-week to-do list //

--- While writing the dates, my mind wandered—reminiscing about something funny: a news parody show. //

--- Skipped a number of dates because of that distraction

--- A few moments later, I realized my mistake, retraced my steps to figure out what I had done wrong and why

--- I realized that "it was because I was not concentrating," and I actually said that to myself—via mumbling //

*** *** ***

8) ***Voice-Note No.:*** <u>Batch 1; No. 8</u>
Date & Time Recorded: <u>Jan 6, 1:06 AM</u>
Audio-Length: <u>11:42</u>
Main Topic/Theme: <u>Various, including: 1) Obama's *A Promised Land* 2) How do/should we talk to ourselves? Answer via my e-filing methods: "What did I mean by this? Why did I do this [e.g./i.e. file something within the subfolder that I saved it in]?" 3) *"Why did I come to the kitchen?"*</u>

Sub-Topic/Theme Summaries:

--- Obama was a lackluster high-school student. Then in college, he transformed; he would have "internal self-dialogues," often thinking deeply about topics such as justice, American "dream"/democracy, freedom, etc. //

--- "He was living a lot, or was too much in his mind"; his friends told him he needed to loosen up. I—and many other people—also do it. //

--- Part of how we communicate with ourselves, is in relation to our research and writing processes—i.e./e.g., while writing academic or professional documents: using a combination of our own findings, as well as those of other previous authors //

--- Part of my own method for preparing or executing the above tasks, involves the use of elaborate electronic-filing systems—"folders within folders within folders." Because of this, we're forced to later figure out: *"What did I mean by this? Why did I save this file within the subfolder in which I saved it?"*
// //

--- A similar situation unfolds via our forgetfulness/distractedness with executing various tasks: "Why did I come to the kitchen?" //

*** *** ***

9) ***Voice-Note No.:*** <u>Batch 1; No. 9</u>
Date & Time Recorded: <u>Jan 6, 2:23 AM</u>
Audio-Length: <u>6:58</u>
Main Topic/Theme: <u>"What was I doing?" AKA Distraction</u>

Sub-Topic/Theme Summaries:

--- Name of the concept referenced earlier—"what was I doing (?)": distraction

--- I need to look up studies about distraction, e.g. in the context of reading

--- In my own personal context of reading, it happens often when I am reading a long text for school or work, as opposed to when I am reading for fun or leisure

--- Reading online is different; it's as if "distraction is part of the experience…it's built in" it is

*** *** ***

Batch 2

1) ***Voice-Note No.:*** <u>Batch 2; 2.1.1</u>
   Date & Time Recorded: <u>Jan 6, 3:39 PM</u>
   Audio-Length: <u>4:14</u>
   Main Topic/Theme: <u>"Tip of the tongue" syndrome</u>
Sub-Topic/Theme Summaries:
   --- Another instance of intrapersonal-comm.: "tip of the tongue" syndrome: "What's that word?! It's on the tip of my tongue!"
      + Editorial note, added during transcription (03/10/21): The word is zeitgeist!

*** *** ***

2) ***Voice-Note No.:*** <u>Batch 2; 2.1.2</u>
   Date & Time Recorded: <u>Jan 6, 10:48 PM</u>
   Audio-Length: <u>7:02</u>
   Main Topic/Theme: <u>Thinking during meditation</u>
Sub-Topic/Theme Summaries:
   --- I realized something during or shortly after meditation: it's hard to "quiet" the mind, to avoid thinking
   --- Analogy: the way our minds work in that regard—always thinking even when we're trying to meditate, might be similar to a computer opening various random files or performing various other random functions, even though you're taking a break and are not using said computer

*** *** ***

3) ***Voice-Note No.:*** <u>Batch 2; 2.1.3</u>
   Date & Time Recorded: <u>Jan 7, 6:11 AM</u>
   Audio-Length: <u>34:48</u>
   Main Topic(s)/Theme(s): <u>1--Paul Thaggard's Cog-Sci text: images, connections, 2--learning-key/ turning-points, and 3--role of emotions in reading and understanding fiction and non-fiction stories</u>
Sub-Topic/Theme Summaries:
   --- Chapter about images: Practical applications of ***<u>images as one of the ways our minds work</u>***. Using images in our learning and teaching: ***<u>for me, images were/are very useful in learning math. E.g., picturing a 1/3 of a circle to understand "a third of X"</u>***
      + Asking myself, "what is this—what does s/he mean?!" Then building a mental image of what's being described. E.g., "a large animal with a trunk and tusks on its face, big legs and a short tail" = "elephant!"
      + Also: 1) Sensory imagery for me during meditation: waves washing over my feet. 2) Images as a learning hindrance: e.g., me with trying to visualize Ken Follet's villages, cities, people, etc. in the medieval ages //
   --- Chapter 7: Connections as another way of using our minds, part of the process of thinking:

--- Key/turning-points in learning: Wrench table 6.10. Very important in the learning process, esp. for teachers/professors, who have to really understand their subject-matter before teaching it. "Aha! That's what the s/he (author) means!" That is intrapersonal-communication.

--- Emotions also help us digest fiction and nonfiction material. Different from images, but with the same resonant effect in our minds. E.g. crying or feeling sad while reading fiction and nonfiction stories.

*** *** ***

4) **Voice-Note No.:** Batch 2; 2.1.4
Date & Time Recorded: Jan 8, 4:24 AM
Audio-Length: 3:10
Main Topic(s)/Theme(s): "To-do" lists

Sub-Topic/Theme Summaries:

--- "To-Do" lists: I use them a lot. Authors such as Atul Gawande have also discussed them. They are a form of intrapersonal-comm.: "What do I need to do today? I need to do X, Y, Z." Also, "what should be the priority here [out of the X, Y, Z list]? 1st most important, 2nd most important, etc."

*** *** ***

5) **Voice-Note No.:** Batch 2; 2.1.5
Date & Time Recorded: Jan 10, 12:38 AM
Audio-Length: 8:45
Main Topic(s)/Theme(s): General Types of Intrapersonal-Comm.

Sub-Topic/Theme Summaries:

---Editorial/Meta-analytical: I'm behind on the recordings, because of my life patterns

--- Distractions: getting distracted by random thoughts during class or conversation; but I crossed out this point because I don't think it qualifies as intrapersonal-comm.

  + Experiment: next time with Ethan—my software-engineering coach, "keep a tally of unrelated distractive thoughts."

  + "There are thoughts in general [in our minds], but every time you're thinking, doesn't necessarily equate to you communicating with yourself!" //

--- Self-encouragement: "Come on, Seif! You can do this!" That is definitely an instance of intrapersonal comm. // //

--- Self-questioning, e.g., "I wonder if I can do that?"

  + I—[/Seif/[Your Name] the primary self, wonder if I—[/Seif/[Your Name]—secondary self-] can do that

--- Corporal states: acknowledging a corporal state to yourself—e.g., "Oh my God, my head hurts!" That's also an example of intrapersonal-comm. //

--- *"Any time you're awake, the question isn't *if* you communicate with yourself at least a certain % of the time; the question is *how*[or when]…any time you're awake, *you will* communicate with yourself at some point, somewhere [,somehow]."*

*** *** ***

6) **_Voice-Note No.:_** Batch 2; 2.1.6
   Date & Time Recorded: Jan 10, 12:57 AM
   Audio-Length: 13:38
   Main Topic(s)/Theme(s): Other types of intrapersonal-comm.; tip of the tongue and meta-analysis; types of thoughts; distractions in your own mind during conversations or other interactions.

Sub-Topic/Theme Summaries:

--- "Tip of the tongue" syndrome happens to me a lot! "What's that word? There's a word I'm looking for!" If you're (your primary self is) referencing your secondary self, as in this instance, then that is intrapersonal-comm. // //

--- Meta-analysis while rehearsing for events such as job interviews, work presentations, etc. "How should I answer this [question]?!" = Intrapersonal-comm. // //

--- "Every thought is not an instance of SC—intrapersonal comm., but every thought referencing your secondary self is obviously a thought." That is an important distinction; thinking in general isn't necessarily intrapersonal-comm. //

--- Types of thoughts and emotions: e.g., missing a loved one. Also not necessarily instances of intrapersonal-comm. // //

--- Two specific instances of distraction from previous day's software-engineering session: 1) Thinking about upcoming job-interview, and 2) Thinking that I had to write an email to someone—"I should remind myself to do that…" //

--- Applying IPC to learning in general, and poetry: "I—[the primary self-] wonder if I—[/ Seif/—secondary self-] can write poetry?!" //

*** *** ***

7) **_Voice-Note No.:_** Batch 2; 2.1.7
   Date & Time Recorded: Jan 10, 11:21 PM
   Audio-Length: 6:59
   Main Topic(s)/Theme(s): Journaling and relation between intra- and inter-personal-comm.; emotions as intrapersonal-comm. instances, including guilt.

Sub-Topic/Theme Summaries:

--- "To-do" lists are a form of us talking to our future selves; "Seif, don't forget to do X, Y, Z." Journaling is somewhat similar; you're also taking to yourself [, recalling what your self did in the past]. However, journaling is also a way for us to preserve our life events for posterity, i.e./e.g. after we die; you're both talking to yourself in the journal, and potentially talking to others in the future. //

--- Many times, when we're talking to other people, we're also talking to ourselves; e.g. the common line in TV/movie dramas: "who are you trying to convince here—me, or yourself?"

     + Intrapersonal-comm. can lead to interpersonal-comm., and vice-versa. In fact, perhaps we're always talking to ourselves, vis-à-vis arranging and screening our thoughts before divulging them?

--- Emotions; particular case of guilt: "I am a bad person" for having done that bad thing. You don't necessarily have to use those or other words; the mere emotion is a de facto intrapersonal-comm.

*** *** ***

8) ***Voice-Note No.:*** Batch 2; 2.1.8
Date & Time Recorded: Jan 11, 8:23 PM
Audio-Length: 11:06
Main Topic(s)/Theme(s): Emotion of guilt continued; my friend's "ethnography of the mind."
Sub-Topic/Theme Summaries:

    --- "Self/[Your Name], you're a bad person!" Related: second-guessing yourself, or blaming yourself later after an argument, a la George Costanza: "The jerk store called, they ran out of you!"

    --- My friend's "ethnography of mind": he discovered that he often speaks to "guest speakers" in his mind. I.E., envisioning himself talking to someone else [in the past or future, or both?]. For me, that isn't really an instance of intrapersonal-comm.

    --- He also realized that it's almost impossible to think without using words, especially in the context of metacognition

*** *** ***

9) ***Voice-Note No.:*** Batch 2; 2.1.9
Date & Time Recorded: Jan 13, 4:16 AM
Audio-Length: 8:16
Main Topic(s)/Theme(s): Self-doubt about learning; theorization about learning ability; self-communication sans self-reference.
Sub-Topic/Theme Summaries:

    --- Self-doubt: me doubting myself vis-à-vis soft-dev and STEM subjects. Answer using theory and empiricism: we all have an ability to theorize. //

    --- You can apply theory and empirical experience to reduce your self-doubt: "Based on past experience, this is my learning style;" but in other words, yes, *YOU* have the ability to learn well, no matter your IQ. //

    --- Can you communicate with yourself without directly referring to, or thinking about your (secondary) self? Yes, it's definitely possible to communicate with yourself without direct reference; e.g., "I wonder if that is possible?" Translation: "I wonder if I can do that?"

*** *** ***

10) ***Voice-Note No.:*** Batch 2; 2.1.10
Date & Time Recorded: Jan 13, 11:34 PM
Audio-Length: 16:09
Main Topic(s)/Theme(s)—Note: Add after listening to entire recording (!): Types of self-communication, including metacognition during interpersonal dialogues, longterm-planning and procrastination, and meditation.
Sub-Topic/Theme Summaries:

    --- Editorial note: v-note recording procedure has evolved at this point; notes are written earlier in a notebook or phone app, before being read later.

--- About talking to my Kean mentor, re: his life-experiences growing up. Mr. DE, in his 70s, African-American, grew up in the "Jim Crow"-era US south: the way questions kept popping up in my mind in reaction to things he was saying—i.e., follow-up questions. Additional observation: outside the context of research, we meta-analyze our thoughts all the time without realizing it. //

--- Intrapersonal-communication is more than specifically thinking or talking to yourself. It also involves: 1)--ordinary (non-academic/research-based) metacognition, 2)--clear and vague emotions, e.g. guilt. //

--- Longterm-planning and procrastination, and vague hesitation in reaction to thinking about upcoming tasks. Also, despite to-do lists, I experience both clear and **vague** emotions of hesitation, re: thinking about upcoming tasks. //

--- Meditation: thoughts wander as you meditate. Need for me to delineate general thoughts vs. thoughts of communication to self. If you catch your thoughts wandering as you meditate, is that an instance of intrapersonal-communication?

*** *** ***

11) ***Voice-Note No.:*** <u>Batch 2; 2.1.11</u>
Date & Time Recorded: <u>Jan 15, 8:51 PM</u>
Audio-Length: <u>24:08</u>
Main Topic(s)/Theme(s)—Note: Add after listening to entire recording (!): <u>Spectrum of knowledgeability, from "know nothing" to amateur, through expert; intrapersonal-comm. via/during intense emotions; suspending and confirming understanding; fields of experience.</u>

Sub-Topic/Theme Summaries:

--- Realization I made about learning processes: knowing a subject **well/thoroughly**—e.g./i.e. as an expert—is different from having a limited amount of knowledge about said subject. [***Editorial Note: Depending on where I'm going with the topic, this realization is rather obvious and most likely isn't useful.***] Related: my upcoming reading of "*30-Second Quantum Mechanics Theory.*" Reading and completing that book doesn't make me a quantum theory/mechanics expert!

--- Above point continued: before starting your learning-process of subject/topic X, it might be useful to ask yourself, "where on the 'knowledgeability spectrum'—so to speak, of this topic, am I?"

--- Realization: a lot of intrapersonal-comm. happens as we experience emotions—happiness, sadness, anger, etc. Implication for learning-process improvement: keep the above fact in mind as you read or study; you're more likely to be distracted when emotional, which interferes with your learning process.

--- Related qn.: "In an average day [or other time-period, e.g., a few weeks/months, etc.], what percentage of the time do we have intense emotions? Intense sadness, intense happiness, etc." Another related question in same context: how much do we communicate with ourselves during these intense emotions?

--- Depending on the answer here, we have to keep it—and the above realization—in mind, as we go about our learning processes. E.g., if I am taking a course—e.g. for continuing-ed.-credits or for working training, and I lose a loved one during that same time-period … [***Incomplete statement completed during transcription:*** *I need to realize that I'm distracted by that sad event; thus, either pay more attention, or take the course later, etc.*]

--- Example of communication with self during intense emotions. E.g., when someone angers you: "I cannot believe she did that to me!" Or, "I feel so sad." Resultant argument: in these instances, you're definitely communication with yourself, albeit unintentionally.

--- The importance of what I refer to as "suspending understanding": i.e./e.g., while reading a novel or academic book, and you don't fully understand the concepts. My proven hypothesis: keep reading, then try to make sense of the combined content later—e.g. by re-reading, or "connecting the dots," etc.

--- That method is harder with STEM subjects, whose topics often build on one another, with the need to master lower-level concepts before continuing onto the advanced concepts. E.g.: need to understand numbers and counting before learning arithmetic—adding and subtracting, etc., and later, algebra, etc.

--- Still, "suspension of understanding" is possible even in STEM subjects: e.g., memorizing formulae and using them to solve problems without really understanding the logic behind those formulae. E.g., memorizing the need to divide both sides of a basic algebra equation, e.g. "$2X = 10$" using a common divisor on each side, until the number of the right cannot be divided further. In the above example, $X = 5$. Later, you can learn how the formulae really work; their logic, etc.

--- "Confirming understanding": you should confirm that the concepts really mean what you think they mean, e.g. using online or other resources and books, etc. E.g., I often look up various vocabulary meanings while writing, as I find myself using specific words that I think fit the context, but I'm not sure.

--- Relation of "fields of experience" with learning processes. Background: old way of teaching the basic process of communication didn't include the above concept (i.e., "fields of experience), which refers to the holistic life-backgrounds of all humans, who constantly communicate with each other. It even applies to children. For instance, kid X whose family is rich, talking to kid Y, whose family is poor. Each of their processing or digestion of information will be influenced by their socioeconomic statuses.
//

--- E.g. of above point, real-life high-school student EA (real name initials), who is very smart. I believe if I had had his advantages at his age, I would probably have performed as well as he does in school.

*** *** ***

12) ***Voice-Note No.:*** Batch 2; 2.1.12
Date & Time Recorded: Jan 16, 1:41 AM
Audio-Length: 8:44
Main Topic(s)/Theme(s)—Note: Add after listening to entire recording (!): Use of visualization and concept-processing using words while reading fiction and nonfiction, academic, and other genres.
Sub-Topic/Theme Summaries:

--- Editorial note, explicated at beginning of v-note: For the following months after this recording, from January to March (end-date of recordings), v-note topics will be related to the books I'm reading, as well as other things I'm learning, esp./e.g., software-development.

--- Paul Thaggard's breakdown of cognitive-scientists' six general types of human-thought representation and computation, i.e.: logic, rules, concepts, analogies, images, and connections

--- Objective of research: application of intrapersonal-communication concepts to the improvement of my own learning processes; essentially trying to figure out, "how do I think?" "How do I think about and/or process the content that I am reading?" Partly, through visualization; I have to visualize what I am reading

--- E.g. of visualization; John Grisham's protagonist in the "Rogue Lawyer" novel, meeting in a smoke-filled diner with someone: how do I process that scene?

--- What about words—what role do they play in my comprehension of the contents that I'm reading? I have to use them to decipher meanings, e.g. in academic reading contexts.

--- Another way to ask the above question(s)—i.e. How do I think about and/or process the content that I am reading (?): As I'm reading these books, what am I telling myself? Answer: I check my understanding of the content, with fiction and non-fiction, I'll exclaim or laugh, etc. I also use visualization. With more prosaic or academic/noncreative books and learning, I'll use words to make meaning. "When [words] get into my brain, they dissolve into concepts—into abstract concepts." Thus, in my mind, I'm not visualizing *words*; rather, I'm processing the meanings of the concepts represented by the various words I'm reading.

*** *** ***

13) *Voice-Note No.:* Batch 2; 2.1.13
    Date & Time Recorded: Jan 16, 2:04 AM
    Audio-Length: 1:16
    Main Topic(s)/Theme(s)—Note: Add after listening to entire recording (!): Semiotics
Sub-Topic/Theme Summaries:

--- I need to dig up my semiotics books from grad school, and perhaps look up some resources online. I just realized that semiotics can/will be useful for me in this research.

*** *** ***

14) *Voice-Note No.:* Batch 2; 2.1.14
    Date & Time Recorded: Jan 17, 12:21 AM
    Audio-Length: 10:14
    Main Topic(s)/Theme(s)—Note: Add after listening to entire recording (!): Intrapersonal-comm. during ethical dilemmas; hypothesis of "brute-force"-style learning.
Sub-Topic/Theme Summaries:

--- Ethics: related to earlier-mentioned concept of our propensity to communicate with ourselves more via/during intense emotions. Relevant example: moments in which you're experiencing cognitive-dissonance between the ethical choice versus the easier choice. In such moments of working through "ethical crossroads/dilemmas" we talk communicate more with ourselves versus in other moments.

--- Learning: e.g., me trying to solve a number of algorithm challenges on FreeCodeCamp. "One of our worst enemies in the learning process is self-doubt; can I really do/learn this stuff?" My new theory hypothesis of longterm brute-force learning: regardless of the intelligence or IQ of a given person, we are all capable of learning any given concept, as long as we're exposed to it for a long period of time. [*__Editorial note:__ further research might be needed to confirm this hypothesis for individuals with learning disabilities. My own anecdotal evidence suggests that there might be limits to what they can learn.*]

--- Meaning of "brute-force" execution; e.g. with hacking passwords: prolonged engagement and attempt at doing something. The key advantage that aspiring polymaths have, is that you can take your time—as long as needed, to learn the various concepts you're trying to learn, regardless of how hard-to-learn those concepts are.

--- Caveat(s) of above hypothesis: it doesn't simply involve passive learning; you have to actively engage with the concept, understand it, practice it, etc.

*** *** ***

15) ***Voice-Note No.:*** Batch 2; 2.1.15
Date & Time Recorded: Jan 17, 4:37 AM
Audio-Length: 10:27
Main Topic(s)/Theme(s)—Note: Add after listening to entire recording (!): "Yearning for coaching"; self-doubt in context of "imposter syndrome."

Sub-Topic/Theme Summaries:

--- Related to learning: "Yearning for coaching." E.g. with me learning algorithms and data-structures: "In your most intense moments of self-doubt, or feeling frustrated, you [at least I, personally, often] yearn for coaching: 'if only I had somebody that could help me figure this out.'"

--- But it depends on your own learning style. E.g., I for one am not afraid of asking for help if/ when I need it, but other people want to figure it out on their own. But regardless, I believe most of us at one point or another—maybe consciously or subconsciously—experience that phenomenon.

--- In my own case, I was able to satisfy that yearning over several months in late 2020, hiring a software-development coach. But what happens if you experience that need; e.g., in a class where you don't get along with the teacher? Or what about outside that class, where your problematic relationship with the teacher isn't necessarily an issue?

--- Self-doubt: relevant reference in Obama's "A Promised Land," specifically in the context of discussing his wife background compared to his. "***Michelle never worried about selling out, because growing up on the south side [of Chicago] meant you were always at some level an outsider. In her mind, the roadblocks to making it were plenty clear. You didn't have to go looking for them. The doubts arose from having to prove no matter how well you did that you belonged in the room. Prove it not just to those who doubted you, but__ to yourself__***" (emphasis by this author). This reference to self-doubt is in the context of "impostor syndrome," unlike my own reference to challenges of learning.

--- Further explication of above discussion of self-doubt: "it [the above discussion] goes to prove that…the enemy here [i.e., in the contexts of learning or career/general success and fighting impostor syndrome] is not other people many times; it's yourself."

APPENDIX B

Voice-Note Recordings Summary-Transcript 2; (*Total: 16 V-Note Summaries*)

Batch 2.2

1) **Voice-Note No.:** 2.2.1
Date & Time Recorded: Mon, Jan 18, 2021; 7:01 AM
Audio-Length: 1:19
Main Topic(s)/Theme(s)—Note: Add after listening to entire recording (!): Recap
Sub-Topic/Theme Summaries:

--- Recap: self-doubt as discussed by Barack Obama in his "A Promised Land" autobiography.

*** *** ***

2) **Voice-Note No.:** 2.2.2
Date & Time Recorded: Mon, Jan 18, 2021; 7:05 AM
Audio-Length: 11:09
Main Topic(s)/Theme(s)—Note: Add after listening to entire recording (!): Knowledge and self-assessment; (self-)motivation.
Sub-Topic/Theme Summaries:

--- Checking and re-checking ongoing knowledge.
--- >> Checking initial knowledge
>> Conscious and unconsciously
Ø "How much do I know about topic X?"
--- Research and learning are two examples of contexts in which we perform that self-assessment
--- Are there are instances in ordinary day-to-day life?
--- Low self-motivation as a big impediment to learning: how can self-communication help with the alleviation of this impediment?
--- "Carrot and stick" system in life; we're forced to work to earn a living. "If you're not motivated to learn, you will not learn." Learning is hard. "*How motivated am I, on a scale of 0 to 10?*"
--- How does self-comm. help with self-motivation? Partial answer: it helps to first figure out the fact that you're indeed unmotivated, that you don't *feel* motivated.
--- One way to deal with that motivational deficit is: "Ok, so I don't feel motivated. Well, suck it up, self (/Seif)! I need to learn this, for my own good."

*** *** ***

3) **Voice-Note No.:** 2.2.3
    Date & Time Recorded: Mon, Jan 18, 2021; 7:57 AM
    Audio-Length: 6:16
    Main Topic(s)/Theme(s)—Note: Add after listening to entire recording (!): Meta-analytical review and plan of study;

Sub-Topic/Theme Summaries:

---   Use of Thaggard's book
---   The "so what" of the study; Google-Scholar articles: I'll be searching for articles/resources about cognitive science and learning styles
---   To help improve your own learning processes via cognitive-science principles, how well does Google-Scholar help you?
---   Cog-sci articles tend to mostly use Bayesian statistical-analysis theory; but regardless of one's level of knowledge of such advanced concepts, can Google-Scholar-sourced articles help? How, and/or how much, exactly?

*** *** ***

4) **Voice-Note No.:** 2.2.4
    Date & Time Recorded: Tue, Jan 19, 2021; 6:53 AM
    Audio-Length: 12:01
    Main Topic(s)/Theme(s)—Note: Add after listening to entire recording (!): Checking emotional state and self-motivation; critical self-questioning of ideation; reading-comprehension as example of checking understanding.

Sub-Topic/Theme Summaries:

---   "Similar to driving," checking emotional state and self-motivation before checking understanding of knowledge
---   Critically questioning ideas; "I am a man/person of ideas": "Hm…is that idea possible/viable?"/Is that a *good* idea?
    +   This in essence is a discussion between your primary and secondary selves: "Seif (Self 2), I (Self 1) is wondering if this idea—e.g., a personal/at-home-museum—is a good idea!"
---   Another example of checking of understanding: while reading the Ken Follet "Columns of Fire" novel: pausing to make sure I understand who is saying what during the dialogue turns of the book's characters. Thus, reading-comprehension is one learning activity during which we regularly pause to check our understandings. "Am I understanding this correctly?"

*** *** ***

5) **Voice-Note No.:** 2.2.5
    Date & Time Recorded: Wed, Jan 20, 2021; 11:51 PM
    Audio-Length: 10:16
    Main Topic(s)/Theme(s)—Note: Add after listening to entire recording (!): 1)--Need for self-communication before interpersonal-comm. 2)--Reading comprehension 3) Procrastination

Sub-Topic/Theme Summaries:
- --- Point 1:
    - \>\> 1)—"Before communicating with others, you have to communicate with yourself first."
    - \>\> 2)—"To me, reading-comprehension demonstrates part of or a similar process."
        - \>\> *"I feel like that note is self-explanatory…I don't wanna um…I don't know, say much more. I feel like I wrote it clearly and the way I just read it out makes sense. So…I'm not gonna say more about it."*
    - \>\> Quick revelation about some psychologists' work, re: the formation of sentences in our minds before speaking. Before you (/"one") divulge(s) a sentence, which is arguably the basic unit of communication, it has to make sense to you first.
- --- Comparison of ordinary people's communication process to rappers in performance of "free-style rapping," as well as "stream of consciousness," a la President Trump: thinking out loud. When we think out loud, not all thoughts make sense. Our thoughts are jumbled. "Thoughts in process are not the same as refined thoughts, [or]…'thoughts ready to consume.'"
    - \>\> <u>Very important editorial note</u>: The above assertion/presumption of my meaning being obvious, was wrong. I had to listen to my own explication to properly understand the main point in question, i.e., "Before communicating with others, you have to communicate with yourself first."
- --- Reading comprehension as an example of the above concept:
    - Ø Acquiring literacy skills in childhood or early-ed.,
        - o Deciphering basic meanings via alphabet and words, sentences, etc.
        - o wider-scale meanings, e.g. an emerging novel's plotline, etc.
- --- Procrastination: basically/essentially is = laziness. Similar to improving your motivation for learning, overcoming procrastination also requires positive self-talk: "I need to do this for my own good." / "I already agreed to help my wife with this."

*** *** ***

6)   ***Voice-Note No.:*** <u>2.2.6</u>
       Date & Time Recorded: <u>Fri, Jan 22, 2021; 1:11 AM</u>
       Audio-Length: <u>15:01</u>
       Main Topic(s)/Theme(s)—Note: Add after listening to entire recording (!): <u>1)—Quantification of news consumed for knowledgeability appraisal, 2)—Awareness of one's lack or possession of knowledge, 3)—Self-debating and censoring during learning discussions (e.g., at school, church, etc.)</u>

Sub-Topic/Theme Summaries:
- --- Point 1: Quantification of all the knowledge we consume: e.g., via news, on a daily basis:
    - Ø Context: I personally consume a lot of news on a daily basis. Not on soc.-media, but big consumer of news—i.e., in a written format, online.
    - Ø For someone like me who loves learning—as much as I do: how can I harness that news-consumption? I'm guessing that "actionable news" out of the total percentage that I consume is in the single-digit percentage range. No more than 9 or 10%.
        - o One example of news utility is the current COVID pandemic. Infection rates, vaccines, etc.

Ø   But apart from such an example, news-consumption = "knowledge is power," but in the abstract sense.

Ø   In the simplistic way we think about intelligence, most of the news we consume is useless… Or is it? "This is an evolving issue in my mind."

Ø   It doesn't have to be a black or white issue.

     o Over time, you can harness the news to your advantage. But "there is a deliberate process you have to put yourself through, I feel like." E.g.: with this book's research, I can perhaps utilize the news I consume. But it has to be a systematic periodic review. "What have I been learning via the news? What are the implications with various knowledge-domain subjects?" E.g., in the context of political-science or in general, social-science.

   --- Point 2: Vague or rough knowledge of enormity of general or specific knowledge you don't know.

Ø   In general overall, or by subject- or topic-matter. E.g.; random example: fish. What are fish? How do you classify them in the biological taxonomy? Vertebrae/invertebrate, genus, family (?), etc.

     >> <u>Editorial note</u>: Experiencing "tip-of-the-tongue" syndrome, re: "taxonomy."

---For you to become smarter, you first have to grapple with how much you don't know. E.g., in my case, apart from the little I learned in primary school about fish—e.g., the parts of a fish (fins, gills, scales, tails, etc.); they're vertebrae and are aquatic—I don't know much else!

Ø   Explication/detail: go to a physicist and ask them about human-comm. theories; they'll probably look at you blankly or puzzled—or annoyed—LOL.

---Conclusion of point 2: first, know the limits of your knowledge, thus know what you need to learn more.

   --- Point 3: Self-debating and self-censoring during studying and other similar sessions or activities:

Ø   Based on the experience I'd just had in Torah study in the Young Friends caucus, and from my experiences back in the day in college.

---Over time, you improve your art of knowing/deciding what to say and what to keep to yourself. Self-debating for the purposes of wanting to share gets in the way of your listening. If as you're listening while forming thoughts for pontification, then your listening and learning will be negatively affected.

---Active listening: checking for shared understanding through paraphrasing, digesting info by thinking out-loud… But again, it's an art form. Essentially however, self-communication is what you're doing in those little debates of "*should I or should I not speak up?*"

*** *** ***

7)   ***Voice-Note No.:*** <u>2.2.7</u>
    Date & Time Recorded: <u>Sun, Jan 24, 2021; 3:14 AM</u>
    Audio-Length: <u>15:53</u>
    Main Topic(s)/Theme(s)—Note: Add after listening to entire recording (!): <u>1)—Curiosity and acting on it or not, 2)—Self-coaching, 3)—Porpora's "How Many Thoughts" papers.</u>
Sub-Topic/Theme Summaries:

--- 1: Curiosity, and choosing to or not act on it. Regardless of the benefit of learning. Explication: part of how we learn is via curiosity. E.g., random example: *"I wonder what the farthest distance is that migratory birds can fly?"* You can then either look it up if you really want to, or not. Or: Having to look up or study something because of an assignment at work or school, etc.

--- 2: Self-coaching, e.g., through time-management. E.g., my 40-min resolution [i.e., for watching at least 40 mins of MOOC videos per day]. Another e.g.: my curiosity about number names between trillion and infinity. Explication: after you realize that you're very inquisitive by nature, it's up to you to self-coach.

---Explication of number-names between a trillion and infinity. This curiosity-instance had happened a few days prior, triggered by something I was reading. Thus, I Googled the question, and found the answer.

---Conclusion of point 2: if you love learning, it's up to you to nurture the love and follow it up with discipline and self-coaching, time-management, etc.

--- 3: Porpora's "How Many Thoughts" papers. Explication: for the most part, this point isn't connected to the ones above; then again, it kind of is. The way Doug and his opponents wrote their papers, is emblematic of how knowledge is discovered and expanded, and it is connected to the above discussion of curiosity. Curiosity leads to our searching, theorizing, formal research, etc.

*** *** ***

8) *Voice-Note No.:* 2.2.8
Date & Time Recorded: Sun, Jan 24, 2021; 3:31 *A*M
Audio-Length: 8:47
Main Topic(s)/Theme(s)—Note: Add after listening to entire recording (!): 1)—"Weirdness" of human beings, 2)—Learning and/or working despite ongoing problems in your life or distractions in general.

Sub-Topic/Theme Summaries:

--- 1: Quote by me: "Human beings are [SO] weird!" Explication: Our behavior, apparent thought-processes, etc. Relation to self-communication and learning: it's notable, given the fact and way in which I say that to myself.

---Further explication of point 1: E.g., at Walmart in line and someone cuts ahead or otherwise does something bizarre, causing you to say that. *"We all say a variation of something like this in such moments."*

---This instance of self-comm. (SC) isn't quite related to learning, but this research isn't limited to the learning methods application.

--- 2: Learning and/or working despite ongoing problems in your life (connected to point no. 3 below):

+ 3: Working and/or learning against distractions overall, … in your mind or around you [or] your surroundings.

---Explication of both points 2 and 3: regardless of our class, society, etc., we all have problems in our lives. "The question is, how do we plod on, chug on, keep going…in our work and in our learning process?"

---Further explication: example of students in college or professional school. College, medschool, etc. You're there, and you have problems in your life going on. E.g., you lose a loved one. How do you keep learning effectively while quieting the mind? For you to suspend yourself from being preoccupied by those other thoughts?

---The above questions are open and unresolved questions. Part of the answer vis-à-vis getting over problems in your life is self-coaching, per my notes of the 22nd.

+ "You have to tell yourself, '*alright Seif/John/Mary/etc., look ok, I have stuff [i.e.,* meaning problems] going on right now, my parents are going through a divorce, I don't know…I just lost my cat who I loved a lot…very much or whatever …I have those problems, but I need to just…concentrate…like, push those out of my mind for now and just concentrate…I don't have a choice; I'm trying to do this for my success…so I have to…push *those thoughts out of my head for now.*" It's ok to grieve, but that should not then stop you from doing your work.

---"It's important that we teach ourselves to do that [i.e., **_learn resilience_**], I believe" if we're going to learn or work effectively. "You don't want to be the person at work with problems at home, and then you yell at your colleagues." Neither would you want others to do the same—i.e., "coming to work with their problems."

*** *** ***

9)   *Voice-Note No.:* <u>2.2.9</u>
     Date & Time Recorded: <u>Mon, Jan 25, 2021; 1:04 *A*M</u>
     Audio-Length: <u>9:49</u>
     Main Topic(s)/Theme(s)—Note: Add after listening to entire recording (!): <u>1)—Free will vs. determinism, 2)—Self-coaching through the long and lonely road of learning.</u>

Sub-Topic/Theme Summaries:

---   1: Free-will vs. determinism: inspired by the relevant journal entry on my website.

     ---Explication of point 1: in undergrad, one of my profs liked asking us about our opinion of that debate. What's the biggest influence on our lives—free will, or fate?

     ---Further explication: if/when we're in classes or other venues discussing such topics/debates, what's our cognitive-SC process? "*I can think out loud, or…before stating [my opinion about the topic at hand,] I have to first…rehearse it in my mind [or] think about it, [then] say, [e.g.,] this is what I think: …e.g., I don't think free will is absolute…[neither is fate.]*" Regardless, I will first have to say that to myself before saying it out loud.

---   2: Self-coaching through long, hard, and lonely road of learning. Indeed, the road of learning is long and arduous.

     ---Explication: part of the way the analysis [for this research] bears out, the way I come to the insights for the research, is via repetition. [Thus my revisiting of the concept of self-coaching.] In my learning of soft-dev, I often reflect on the fact that, that stuff is hard. Especially algorithms and data structures.

---Given the above fact/consideration—i.e., that CS concepts are hard, you have to do a lot of self-coaching as you're learning the topics. You have to remind yourself, "alright Seif, remember, this is for your own good. [It] is gonna pay off in the long run. Come on, you can do this, Seif. You can do this." Regardless of the fact that it's a long arduous road.

*** *** ***

10) *Voice-Note No.:* <u>2.2.10</u>
    Date & Time Recorded: <u>Mon, Jan 25, 2021; 7:00 PM</u>
    Audio-Length: <u>12:37</u>
    Main Topic(s)/Theme(s)—Note: Add after listening to entire recording (!): <u>1)—"Brute-force prolonged learning" 2)—Research-design tailoring through SC 3)—Messiness of that tailoring through SC process</u>

Sub-Topic/Theme Summaries:

--- 1: "Brute-force prolonged learning"; stats and content-analysis: You can learn anything, regardless of how hard it is, if you keep at it for a long time... and [using] a triangulation of methods; teachers, study-partners, etc.

---Explication of above point: it's a firm belief of mine, and it's bearing out with my software-development learning. Like most folks, stats—anything to do with math—was for a very long time challenging for me. I had and to some extent, still have, math anxiety.

---The starting point for that "phobia"—for lack of a better word—was in the 2nd grade when the teacher at the catholic boarding school would inflict corporal punishment on us for not learning the multiplication table. Ever since, math for me was something to be feared, something hard, not fun, etc.

---Turning point: grad school; passing the stats course was the spark for the fire of math love, so to speak. *"Ok...I can do this."*

---Still, stats was until recently a methodology that I try to avoid. Lately though, I've fallen in love with stats. The breakthrough: the text by Wrench et al., *"Quantitative Research Methods For Communication,"* especially the table from chapter 6, describing appropriate measurements for different variables (nominal, ordinal, interval/ratio), research goals (differences or relationships), and variable numbers (two or more variables or groups of variables).

---And in recent project-management study, voila, I'm enjoying myself while using statistical content-analysis!

---Relation to above point(s) to SC: you have to keep telling yourself, "just keep going, you'll get this sooner or later."[You have to often use] suspending of understanding, ...stick with it, [and use] self-coaching. Much of this voice-note's content might be repetitive, but part of the way you get to insights is through repetition. E.g., Owen's (1984) thematic analysis: repetition, recurrence, and forcefulness.

---Conclusion of above point (1): brute-force learning involves a lot of SC.

--- 2: Research-design: e.g., with Te's project. Using methodologists' systems for your own studies involves a lot of SC, to tailoring the methods to your own research-purposes.

--- 3: Messiness of the that process of research-design tailoring through SC. Between all the triangulation, hammering out of details, using some principles and ignoring others, the process is very messy!

*** *** ***

11) *Voice-Note No.:* <u>2.2.11</u>
Date & Time Recorded: <u>Tue, Jan 26, 2021; 1:59 PM</u>
Audio-Length: <u>10:52</u>
Main Topic(s)/Theme(s)—Note: Add after listening to entire recording (!): <u>1)</u>—<u>The vague or abstract and mysterious digestion…of knowledge</u>

Sub-Topic/Theme Summaries:
--- 1: The vague or abstract and mysterious digestion, processing, connection, interconnection, etc., of knowledge. Explication: H's concern about retaining all the knowledge she was soaking in, in our Master's program: *"Will I retain it?"*
---Explication: my common experience of "tip-of-the-tongue" syndrome. In those moments, you clearly have the knowledge in your mind, and you know what the concept is. But there is a mental block between your knowledge of the concept and the specific [name of the concept].
---Further explication 1: while we often have clear access to the knowledge in our minds, most of our education is turned into a vague abstract knowledgeability or collection of knowledge. We're educated in general, but we often perform poorly in contexts like that "Are You Smarter Than a Fifth Grader?" show.
---Further explication 2: the physicists who—according to Quora—can't answer the lower-level physics-topics questions. They are so steeped in their higher-level knowledge, their brains/minds can only hold a certain finite amount of knowledge in the short-term-accessible memory. "You have this mushy soup of all the things you've learned throughout your life." Sometimes, some things are at the top of that "soup," easily skimmable, [but]…you won't be able to retrieve most of the knowledge."
--- 2: Is it worth it? For self-taught individuals, there is a big issue with self-doubt and impostor syndrome. This question ("is it worth it?") is also related. The question is mostly rhetorical. Of course it's worth it. But that aforementioned mushy soup of knowledge is most often not useful. Regardless, education helps out most of us well; the benefits of education are clear.
--- 3: Obama's mum about to die, [in] chapter two [of his "A Promised Land" memoir]. Explication: as aforementioned, when we cry while reading a sad book [or express another emotion], that is a form of SC. Not quite you speaking quietly in your mind, or speaking out loud, but that emotion is a form of SC.
---Further explication: often/sometimes, the emotion is vague; you might be experiencing a mixture of two or more emotions. You might be able to able to divulge the emotion, but [often,] it's just that vague feeling in your mind. In the case of Obama's mum about to die, I wept.
---Part of the cause of my sadness was the fact that his mum was dying of cancer, and I've been through a similar experience—i.e., having a mum afflicted with cancer. Thus, this is an example of my aforementioned description of SC via emotion.

*** *** ***

12) ***Voice-Note No.:*** <u>2.2.12</u>
Date & Time Recorded: <u>Thu, Jan 28, 2021; 5:07 *PM*</u>
Audio-Length: <u>19:11</u>
Main Topic(s)/Theme(s)—Note: Add after listening to entire recording (!): <u>1: Bending, break-ing, or making your own rules, 2: Journaling; similar to "to-do"/notes-to-self, as a form of SC, 3: Wonderment and exclamation in Luganda, in reaction to Ken Follet novel, 4: Porpora's *Thoughts* and *Barry Paradox* papers, 5: "For now, just take it on faith," 6: Breaking the rules; poetry.</u>

Sub-Topic/Theme Summaries:

--- 1: Bending, breaking, or making your own rules: e.g., about how amateur coders/programmers should learn by building apps. Explication: everyone learns differently. Learning is inevitably intertwined with the concept of stylistics. Note: that is a layperson's, not a professional educa-tor's opinion or evaluation. E.g., the opinion or myth that some people are visual learners

---Further explication/example: my preference of real books over PDFs. Or math. For me, it was very helpful to visualize fractions, so as to learn that concept well.

---Many software-developers assert that building apps is better for you learn, versus endless learning of concepts. But my argument is: how am I supposed to build apps, etc., before really learning the concepts first? Thus, in that particular case, that's my personal opinion.

---Relation to SC. Each of us has to play out that personal debate in our minds. You know yourself best vis-à-vis personal preferences.

--- 2: Journaling; similar to "to-do"/notes-to-self, is a form of SC. It is actually you talking to your future self. E.g., random example of a to-do-list (TDL) note to self. "1) Read XYZ book, 2) Go do XYZ task." I'm telling Seif of the future to go do those things. Journaling is similar.

---Further explication of 2; journaling, etc. This point isn't related to learning—i.e., the ap-plication I'm focusing on, re: SC. Mostly related to general SC.

--- 3: Wonderment and exclamation in Luganda, in reaction to Ken Follet novel. Explication: by "wonderment," I'm referring to the good "pictures" painted by Follet of life in the dark/medieval, pre-Elizabethan, Elizabethan, and other eras in England and Europe.

---"Columns of Fire" is about the transition from the older members of the House of Tudor to that Elizabeth I, and from Catholicism to Protestantism. Most of the wonderment is in English. However, every once in a while, I express said wonderment in Luganda, using a common Ganda wonderment expression, which directly translates as "mum, lady!"

--- 4: Porpora's *Thoughts* and *Barry Paradox* papers; context: in ref. to my notes of the 26th. Abstract processing of ideas, knowledge, etc.

---Those papers discuss very complex philosophy of physics topics. While reading them, one feels as if, s/he's not really understanding said topics. My plan: to revisit the papers before I end these verbal-note recordings, to test how the mind processes knowledge over time. In other words—for instance: do our understandings get better/clearer?

>> <u>Editorial note, 04-26-21</u>: I haven't yet gotten a chance to do that, but I will asap before I finish my analysis in the book's chapters, and I will report on the results accordingly.

---Further explication of point 4: Deleuze and Guattari's book didn't make sense either, and I highly doubt whether I will eventually understand it, at least without other online resources' expositions. In any case, the instructor—Brent—advised us to also revisit that book in the future, to evaluate our understandings over time.

---Relation to SC: my understandings of whatever I'm reading or learning are usually informed/constructed via SC. *"Do I even understand what s/he's//they're trying to say?"* There are two parties within you, namely the primary and secondary self. The primary self is asking the secondary self, *"do you even understand this (?),"* [or] *"what's going on here?"* Usually, this process is happening by default or reflexively. But as I am doing this study, I realize that that process is playing out. It's very clear.

--- 5: "For now, just take it on faith..." Explication: David Malan often says that; he is the instructor of Harvard's CS50 course, a MOOC on Coursera. That course—along with other resources online—is how I'm learning software-development. He says that phrase as an implicit promise that he'll explain the meanings/details later. It stood out to me, because it's very similar to my own "suspend understanding" maxim/concept.

---Further explication of 5: E.g., with Doug's above-mentioned papers, and Douglas Hofstadter's *"Goder, Escher, Bach."* With the learning involved in those two works, the meanings won't be clear from the outset. One of the tools that help me learn is to say to myself, *"hold on, don't panic; suspend understanding, the puzzle pieces are gonna make sense later. In the meantime, just try to keep concentrating, reading, learning, and later on you'll make the connections."*

---Clearly thus, I am not the only one with that thought. Further explication: Malan is talking to us and himself that way because he performing a formal lecture. But regardless, that "suspension of understanding" concept is clearly not just my own. It seems to be a useful technique vis-à-vis learning.

--- 6: Breaking the rules; poetry. Explication: I plan on using that strategy (breaking the [learning] rules) to learn poetry. You can use as many resources from online as possible, but along the way, "you'll have to make your own rules for sure."

*** *** ***

13) Voice-Note No.: <u>2.2.13</u>
Date & Time Recorded: <u>Sat, Jan 30, 2021; 4:24 *PM*</u>
Audio-Length: <u>23:10</u>
Main Topic(s)/Theme(s)—Note: Add after listening to entire recording (!): <u>1) What is the average attention-span ability of the average Joe/Jane? 2) One possible answer: via learning activities that hold our attention. 3) Explaining things to my mum; sub-point: arranging thoughts for explanation and/or during arguments etc., outside of formal contexts. 4) Sense of self, 5) Related to self-doubt and other concepts I've discussed before. Context: as a professor. 6) Sense of time and place as well.</u>

Sub-Topic/Theme Summaries:
--- 1: What is the average attention-span ability of the average Joe/Jane? Explication: using a number of metrics including the controversial IQ and intelligence in general. Do some people have superior focus? Can they sit through a 3-hour lecture and be alert the whole time their thoughts starting to wander? Or, do most of us in a three-hour lecture inevitably wander off mentally? Further explication: regardless of the answer and negative influence on learning ability, what are some of the ways we can overcome attention-span shortcomings?

---Further explication of above point: I will try to provide an answer to the above question in my book's analyses.

--- 2: One possible answer: via learning activities that hold our attention. One example for me: reading French news out loud. Explication: I have been trying to learn French for a very long time, and I am currently at an advanced beginner or intermediate level. One of the ways I improve my French is via Radio France Internationale (RFI) and France 24 television. However, I often find myself trailing off via attention span. Not only does that happen with listening and watching broadcasts, but also with **reading** French news stories. In this context—unreliable attention span via reading, I've found over time that it helps when I read the stories out loud.

--- 3: Explaining things to my mum; sub-point: arranging thoughts for explanation and/or during arguments etc., outside of formal contexts. Explication: a lot of us have trouble connecting with our parents. Being/getting "on the same page" with them, regardless of their education levels. One of my challenges with my mum: our starkly different levels of education. She's very smart, but unfortunately didn't go far in her formal education.

---Further explication: despite her limited education, she can understand things clearly if you patiently explain them to her, if you **connect** with her. But **that** is the challenge; **how** do you "connect" effectively with her?

---Sub-point: arranging thoughts for explanation and/or during arguments etc., outside of formal contexts. Explication: despite my advanced education, I hate the concept and activity of debate. Why? "I am not quick with words" (or 'quick-witted,' I guess). I am also not big on confrontation. I can do it if I have to, but I tend to be a conflict-avoider.

---Further explication: you often know you're right in the argument, ethics dilemma, or vis-à-vis the conflict you have with someone. The other party often has the wrong argument, clearly. And yet, it takes a great effort for you to arrange the words [in your mind], to be able to respond or rebut arguments, etc. Even in formal contexts—e.g., Doug responding to that critic, it takes a great deal of effort and time.

---Final point in above point's context. Nowadays, we're lucky to have Google. We often don't realize that it's not necessary to argue; you just have to Google the topic/question, etc.

--- 4: Sense of self; this concept is probably related to symbolic interactionism and pragmatism.

--- 5: Related to self-doubt and other concepts I've discussed before. Specific context here: as a professor, I have no choice than to really immerse myself in knowledge to be competent. Explication: as a professor, one of my worst nightmares is being in that state/situation where you haven't prepared well enough, and you find yourself in that "deer in the headlights moment" when a student asks you a question and you can't answer it.

---Further explication: It's ok if it's a philosophical [/"grey area"-related] or debate-related question, or if you in fact prepared well, but the question is about something not in the relevant text chapter(s). But overall, for me—as an instructor in the context of teaching, I believe I have no choice but to be steeped in knowledge, especially the knowledge I'm supposed to teach. Conclusion: in general, self-doubt is related to the general concept of a sense of self.

--- 6: Sense of time and place as well. Learning doesn't happen in a vacuum. You learn things in particular times and places: at school, work, from your childhood, through your late adulthood years. Explication/implication (partial-relevance) for me: as I am trying to retrieve a piece of knowledge, e.g., "*I come across a factoid or question I don't know and I'm trying to answer it in my head, ...a big part of the clues will come from*" my recollection of the time-period in which I might have learned the content, e.g., grad-school or undergrad, etc., and try to "connect the dots," construct the answer.

---When that happens, your memory is transporting you "across the plane of time." And inevitably your memory is also connected to a location, where you learned that knowledge. However, as I have already mentioned, longterm comprehension of knowledge, and making connections between various knowledge-domains in your mind, is vague and abstract.

---Further explication and preview of point-to-come: knowledge is also tied to people, i.e., the people who taught you or helped you gain your knowledge. Despite the fact that some knowledge is tied to time, place, and people, much/most of it is an unclear collection of various facts, topics, subjects, etc.

*** *** ***

14) *Voice-Note No.:* <u>2.2.14</u>
   Date & Time Recorded: <u>Sun, Jan 31, 2021; 9:49 PM</u>
   Audio-Length: <u>8:20</u>
   Main Topic(s)/Theme(s)—Note: Add after listening to entire recording (!): <u>1)--Ideas and/or knowledge and their confluence with people in our minds. 2)--Creativity through making and bending, breaking, and remaking etc., your own rules. 2)--Suspension of understanding. 3)-- Break-up into bite-size pieces, hope/optimism, and the knowledge that as a lifelong-learner, time is your best friend. 4)--.</u>

Sub-Topic/Theme Summaries:

---   1: Ideas and/or knowledge and their confluence with people in our minds and later, in our writings. E.g., the Wright brothers or Doug, my mentor.

   ---Explication: its hard to separate ideas from their discoverers or inventors. I might think about ideas and knowledge, but my mind will inevitably also think about the inventors/discoverers of said ideas and knowledge. Regardless of their ideas, they are fellow human beings, and I relate to them as such.

---   2: Creativity through making and bending, breaking, and remaking etc., your own rules. Connected to the latter, as well as the one written/recorded on the 27th.

   ---Explication: one technique to help myself learn better is through bending rules, and making my own rules. First of all, what **is** creativity? Regardless, what does this topic have to do with SC? My assertion: a lot. In context of SC, metacognition, and improving our learning methods, creativity has a big part to play. Via SC, "hm...how does this work (?),"/"how can I make this work (?)," we're engaging in a creative process.

   ---Further explication: one can also argue that journaling is creative activity or output. Poetry is also a form of creativity, regardless of whether you're writing the poem to yourself or others.

   ---Most of us that write journal entries or poetry aren't famous. There's always a chance that your writings will become famous later. But primarily, that journal-entry or poem is for your own consumption.

*** *** ***

15) *Voice-Note No.:* 2.2.15
   Date & Time Recorded: Mon, Feb 1, 2021; 11:23 *PM*
   Audio-Length: 10:19
   Main Topic(s)/Theme(s)—Note: Add after listening to entire recording (!): 1)--"Good job Seif."
   2)--Similar to comp-sci/soft-dev concept, "sanity check."

Sub-Topic/Theme Summaries:

   ---   1: "Good job Seif," after finishing a task. A psychological trick I use to reward myself after
         successfully completing a task (quantitatively, not necessarily qualitatively—i.e., as long as
         it's **done,** regardless of quality).
         ---That psych-trick is especially important when you're doing favors or work for others.
         Metaphor: when you present the cake to them, all they see is the finished product, not all the
         hard work that went into it.

   ---   2: Similar to comp-sci/soft-dev concept, "sanity check." The original concept in soft-dev:
         executing a print command to check if the code is working correctly, or using said command
         to debug your code.
         ---Explication: It can be helpful in our work in general for us to borrow that (sanity-check)
         concept and tailor/apply it to the different tasks we're executing. For me, TDLs help in a way
         similar to sanity-checks, to help me stay well-oriented and level-headed, knowing well what
         is done and what remains undone.

*** *** ***

16) *Voice-Note No.:* 2.2.16
   Date & Time Recorded: Thr, Feb 4, 2021; 12:59 *AM*
   Audio-Length: 28:17
   Main Topic(s)/Theme(s)—Note: Add after listening to entire recording (!): 1)--Imagining what
   others are thinking and/or saying about you.

Sub-Topic/Theme Summaries:

   ---   1: Imagining what others are thinking and/or saying about you in your presence now or before,
         or in the future, in your presence and absence. Explication/recap: the focus of the book is SC,
         in the context of improving our own learning methods. You can't just stick to the application
         of learning-method-improvement, and the concept is tied up with that of metacognition and
         cognition in general, and *selves*. Thus, I will also focus on other areas of life in addition to
         learning-method improvement.
         ---At any given time, we all most of wonder what others think of, or say about, us. Random
         example: "boy meets girl or in LGBT context, boy meets boy;" a love story. The guy falls in
         love, then starts to wonder, "I wonder if she thinks about me" / "I wonder what she thinks
         about me" / "…what she tells her friends about me."
         ---Unless you're self-employed, don't all of us want to know what our bosses think of us?
         And you're self-employed, you have employees, suppliers, etc. Regardless most of us—in the
         context of business or work—also wonder what others think of us. Relation to SC: your *self*
         has a *primary* and *secondary self* (my terms). The primary self is me—my name is Seif, 36
         years old, etc. But it also helps to "get out of your own body/mind" to imagine, via the two

selves (primary and secondary): "hm, I wonder what she thinks about me?" That is a clear instance of SC.

--- 2: Suspension of understanding: E.g. of David Malan, "just trust me for now," as he introduces concepts to students that have never done comp-sci. This phrase is similar to my own "suspension of understanding" concept. In this context, related to active listening. Unlike Malan's students—i.e., who aren't familiar with comp-sci basic concepts, you understand your interlocutor's points. However, you're also suspending many of your own thoughts, ideologies/beliefs, knowledge, questions, etc. You're setting them aside to first listen well to the current incoming message.

---Primary reasons for your suspension of understanding and/or active listening. Seeking of "meeting of the minds," and looking forward to new pleasant or useful facts, analyses, etc. When you listen actively, you learn things you didn't know. You encounter different angles/ perspectives to old or already-known knowledge, etc.

---Relation of above point to SC: If I am reading a book and I don't understand the concepts therein, and I say to myself, "*Ok, I'm going to suspend understanding for now, I don't have to understand each and everything that I'm reading in this book. First, let me read, and eventually it's going to make sense.*" That's you talking to yourself.

---In the case of David Malan, he is not talking to himself; rather, he's lecturing to students. But again, the point stands. Another example: me reading *Goder, Escher, Bach*. You can't shouldn't say: "Oh well, I don't understand it all, so I'm going to try understand at least a chunk of it."

--- 3: Break-up into bite-size pieces, hope/optimism, and the knowledge that as a lifelong-learner, time is your best friend. Similar expression with potential utility and/or similar proverbs etc., [to] "*time is a life-long learner's best friend.*"

---Explication; application of above concepts to autodidactic-learning of software-development and other topics. How to use these techniques: first, you have to (/can) tell yourself, "*Ok, let me break this [soft-dev, French, etc.] into bite-size pieces for me to learn this better.*" Second, realize and admit that yes, learning is hard, but you have to nurture hope and optimism, telling yourself "*you know what, I'll get this; I'm gonna learn this.*"

---Further explication: sometimes, you're studying for an exam or otherwise have a deadline. But if you're life-long-learner, you have to realize that you have the whole of your life ahead of you to do what you love doing, i.e., learning. You don't have to learn *everything* today or *right now*.

--- 4: Me composing a question to my mum in Luganda and English. "*Wawulidde* about Kabushenga retiring from the New Vision Group (NVG)?" Clarification: NVG is a Ugandan print-news and general media-company. Explication: I caught myself composing the above question to my mum because of the current research. I try to execute metacognition on my own thinking processes.

---Further explication. The above words ("*Wawulidde* about Kabushenga retiring from the New Vision Group?") were the exact words that were forming in my mind, and I was planning to text my mum that question. I wasn't thinking about the topic in the abstract. With this research, I am trying my best to better understand human cognition and memory.

---Further explication; related question: what's the percentage break-down of our thoughts—i.e., by type (memories, future plans, general abstract thoughts, images, etc.). You can divide time into two basic parts: moments before now, and moments after now. And "now" keeps changing. Thus, in your mind, you can think about the time that you've already lived, or you can imagine the future. [Editorial Note: Interesting how the "present/now" is almost none-existing, or is very fleeting, in this breakdown!]

---Further explication: I don't think the above concept/practice (/concept-practice) is indeed an instance of SC, as I wasn't composing that message for my own consumption. [Editorial Note: The current self/Seif disagrees strongly with the self/Seif that made the preceding judgement! That instance is both one of SC, and interpersonal-comm.] For instance, a typical pure-SC instance is one saying "I wonder how that works." But in the above case, I was composing a message for my mum's consumption.

---Further explication: the above analysis might be a breakthrough or insight in the question of "what counts and/or doesn't count as SC?"

APPENDIX C

Voice-Note Recordings Summary-Transcript 3; (*Total: 24 V-Note Summaries*)

Batch 3.1

1) *Voice-Note No.:* <u>3.1.1</u>
   Date & Time Recorded: <u>Sat, Feb 6, 2021; 4:58 AM</u>
   Audio-Length: <u>9:24</u>
   Main Topic(s)/Theme(s)—Note: Add after listening to entire recording (!): <u>"Different folks, different strokes" for learning styles; Metaphors and analogies: what are they?</u>

Sub-Topic/Theme Summaries:
   --- 1: "Different folks, different strokes" for learning styles. Metaphor: learning to code is similar to learning to use a fork and knife, and building software projects is similar to using newly-acquired fork and knife skill on a meal of beef or chicken.

   ---Also, if a student has math-anxiety, how do you expect them to solve mathematical problems [before somehow alleviating the math-anxiety]?

   ---Metaphors and analogies: what are they? Thagard differentiates the two. An analogy is under the umbrella of a metaphor. An analogy is closer in meaning to a simile. And a metaphor is a story or situation that stands in for a different situation or concept, etc. E.g., "this job drives me crazy, coz I keep putting out fires all day long." Obviously, my meaning isn't literal, re: "putting out fires." Rather, there are always disasters that I have to keep mitigating with clients, colleagues, etc.

   ---My explanation to G using the fork and knife learning metaphor demonstrated the validity of Thagard's synthesis. Analogies and metaphors help makes sense of things within our minds, but it also helps us while explaining concepts to other individuals.

*** *** ***

2)   ***Voice-Note No.:*** <u>3.1.2</u>
       Date & Time Recorded: <u>Sat, Feb 6, 2021; 7:08 PM</u>
       Audio-Length: <u>7:22</u>
       Main Topic(s)/Theme(s)—Note: Add after listening to entire recording (!): <u>"What's the best way to learn?"</u>

Sub-Topic/Theme Summaries:

---   Debate w- G; what's the best method of learning? "Learn by doing," or get the basics first, then dig in? My own pref.: the latter (get the basics first, etc.).

---   We retain a lot more than we think/know we do. The information/knowledge is stored in the back of our minds. Paradox: for you to use that stored-away knowledge, you have to kind of re-learn it. Which frustrates us, making us think we're not learning. But my hypothesis: it's there, the key to really memorizing it is how to harness it.

---   Often, when we go over something we've already learned, we discover new angles to that knowledge; new angles/ways of looking at that topic/subject, etc.

---   How can we relate the above revelation to intrapersonal-comm.? That debate w- G—re: the best learning methods—is similar to the debate we always have in our minds: "what is the best way for me to learn this stuff?"

*** *** ***

3)   ***Voice-Note No.:*** <u>3.1.3</u>
       Date & Time Recorded: <u>Tues, Feb 9; 12:03 AM (Recordings for 3 days.)</u>
       Audio-Length: <u>28:43</u>
       Main Topic(s)/Theme(s)—Note: Add after listening to entire recording (!): <u>1.--"What was I thinking?" 2.--Continuous self-assessment of learning abilities. 3.--Self-assessment and empathy in ref. to ethics 3--"Finish what you start." 4--"Hm, I wonder how that works."</u>

Sub-Topic/Theme Summaries:

---   "What was I thinking?" Left a message app on my phone to go the browser, Safari, then had to ask myself that question. Something that happens to many of us, same as, "why did I come to the kitchen?" Mindfulness can probably help us in that regard; i.e., avoiding absent-mindedness.

---   Continuous self-assessment and sense-making of learning methods and abilities, e.g. with my thinking of myself as a (no offense meant here!) special-needs learner. First time I made that reference was with soft-dev coach. If I think of myself as a special-needs learner, then I won't feel bad when it takes me a long time to learn something.

---   We tend to be our own worst critics—we're very hard on ourselves in the various endeavors we partake in.

---   Similar to above concept: self-assessment, re: math. As in, "I can't do arithmetic, but I can do quantitative-analysis." I for one tend to be very deliberate about my learning-ability self-assessment. "The first step to solving a problem is to realize you have a problem."

---   Active and clear self-assessment and empathy in ref. to ethics; e.g. deciding to reply or not an email. "Do unto others…": case in point, instance that happened to me earlier, re: receiving an email from someone requesting help. At first, I didn't want to reply, but I remembered my own feelings in similar situations—i.e., as the requesting party. Thus, I replied and regretfully told the person I can't help them.

--- That is another example of how we communicate with ourselves. The discussion has echoes of other similar concepts, e.g. id, ego, super-ego, or primary and secondary needs, per Doug Porpora.

--- That above is an example of an internal dialogue.

--- "Finish what you start." Context: there is so much material online for learning. The problem however: I often come across fascinating material, bookmark or print it out, but don't finish it. Hence my stern reminder to myself, written and hung on the wall: "Finish What You Start."

--- Instead of looking for more resources, how about first finishing the ones you've already started? You don't even have to do the homework assignments—you can just audit them, watching, the videos.

--- Curiosity about facts or concepts leading to self-questioning: "Hm, I wonder how that works." What's the connection between that, as in—for those of us in knowledge-heavy fields, "what's the difference between that [amateur] general curiosity and professional-level focused curiosity?"

--- Reaction to above point: There are some professions—such as mine—that require an inquisitive mind/personality. However, can other people—i.e., without that inquisitive personality do those jobs? Also, how much of our professional research results from that day-to-day curiosity? Are our serious theory-based questions enriched by those "trivial" curiosities?

--- My own answer: my professional and general/amateur curiosities merge. Good example: this book's research. Another example: my dissertation. The topic arose from a morality question I had asked myself before, how can humans—the most intelligent of animals, commit such atrocities?

*** *** ***

4) *Voice-Note No.:* 3.1.4
Date & Time Recorded: Sunday, Feb 14; 1:44 AM
Audio-Length: 13:22
Main Topic(s)/Theme(s)—Note: Add after listening to entire recording (!): 1—Curiosity about % of human thoughts that are memories, etc.; 2—Classification of thoughts into 3 main; categories; 3—Realization that the knowledge you consume via reading is interconnected; 4—Mnemonics and other memory-tricks.

Sub-Topic/Theme Summaries:

--- Curiosity about % of human thoughts that are memories, etc. Tried Googling it and got no good results. That highlights the problem of Googling stuff.

--- Logic-based postulation: You can probably classify thoughts into 3 main categories. 1—abstract collection of lived experiences; 2—actual memories of lived experiences; 3—the future that you haven't lived through yet. Our sentience means that time is a primary part of our experience for us. All our knowledge can be [probably] be categorized as belonging to time *before* a given present moment—e.g., the time and date listed above, and the time *after* said given moment.

--- Realization that the stuff you read is interconnected; e.g., me being curious about "Farmer's Almanac" decision vs. "weather-predicting" groundhog. As I was reading a story about that groundhog, I became curious about the "Farmer's Almanac's" precision.

--- That might be a good example of how we make connections between different types of stored-away knowledge (in our brains). These connections happen spontaneously depending on context; but one of the situations that causes that arising of connection is while we're reading. In this case, the relevant concept was weather prediction; as I was reading about the groundhog, it triggered curiosity, re: precision of the "Farmer's Almanac" vis-à-vis weather-prediction.

--- I actually looked it up; both weren't accurate in their predictions. Modern meteorologists point out that beyond three months, we can't predict weather that well.

--- Mnemonics and other memory-tricks. Context: academic environments. Memory and its role in learning for me personally. Personal quote from day-to-life personal/familial relations: "if I had the same good memory in class, similar to that I have for other non-learning past events," I would be very smart!

*** *** ***

5) ***Voice-Note No.:*** 3.1.5
Date & Time Recorded: Monday, Feb 15; 3:58 AM
Audio-Length: 12:59
Main Topic(s)/Theme(s)—Note: Add after listening to entire recording (!): 1--E-filing system on my PC; 2--Self-counseling about need for patience; 3--Curiosity: humanzee Wikipedia article; 4--Suspending understanding.

Sub-Topic/Theme Summaries:

--- E-filing system on my PC; keeping myself organized, remembering what task I was doing or where I ended with a particular task, etc. The way we organize e-files in particular e-folders, and marking where we've ended while reading books, etc.; all those are ways in which we communicate with ourselves.

--- Self-counseling about need for patience after choosing a learning strategy. E.g., while coding on FreeCodeCamp. The strategy that I chose for myself to learn soft-dev is to first really learn the concepts before attempting to build actual projects/apps, despite relevant pop-conventional wisdom asserting the validity of "learning by doing." But sometimes, I doubt that strategy; and in such moments of self-doubt, I have to self-counsel about need for sticking to the chosen strategy.

--- Curiosity: humanzee Wikipedia article. Is curiosity part of how we improve our learning methods? The answer: "yes." Relevant example for me: Googling if it's possible for humans and great apes to cross-breed. Google result: Wikipedia article about "humanzee." A good example of self-triggered curiosity.

--- Suspending understanding; similar to suspending visualization while reading a fiction or non-fiction book with authors' descriptions of scenes or things. The metaphor for such situations is: it is akin to drawing a sketch of a suspect's description as the witness describes said suspect.

*** *** ***

6) *Voice-Note No.:* <u>3.1.6</u>
    Date & Time Recorded: <u>Wed, Feb 17; 3:09 AM</u>
    Audio-Length: <u>13:18</u>
    Main Topic(s)/Theme(s)—Note: Add after listening to entire recording (!): <u>1)—Constant anticipation while reading fiction and [creative] non-fiction; 2)—Can that sense of anticipation be applied to academic and applied-setting learning (?); 3)—Metacognition; 4)—NY Times paywall and the negative influence of poverty on an aspiring polymath.</u>

Sub-Topic/Theme Summaries:

--- Constant anticipation while reading fiction and [creative] non-fiction—as in, the plot of the story, re: background, tension buildup, climax of tension, and denouement: "What's gonna happen [next]?"

--- Can that sense of anticipation be applied to academic and applied-setting learning, treating it like a story whose end you're anticipating? This might be a good (self-directed) psychological trick. Example of a (self-directed) psychological trick: telling myself, "just write one sentence for that paper—just one sentence or paragraph, then you can have a treat!"

--- Implementation of that psychological trick is paradoxically easy and hard at the same time. E.g. with research, (arguably,) in regard to findings of research, a well-done study should have a tabula-rasa of sorts—you shouldn't know what you'll find in the end. You should go into the field or lab…expecting w0hatever [or nothing in particular, vis-à-vis the results you'll eventually find]. The results might surprise you, regardless of what your hypothesis is.

--- Emphasis: I hope the connection is clear between the points made throughout these voice-notes, and the topic of the book's research, i.e. intrapersonal-communication, and applying that concept and practice to our learning methods.

--- Points such as the above are epiphanies or realizations that I make in my mind while reading, thinking in general, and learning. For the purposes of the book's research, I pause and take note of such potentially-insightful realizations. And both points are (arguably) good examples of how we talk to ourselves. Another example of SC: when you catch yourself thinking (while watching a movie, etc.): "Oh my God, I'm 'on the edge of my seat'!"

--- In my case, my thinking process in such instances involves meta-cognition, thinking about how I think and communicate with myself, and applying the resulting realizations in my research.

--- NY Times paywall and the negative influence of poverty, class, race, etc., on an aspiring polymath. Explication: NY Times and other top national daily newspapers' websites have paywalls. One related question in this context, how easy is it for a poor person, … or a disadvantaged person to implement their goal of becoming a polymath? Ans.: it's not easy for a poor person to cultivate that trait. Apparently, knowledge is a commodity. Sure, there are ways for one to learn without paying. But, it's definitely harder, compared to more privileged members of society.

*** *** ***

7) *Voice-Note No.:* <u>3.1.7</u>
    Date & Time Recorded: <u>Sat, Feb 20; 12:22 AM</u>
    Audio-Length: <u>21:34</u>

Main Topic(s)/Theme(s)—Note: Add after listening to entire recording (!): 1)—Social-awkwardness and loneliness as a result of one's love of learning; 2)—Latter point includes social-judgments of pretentiousness, arrogance, etc.; 3)—Obama's book: from the NY Times article promoting the book.

Sub-Topic/Theme Summaries:

--- Main point and digression for emphasis:

>> Social-awkwardness and loneliness as a result of one's love of learning.

>> Digression/recap/emphasis: my goal with these notes is to continuously clarify the over-arching goal of the autoethnographic data. We communicate with ourselves; so what?

>> Utilization or application SC on the concept/practice of improving our individual learning methods.

--- Back to main point above (social-awkwardness and loneliness as a result of one's love of learning): Even though learning and knowledge are good concepts, they aren't as valued in society as they should be. E.g.: of XX's story; how I told him I'm doing a Ph.D. and his response: "another damn degree?!" From a young age, we hear these derogatory words such as "nerd" and "geek," about these weird awkward kids—or bookworms.

--- That element described above is a sentiment "made up of"—or which triggers—different emotions. [Part of] The way you conceptualize it in your mind is through emotions. And those emotions come from reactions we have when we're ostracized for our love of learning; e.g., how I felt after XX's reaction to my pursuit of the Ph.D.

--- That above—that conceptualization and thinking or feeling of that sentiment—is an example of SC. SC isn't just verbalized or clear or spoken-out-loud thoughts. Self-directed abstract thoughts themselves are a kind of SC. And emotions—e.g. what I felt in reaction to XX's ostracization—are a form of SC.

--- So what? It means that we should strive to develop our self-confidence; "this is who I am; I love learning, and [/but] society looks down on people that love learning. But I am who I am…" Otherwise, you'll feel like you're doing something wrong, loving to learn.

--- Above point includes social-judgments of pretentiousness, arrogance, etc. This particularly hurts; as in, is very hurtful to an aspiring polymath or a lover of learning. Example of story of professor at XX institution, who was worldly and highly-experienced, and had lots of questions. As a result, people started sighing and rolling their eyes, etc.

--- Accusations of being pedantic. Also, "mansplaining"—i.e., sexist/misogynistic pedantic behavior is a real vice. However, regardless of intention/personality/etc., there are often judgements of being a "know-it-all," etc.

--- Implication: "sometimes, you gotta shut up…it's ok to share, e.g. in class, but it has to be relevant, not because you love hearing the sound of your own voice," etc. "Sharing is ok, but don't overdo it; know when it's appropriate," etc.

--- Obama's book: from the NY Times article promoting the book. "*And much like the way that earlier book [i.e., Dreams From My Father] turned the story of its author's coming of age into an expansive meditation on race and identity, so a "A Promised Land" uses his improbable journey from outsider to the White House, and the first two years of his presidency as a prism by which to explore some of the dynamics of chance and renewal that have informed two and a half centuries of American history. It attests to Mr. Obama's own story-telling powers and to his belief that, in these divided times 'story-telling and literature are more important than ever,' adding that 'we need to explain to each other who we are and where we are going'"* (my emphasis via underlining).

--- In context of SC: How do I explain to you who I am, before I know who I am to myself? In this book, I utilize a good amount of SI theory, which deals extensively with the concept of "selves." E.g., Norbert Wiley's "The Semiotic Self." Before explaining to others who you are, you have to first really know who you are. Also related to earlier statement(s) about the need for developing self-confidence.

--- Related somewhat: LGBT folk who have to [first admit to themselves their sexuality, and then] tell the public, "love me or hate me, this is who I am."

*** *** ***

8) ***Voice-Note No.:*** <u>3.1.8</u>
Date & Time Recorded: <u>Tues, Feb 23; 12:44 AM</u>
Audio-Length: <u>18:14</u>
Main Topic(s)/Theme(s)—Note: Add after listening to entire recording (!): <u>1)—Role models (male); 2)—NY Times one about Obama and his journey as an author; 3)—Implication of that fact for purposes of the book; 4)—The messiness and grind of the journey; i.e., deciding whom to emulate, how, and why.</u>

Sub-Topic/Theme Summaries:

--- Role models (male): my dad, Mr. Everett, Pres. Obama, and Doug. Recap of book's research topic: intrapersonal-/self-comm., and applying it to our learning methods and our general lives. I (often) communicate with myself and say, "these are my role-models." Specifically:

>> Obama: e.g., the recent NY Times article about his writing process

>> My dad and Mr. Everett: I also learn from them. And when you learn from someone, you automatically improve your learning process.

>> E.g., Mr. Everett: I've learned a lot from him vis-à-vis values.

>> As I mentioned earlier, learning is often ostracized by society. Antidote: you have to develop your self-confidence. And Mr. Everett helps me with that.

>> Obama also helps via his life-stories, i.e. in the context of getting to your real self; your role in the world, etc.

>> My dad: our relationship is complicated, but he is a great role-model, given the challenges he faced, and the success he has achieved regardless.

--- "I'm bathing / 'drinking in' this article;" i.e., the NY Times one about Obama and his journey as an author:

In high school, Mr. Obama says, he and a "roving pack of friends" — many of whom felt like outsiders — discovered that "storytelling was a way for us to kind of explain ourselves and the world around us, and where we belonged and how we fit in or didn't fit in." Later, trying to get his stories down on paper and find a voice that approximated <u>the internal dialogue in his head</u> (this author—S. Sekalala's—emphasis via underlining), Mr. Obama studied authors he admired. "As much as anybody," he says, "when I think about how I learned to write, who I mimicked, the voice that always comes to mind the most is James Baldwin. I didn't have his talent, but the sort of searing honesty and generosity of spirit, and that ironic *sense of being able to look at things, squarely, and yet still have compassion for even people whom he obviously disdained, or distrusted, or was angry with. His books all had a big impact on me."*

- --- Implication of that fact—i.e., as underlined—for purposes of the book: first, be it for writing or discovering your true self, you have to find your voice. Notable postulation: when you're trying to find your voice as an author, where do you draw the line between the inner and outer voice? The outer voice—i.e., for public consumption—is sieved; cleaned up, refined. Inner voice is raw. I have to think about this some more; what's the difference between the inner voice and the outer voice?
- --- The messiness and grind of the journey; i.e., deciding whom to emulate, how, and why. This also highlights the challenge of finding your true self. Apparently, this is also related to the above point about Obama's journey of finding his true self and his authorial voice. However, the current point arose independently in my mind. In other words, I wasn't actively thinking about the Obama point above.

*** *** ***

9) *Voice-Note No.:* <u>3.1.9</u>
   Date & Time Recorded: <u>Wed, Feb 24; 4:04 PM</u>
   Audio-Length: <u>10:12</u>
   Main Topic(s)/Theme(s)—Note: Add after listening to entire recording (!): <u>1)—Using emails— especially, as well as texts to improve my writing continuously; 2)—Self-questioning about one's own creativity and desire to learn; 3)—Debate about reading and/or writing problems based on socialization with English [or any other relevant language for literacy purposes]; 4)—Perfectionism with writing.</u>

Sub-Topic/Theme Summaries:
- --- Proof-reading your writing in various contexts as well as in email and text-correspondence.
  - \>\> Interesting how ordinary folks don't have to have higher-ed to suss out someone's tone(s) in writing.
  - \>\> Using emails—especially, as well as texts to improve my writing continuously.
- --- Self-questioning about one's own creativity and desire to learn, including using the above tool, i.e., using regular email and text-writing as ways to improve writing. In other words, second-guessing such personal learning tricks.
  - \>\> What are the various mid-points while learning using self-invented tools and tricks?
  - \>\>\> E.g., using logic to decide or figure out the best or correct number of words in a draft of a paper, email, etc.
  - \>\> Conclusion: on one hand, it's healthy to doubt ourselves. However, you shouldn't get stuck in "analysis-paralysis syndrome." At some point, you need to just get started. "Decide on a course of action," then implement it.
- --- Debate about reading and/or writing problems based on socialization with English [or any other relevant language for literacy purposes].
  - \>\> Confusion with words such as we're vs. were.
    - \>\>\> And too vs to.

>> Explication: e.g., with Ebonics. If I speak using "wrong"/irregular grammar in my day-to-day life, will I write in a similar style?

>>> In my own experience with students, people tend to write the way they speak. E.g., leaving out the letter "d" while writing past tense words.

--- Explication: the above point demonstrates one outcome of the process of translating our thoughts into words. If I usually construct my sentences—i.e., in spoken-language contexts—in a certain style, why should I write differently?

--- Perfectionism with writing. "Beating yourself up" about grammatical errors, but also second-guessing your word-choices after sending an important email.

>> Explication: I often write important emails, then after sending them, read through them again, and catch typos or grammatical-structural errors. And that really upsets me. Another uncommon problem for me in this context: second-guessing word-choice. I tend to be decisive. I'm lucky to have an above-average writing capability. Thus, I try to utilize it well in such contexts. [Editorial note: Some instances—1 in particular, an email to Dr. Best back when I was a grad-asst. in the English dept. and she was the chair—stand out.]

*** *** ***

10) Voice-Note No.: 3.1.10
Date & Time Recorded: Sun, Feb 28; 1:46 AM
Audio-Length: 17:00
Main Topic(s)/Theme(s)—Note: Add after listening to entire recording (!): 1)—Example of benefits of "suspending understanding": the two French conjugation charts. 2)—Obama's "A Promised Land" and Ken Follet's "A Column of Fire" are both related to the book of Exodus. 3)—Listening to your mind forming sentences, including during prayer. 4)—Resultant discussion question: is this really interpersonal comm.? Ans.: yes.

Sub-Topic/Theme Summaries:

--- Example of benefits of "suspending understanding": the two French conjugation charts. Before, I didn't like the "recipes" chart; now, I like it. Explication: that "recipes" chart is more succinct compared to the other chart, yet does a good job of explaining the general formula of French conjugation.

>> Somewhat related to quant vs. qualitative analysis. Quant might involve math and is thus more feared, but it might in the end be easier than qualitative, given the fact that you have to do much less explication vis-à-vis results discussion.

>> Also, example of cross-discipline training efficacy: French conjugation and the logic of PC programming.

>> Clarification: Not just French conjugation, but in general, rules of grammar of any other language (including English). Background: my English-undergrad "Structures and Origins of the English Language." The point: grammar rules (of any language) are very formulaic.

>>> Postulation: Thus, given the fact that "the brain is a muscle," if you train yourself regularly with learning or practicing such formulaic or intellectually-challenging cognitive tasks.

>> Meta-analytical point: I'm actually not sure if that is what I meant by the point summary. Thus, "I should have written it better!"

--- Obama's "A Promised Land" and Ken Follet's "A Column of Fire" are both related to the book of Exodus. I realized that connection. I wonder how many such connections we miss regularly!

>> Quick pause-point for explication: all the notes in this transcripts are made in reaction to the realizations I make in my metacognitive day-to-day basic analyses.

--- Explication of above point: Follet's title is a direct reference to the column of fire that led the Israelites to Canaan during the night, and Obama's "Promised Land" is metaphorical in nature, alluding to a better, more-perfect-union America.

>> Further explication: this is an important point, as we tend to gloss over things often. But every once in a while, when you pause and process/digest these connections, it's a good feeling. In this case, I happened to be reading two random books, one fiction, the other non-fiction, by two separate authors. And yet their titles apparently both reference the biblical book of Exodus.

--- Listening to your mind forming sentences, including during prayer:

>> First realization: there are multiple voices competing, then combining to choose an appropriate word.

>>> I pray often, despite my past and present religious-belonging ambiguities. E.g., dear Lord, please help me achieve: _____.

>>> The process that takes place in this context is the concept/act of verbalization. Use of a different context/example for demonstration: sometimes, you feel an emotion but are unsure of what exactly that emotion is—how to label it. E.g., someone embarrasses you, and in that moment you feel a combination of many different emotions, i.e. anger, guilt, etc.

>>> Then later, you can pause and name the different emotions you're feeling.

--- Explication of above point, re: praying. E.g., you have a sick relative, and want/need to pray for them (the relative) to heal: you have to form that sentence in your mind.

>> Perhaps the challenge is presented when you're praying for a number of different things. In fact, the challenge of verbalization isn't only applicable to prayer. For instance, I wrote down these main points earlier, but I still have to explicate them somewhat extemporaneously—forming specific sentences as I go along, while explicating the individual points.

--- Continuation of above point: thus, similar to the process of me explicating this transcript's main points, we have to form the specific sentences for our needs during prayer.

--- Resultant discussion question: when we form sentences in our minds during the above various contexts—i.e. prayer, discussion, etc., is it a form of intrapersonal communication? And for me, the quick answer is "yes."

>> Part of the application of intra-personal communication is for the contexts of inTER-personal-comm. Before communicating with others, we first have to first clarify to ourselves what we're really feeling, what we mean, what we want, etc.

--- Conclusion: thus with prayer, we're talking to G_d technically. But first, we have to first clarify to ourselves what we want/need exactly, from G_d.

*** *** ***

11) **Voice-Note No.:** 3.1.11
 Date & Time Recorded: Thur, March 4; 1:12 AM
 Audio-Length: 24:28
 Main Topic(s)/Theme(s)—Note: Add after listening to entire recording (!): 1)—Meta-analytical discussion of the challenge of record-keeping or tracking of voice-note recordings; 2)—NY Times article by Michael Sims: "Darwin's Dim View of the Second Sex."; 3)—Slowly re-reading emails or other documents written materials before sending or submitting them.; 4)—Take-away from the latter point (no. 3); 5)—Key take-away for intrapersonal-comm. and learning from previous point (no. 4).

Sub-Topic/Theme Summaries:

- --- Meta-analytical discussion of the challenge of record-keeping or tracking of voice-note re-cordings. Implication: the possibility of occasional secondary or tertiary challenges with SC vis-à-vis, despite—or, as a result of—the use of tools such as to-do lists, etc.

- --- NY Times article by Michael Sims: "Darwin's Dim View of the Second Sex." Despite Darwin's sexism—especially in regard to intellectual abilities, he thought very highly of one particular female scholar—probably a sister of one of his colleagues.
  - >> Explication: that article stood out to me because of that scholar's—i.e., the lady—passion for learning.
  - >> Long quote from article:

"Her second published work, "On Female Education," a defense of her own passion for learning and a critique of the expectation that her education would end when she reached adulthood, came out when she was 20. Her most famous book, "Illustrations of Political Economy," dramatized human stories featuring the economic theories of Adam Smith, Thomas Malthus and James Mill. Didactic, at times simplistic, it was lucid and accessible and considered a step forward for progressive, human-centered economics."

- --- Explication: the reason that article—and that quote in particular—stood out to me, was the fact that her political-economy book *"dramatized human stories featuring the economic theories of Adam Smith, Thomas Malthus and James Mill... it was lucid and accessible and considered a step forward for progressive, human-centered economics."*
  - >> "I love the fact that [she] took these hard-to-understand theories and dramatized them… in fact, I think I'll look for this book at some point."
- --- Another long quote:

"Her most recent success was "Society in America," based on two years of travel during which she was feted by artists and legislators. She visited President Andrew Jackson and lodged with the former president James Madison. Both owned slaves, but Martineau did not shrink from portraying the horrors of slavery. She had been writing about it, as both a moral sin and an economically inefficient system, since early in her career. She would have found common ground on this topic with Darwin, a passionate abolitionist who had witnessed slavery's abuses from Africa to Brazil. His theories about natural selection were motivated in part by a desire to undermine racist notions promulgated by scientists.

They compared writing methods. Several of Martineau's books grew out of her detailed travel journals, which was how Darwin had constructed his own book about his voyage around the world. Martineau was said to require little revision for the many pages that flowed from her pen. Darwin thought her invincible and seems to have expressed this idea. Not at all, she replied; a few consecutive hours of hard work tended to exhaust her. Darwin felt the same."

--- Also fascinating and ironic (!) that Darwin was an abolitionist, yet his work was later misused extensively by racist scholars and elites. [Editorial note: apparently, I was thinking of the fact that his work **inspired** social-Darwinism.] In fact, apparently, part of the reason he discovered and popularized the theory of natural selection, was to debunk racism!

--- Key word I use while explicating my interest in that article: "resonate/resonation;" those quotes above—and the article in general—really resonated with me.

>> When you read an article and it triggers emotions of enjoyment from/within you, that is arguably a form of intrapersonal-communication. [Editorial note: I don't think I'm certain about this point. Perhaps one way to think of it is that the mind and body are automatically communicating with themselves via hormones and nervous-system signals—i.e., by producing those emotions. But if you pause and notice those emotions, then I suppose you are indeed communicating with yourself!]

>> In my case, the resonation was marked enough to compel me to save the article for these notes!

--- Slowly re-reading emails or other documents written materials before sending or submitting them. And realizing later that your eyes were moving too quickly, in tandem with your mind's intended meaning, but in that revision or "spruce-up" process, you skipped one or more words [Editorial note: my meaning is a little unclear here; I believe I meant to say, "[omitted writing]" one or more words."]

>> The debate between training myself to send after the first draft, vs. revising till ready. Winner: the latter!

--- Further explication of above point. I usually write important emails to various people—i.e./e.g., to request for help or resources of some kind. Apparently, in many of those instances, I have already formed the sentences in my mind. But while writing the email, I might omit some words. Then during the revising process—trying to catch mistakes or such omissions, because I already know what I'm trying to say in my mind, I read fast, using the meaning stored in my mind, vs. reading the actual words on the screen/page.

>> Then later, while reading the already-sent email, I realize that "aw shucks (!)," I left out some words.

--- Take-away from the above realization. Slow down; breathe; write (the email/document/etc.) carefully, and then proof-read it. But as you read through the draft, keep in mind that you might be "reading" or thinking about the words you've already composed, instead of reading the actual words you wrote in the draft.

--- Key take-away for intrapersonal-comm and learning: That process above is one of you communicating with yourself indeed! It is also a good example of how we can improve our learning methods; first, you necessarily have to communicate with yourself—e.g., as I do in the above point. Then you have to implement the process-improvements you've decided.

>> Key quote from "an unnamed source" (my software-engineering coach): "It's important to have a good intrapersonal relationship with your *self.*

*** *** ***

12) *Voice-Note No.:* <u>3.1.12</u>
Date & Time Recorded: <u>Sun, March 7; 7:42 PM</u>
Audio-Length: <u>17:47</u>
Main Topic(s)/Theme(s)—Note: Add after listening to entire recording (!): <u>1)—Use of "XX"</u> <u>while typing a paper, email, etc. as a placeholder; 2)—Nature of thoughts; 3)—Is it possible to</u> <u>have a "Eureka!" moment…?; 4)—Minor worry: that I'm taking on too much.</u>

Sub-Topic/Theme Summaries:

--- Use of "XX" while typing a paper, email, etc. as a placeholder for a word I can't remember or choose for my meaning.

--- Explication of above point: as I'm writing various documents/messages/etc., I often get stuck in the process at some point. I know what I'm trying to mean, but I can't pick it easily from my vocabulary in my mind.

>> We all have "databases" of vocabulary and knowledge in our minds, but we're often caught up in those moments when we know what we're trying to say, but don't know how to express it. Usually, we eventually manage to express that hard-to-express word or concept.

--- Related to above point. George Costanza's situation in Seinfeld, after his argument with a colleague, and Costanza realizing a good comeback he could have retorted to said colleague after the colleague insulted Costanza.

--- My use of the placeholder "XX" has arisen over the past few years. After undergrad school, probably during the PhD program. I use it because I don't want to lose momentum with my overall meaning or message, despite my temporary vocabulary-block for that particular missing word or concept.

>> Later, I can go back and look at the surrounding words and consider the context, then pick a word. It doesn't have to be a "perfect" word vis-à-vis expressing your meaning.
>>> Our minds work in a very fluid way, so you might not manage to get that "perfect" word. Just pick another one, and move on

--- Nature of thoughts: how or why do they stream into my mind the way they do? [<u>Editorial</u> <u>note: I believe this refers to what other scholars have named the "stream of consciousness."</u>]
>> If my current thought is "Z," can I trace my previous thoughts "Y," "X," etc.?

--- Also related to a conversational phenomenon I experience in my day-to-day life, where my mum and I, or other interlocutors ask ourselves at some point after talking for a while, "how did we get to this topic?" Then we retrace our conversation steps, so to speak.

--- That practice—i.e., keeping track of your thoughts—can also be applied to the improvement of our learning methods. E.g., you're reading a book or trying to follow along with an instructional YouTube video, you might get distracted. It might help to ask yourself in that moment—and/or other such moments: "why [and/or how] did I get distracted, or why [and/or how] do I always get distracted—what leads up to it?"

>> If you figure that out, it might help you pay better attention to what you're doing, including learning.

--- While listening to French news on Feb 19, I wonder: is it possible to have a "Eureka!" moment where when knowledge clicks all of a sudden, or do we have to keep learning little by little? Explication, via the example of learning French: ever since high school, I've been learning French, and I am at an intermediate level or working-knowledge level.

--- Further explication: supposing I currently understand around 60-80% of French news or dialogues etc., how did I get to that knowledge level—did it click [at some point] all of a sudden?

>> Self-debate: that doesn't sound like it makes sense, and it's a rather minor or insignificant question. Sure, it might be possible to have that "Eureka!" moment when all the knowledge clicks into place all of a sudden, or, you'll keep learning little-by-little. But it seems insignificant.

--- Conclusion of above point: we have to keep learning little-by-little, unless the subject-matter is very easy for you. Most of the topics/subjects/concepts/etc. that we learn take time to learn. Sure, that "Eureka!" moment might come to us, but there's only one way to arrive at it—keep learning over time whatever it is you're trying to learn.

--- Also, what do I mean by "when knowledge clicks?" This is mostly referring to making connections between the things you've learned, vs. the actual process of learning per se. E.g., with a subject like chemistry. There are concepts in chemistry with which you'll benefit if you've learned some related concepts in physics and vice-versa.

>> Thus, one day in a biology/biochemistry class, you might find yourself learning a particular concept easily, then you think about it and realize it's because you learned the periodic table.

>> But regardless, in general, you have to learn concepts incrementally.

--- Minor worry: that I'm taking on too much; for instance, with plans to start learning Chinese, Arabic, and Hebrew.

>> Reason for "minor" designation: "with good planning, almost anything is possible."

--- Explication of above point: you don't want to "bite off more than you can chew." But as I say in that self-coined quote, if you plan well, you can pull it off.

>> E.g., the way we learn so much—via numerous classes—in college or high school. A particular example comes to mind, of this one bright young man I know who takes a bunch of AP courses in high school. But that is possible because of the way those courses are scheduled.

>>> Or my example from undergrad, from the semester in which I took a total of 21 credits—i.e., 7 classes—and did well.

--- Conclusion of above point: relevance to improvement of learning methods is obvious. However, in regard to intrapersonal communication in general, apparently, worrying—i.e. the way I do above about taking on too much—is also a type of intrapersonal-communication.

*** *** ***

13) **Voice-Note No.:** 3.1.13
Date & Time Recorded: Tues, March 9; 9:17 PM
Audio-Length: 20:10
Main Topic(s)/Theme(s)—Note: Add after listening to entire recording (!): 1)—A general typological catalogue of my thoughts: what are the different types of thoughts that run through my mind regularly?; 2)—Spontaneous vs. gradual **success** with my research and other works; 3)—Concept of "work-per-sitting."

Sub-Topic/Theme Summaries:

--- Meta-analytical discussion/recap of voice-note recording methodology.

--- A general typological catalogue of my thoughts: what are the different types of thoughts that run through my mind regularly? Answer:
[Editorial note: this Wikipedia article is or might be helpful: https://en.wikipedia.org/wiki/Outline_of_thought]

>> 1—Memories and other time-related conceptions, i.e. past, present, and future (e.g., wondering about the future);

>> 2—Ideas and creativity both general and work-related;
>>> General: E.g., I might be walking by an empty field—i.e., a plot of land, and I picture a house or park/garden/etc. on that land.
>>> Work: I love Christopher Nolan's Batman trilogy movies so much. So, as a communication and media scholar, why not just study it? Of course, this is simply a germination of a raw idea—lots of studies have probably been done already about that trilogy. Still, as long as you have a good "so what," there's always room for more!

>> 3—Sense-support, meaning the five senses: i.e., working cognitively with one or more of the five senses.

>> 4—Metacognitive-related.

>> 5—Other. E.g., various thought-types that combine elements of some of the above points.

--- Similar to the question of spontaneous vs. gradual understanding (see note no. 3.1.12), a question of spontaneous vs. gradual **success** with my research and other works.

>> In other words, e.g., publishing a paper or book that earns you notoriety immediately in your early career years, vs. earning that notoriety because of the totality of your body of works over the course of your career.

--- Explication of above point: that question occurs in my mind from time-to-time. Another way to pose the question: can/will I achieve success from one particular project, or will my success come from the combined body of my work.

>> The point is mostly relevant to intrapersonal communication (/SC), vs. both SC and learning.

>> However, I can apply it to my learning methods by saying: "you don't have to understand this (particular concept/topic/etc.) right now, you can understand it gradually, and in turn, you can get rewarded eventually."
[Editorial note: that I suppose is my approach to software-development.]

--- Concept of "work-per-sitting": be lazy before the sitting, then push yourself hard during the sitting.

>> A self-invented modus of working: e.g., if I have to read a total of six chapters, I can tell myself, "ok, in this sitting (and "a sitting" counts as a certain block of time on a particular date), I have to read a minimum total of three chapters."

>> On a day-to-day basis, there are numerous tasks of *minor importance** that I have to attend to, e.g., sending emails, writing my "to-do" list, etc. *Then again, these tasks are important in the long run—especially the "to-do" list a meta-task of sorts*. In any case, beyond those minor tasks, for the more important tasks, it is important to set goals vis-à-vis amounts of work to be completed within specific amounts of time. "And before I complete this task, I am not going to get up [i.e., from my seat, until after finishing the task(s) at hand]."

--- Application of above point to general intrapersonal-comm. and learning:

>> SC: When you think about the tasks you have to complete on a particular day, that's you communicating with yourself.

>> Learning-method improvement: I developed that concept of "work-per-sitting" sometime during my Ph.D. program—around mid to late into the program. And it can be very useful for learning-method improvement. However, this particular method is just one of my preferred methods—everyone has different working styles and preferences. E.g. some folks might prefer working little-by-little throughout the day, a method to which I can also relate. In any case, this is a prime example of how we can use SC to improve our learning—and general **working**—methods.

***

Batch 3.2

1) *Voice-Note No.:* 3.2.1
Date & Time Recorded: Thur, March 11; 7:55 PM
Audio-Length: 27:38
Main Topic(s)/Theme(s)—Note: Add after listening to entire recording (!): 1)—Creativity and productivity from dynamism of rotating tasks; 2)—The basic 13; 3)—Previously covered: are lists (bullet/number) and check- or outline-, etc. underutilized?; 4)—"Sanity-Check" from PC-sci or soft-dev: usable in general life as well.

Sub-Topic/Theme Summaries:

--- 1--Creativity and productivity from dynamism of rotating tasks: i.e., do X today, Y tomorrow, etc. Explication:

>> First, a side-note: to be covered later: a mind-based vs. written "to-do" list (TDL). The general rough lists we have in our minds of our necessary tasks at home, work, etc., on a day-to-day basis.

>> That^ mind-based TDL is a form of SC.
>>> However, for me at least, the process of composing and working through the mind-based and written TDLs is messy. Regardless, I tend to be more productive when I switch up my tasks from within the day or from day-to-day. E.g., do X in the AM and Y in the PM, versus X all day long, and/or do X today and Y tomorrow, instead of X on both days.

---Another notable fact about TDLs. For me (personally), my TDLs are often in random order. In other words, instead of listing them in order of importance for instance (e.g. X is most important and Z is least important), I will instead list them randomly with either Y or Z as no.-1, and then the other two.

>> I thus have to decide how to prioritize those items—to figure out that X is indeed the most important one—and/or has an attached impending deadline, thus I have to do it first, etc.

---Conclusion: regardless, beyond such prioritization considerations, again—for me personally at least—I find that dynamism of switching between tasks helpful.

--- 2--"The basic 13." Quick (re-)introduction: "The Basic 13" (TB-13) is a collection of important algorithm challenges, curated by a software-development "bootcamp"-school called "The Coding Dojo."

>> It's purpose: for students to master how to solve those challenges in less than 3 minutes each, so as to train themselves for more complicated software-development algorithm and data-structure-related challenges.

>> In 2016, I enrolled in that bootcamp, but unfortunately, I had to drop out because of a number of challenges—and simply because I wasn't well-prepared.

>>> However, I got to keep their algorithms book, which is proving to be very useful in my self-tutelage of soft-dev!

---That book contains TB-13 somewhere at/near the end of the first chapter of the book. Reason for including it as a voice-note agenda item: it puts to the test a lot of my personal theories/beliefs/practices in relation to how I can learn effectively.

>> Despite my academic achievements to date, I don't think I'm necessarily "smart"/intelligent. Thus, I try to be very careful as I build my knowledge based, to really understand the nitty-gritty of topics/concepts.

>>> Related example: my learning breakthrough with statistics. I can compare myself to someone who might understand the general theory immediately.

>>> For me however, it took me a while to understand that the key to understanding the subject is to have to ability to know what tests to use for what variable-types (i.e. nominal, ordinal, or interval ratio) and for which general research goal—i.e., differences vs. relationships, etc.

---Thus, for me, that point above demonstrates a key aspect of my learning style; I have to first really grasp the basics before moving on to the more complex stuff.

>> **<u>Important editorial note:</u>** The current point, expressed differently. A lot of students probably don't mind cramming facts to pass exams, regardless of whether or not they really understand what they're studying. My approach is different: I prefer actually understanding the topic—being able to explain it in my own words, versus simply having the ability to pass a quiz about the topic. And I suppose that preference might be problematic for science majors in college for instance, for whom that kind of holistic understanding would take a long time to establish.

---Conclusion; application to SC: I had to earnestly stress the above point to myself, vis-à-vis TB-13.

--- 3--Previously covered: are lists (bullet/number) and check- or outline-, etc. underutilized?

>> First, a brief discussion of the general importance of lists—e.g., checklists in particular, as demonstrated by Dr. Atul Gawande. E.g., with the aviation industry's good safety record.

>> Second, my attestation to the importance of checklists to pilots.

---Essentially, TDLs are us telling ourselves: "Self, you need to do X, Y, Z today!"

--- 4--"Sanity-Check" from PC-sci or soft-dev: usable in general life as well. Related qn.: is PC-sci a good reflection of how our minds work naturally, and/or an arbitrary framework we find useful? E.g., with me and sanity-checks.

>> Explication [underline: editorial note: the following explanation is edited using an online-search for validity, i.e. to ensure that I'm defining the concept correctly]: sanity-checks are a way to ensure that the software-program written so far works as intended.

---I for one tend to utilize sanity-checks—or, processes similar to sanity-checks—in my day-to-day life for non-software-dev tasks. Resultant/related qn.: is the CRUM (Wiki-link) hypothesis a valid one vis-à-vis explaining how our minds work?

>> Ans.: I don't know. Sure, cog-sci scholars have made good progress with this research, but there is a lot we don't know.

>>> For me personally, what do I think? Well, my brain—similar to a PC stores data, i.e. memories, knowledge, etc. My brain also processes or utilizes the data saved, just as a PC does for data analysis, etc. So overall, sure, CRUM is a generally good construct for understanding our minds.

---Conclusion: regardless of that debate—i.e., CRUM's validity or lack thereof, processes such as sanity-checks—used in the world of computing or soft-dev, can apparently be of good use to our general lives.

---And in essence, a sanity-check is SC: "Seif/self, wait a minute—pause; what's going on here? Is this—e.g., this prioritization order of the TDL—working?

***

2) *Voice-Note No.:* 3.2.2
Date & Time Recorded: Sun, March 14; 11:55 PM
Audio-Length: 26:38
Main Topic(s)/Theme(s)—Note: Add after listening to entire recording (!): 1)—Pause, (re-)scrutinize if or as needed; 2)—Not everything is meant to be said; 3)—Realization about productivity: movement is good (!); 4)—Page 326, Grisham's "Rogue Lawyer." "Lol" at end of first (long) para, abt. Rudd counting punches, etc.

Sub-Topic/Theme Summaries:

--- 1—Pause, (re-)scrutinize if or as needed. E.g., an angry email to your boss, etc. Explication: if/when you're experiencing an extreme emotion—not just anger, pause, (re-)scrutinize the action you want to take, and then if it's needed or as needed, implement it.

>> True story: this actually happened to me in real life back in the day, as a grad-assistant at Kean U.; I once sent an angry email to my boss. All I had to do was to first pause, and either not write it all, or written as a draft and saved it without sending it.

---That is an example of SC—i.e., an individual counseling themselves to pause and (re-)scrutinize their intended actions; but it's not directly relevant to improving our learning methods.

--- 2—Not everything is meant to be said. After hearing it in your mind, pause; think it over: is it really worth saying? If "yes," say it. If "no," enjoy it yourself if it's worth enjoying, and smile. If it's not worth enjoying, make peace with it; it's just a thought, nothing more. It's most likely not worthy of an emotional or other reaction.

---Explication: this point is related to the previous point, and it could have been listed as a sub-point of the previous one.

>> However, the previous point is more general, discussing the concept of actions in general, e.g. sending an email. With the current point, we're specifically discussing the act of speaking—but granted, an email is also a mode of communication, similar to speaking.

---Explication continued: the current point can be useful in preventing situations that give rise to the problematic behavior in the previous point, i.e., the email-sending-while-mad behavior. The argument here is thus: this issue—that you're thinking about—isn't worth getting upset about to begin with. It's just a thought, nothing more.

---Important disclosure: the current/above point isn't my own idea; it's a mindfulness practice. You shouldn't become your thoughts, so to speak. E.g., in cases of past traumas: yes, you were hurt badly in the past (e.g., the violent robbery that befell me).

>> But you shouldn't relive it or blame yourself for it, etc.

---Conclusion: clearly, the above points are (arguably) good examples of SC, ways of training ourselves to behave better. As for applying the points to personal learning methods, we can discuss the practice of good listening skills as a specific (sub-)area. E.g., if you're in a seminar or workshop at work, or in a college class, etc.

>> One mistake I used to make in undergrad (and my Master's program) was speaking to much during Socratic or other similar discussions.

>>> Lesson: contribute to the discussion, but don't overdo it. And stay on Topic; if something is irrelevant, don't bring it up.

--- 3—Realization about productivity: movement is good! Just move (like that physics theory)! [Editorial note: I seem to have been referring to Newton's first law of motion.]

>> Personal postulation: action begets more action. If I've just woken up and I'm slow to start, I just have to push myself to get up and start doing one thing at a time, and that gives me momentum.

---Further explication of current point: relatable to writing. If you have writers' block, just push yourself to start writing by using techniques such as freewriting; do some secondary research to help you get started, but just start writing asap.

--- 4—Page 326, Grisham's "Rogue Lawyer." "Lol" at end of first (long) para, abt. Rudd counting punches, etc.

>> Ditto ("lol")—laughing and "this is so funny...", pg. 327, re: Dr. Taslman

---Explication: contextualization of above note via brief summary of the court-scene in question in Grisham's novel.

>> Quick side-note: I don't like Grisham's use of first-person P.O.V-narration vs. Omniscient-narration. But the style tends to suit his stories.

>>> In this case, the narration has a somewhat comical effect, vis-à-vis the irreverence with which the central character—also the story's narrator—describes the overly graphic video that the prosecutor plays for the jury, which depicts the numerous blows the defendant landed on his victim.

>> More explication, re: the humorous nature of this scene in the story.

---Conclusion of point: this is an apt example of SC with myself; in this case, my reaction is humor. But sometimes, I react with sadness for relevant scenes or plots, etc.

---Application of current point to improvement of personal learning methods: should we treat academic learning—both for arts and sciences—the way we treat reading and enjoying stories/novels/etc.?

>> E.g., if you're reading a sociology text, and a particular chapter therein in preparation for an exam, should you perform a psychological trick on yourself, to ask yourself for instance, "I wonder how this chapter ends?" [Editorial note: I suppose this question is related to the concept of gamification of learning (?).]

---Further explication. E.g., if you read an early chapter in that sociology text about socialization, can/should you ask yourself what the author will say in the later/concluding chapters about the concept?

*** *

3) *Voice-Note No.:* <u>3.2.3</u>
Date & Time Recorded: <u>Wed, March 17; 9:35 PM</u>
Audio-Length: <u>30:08</u>
Main Topic(s)/Theme(s)—Note: Add after listening to entire recording (!): <u>1)--Irony: that guy telling me that my internal dialogue is telling me to fail; 2)--Communicating with yourself about novel stories' trajectories.</u>

Sub-Topic/Theme Summaries:

--- 1)--Irony: that guy telling me that my internal dialogue is telling me to fail. Explication: the gentleman I met in Paris, a fellow American who lives in LA. I was relieved to run into him at a Starbucks, fresh off the train from the airport, overwhelmed by being in a foreign city whose language I know very little!

---He helped me a little bit, vis-à-vis getting my bearings. Him and his wife were with a young French lady—an aspiring actress. Apparently, they owned/own some kind of placement agency. Ever since, we connected via LinkedIn.

---Recently, I got in touch with him on LinkedIn to say hello. I also requested him for remote job leads. From here, things got very strange with this gentleman. He got in touch very quickly via my cell-number which I'd just given him (via LinkedIn).

---He asked me what types of jobs I'm looking for, and I tried explaining my academic/research foci; i.e. intercultural-comm. research, etc. Apparently, he didn't understand what my research is about, but in his attempts to get clarification from, he came off very rude and pushy.

---I became suspicious, wondering what his motives were—I even pondered the possibility of him trying to recruit me into a cult! Reason for that suspicion; in addition to his attitude, he sent me a job posting for which I'm not qualified—the position was related to data-science and software-development. And he essentially wanted to at worst—lie, or at best—"fudge" my qualifications. He told me he can "coach" me if I am willing to succeed.

>> In the end, I don't know what his motivations were. Perhaps he would have gotten a finder's fee (?); again, I'm not sure.

---Relevance of the story to this research: in his attempt to give me a pep-talk, he said my "internal dialogue" is telling (/encouraging) me to fail. In response, I told him I'm writing a book about that topic. His response: I am indeed the right person for such a topic (spoken in a negative way, given his assertion that I'm doing the opposite of self-actualization)!

---Conclusion; the irony: this research is actually intended to help us harness SC for self-actualization, yet here was this gentleman claiming that I am using mine for the opposite detrimental effect.

---Further explication: my research is parallel to that of Margaret Archer and her social-reflexivity theory, which is also premised on the utilization of internal dialogue to improve our lives. Similarly, other scholars point out the utility of positive thinking.

>> Perhaps this is what that gentleman was trying to tell me, that I should instead channel my SC toward self-actualization?

>> But even if the response to the above question is "yes," his immoral ethics were muddling his message.

--- 2)--Communicating with yourself about novel stories' trajectories. E.g., me at the end of John Grisham's "Rogue Lawyer": most common feeling or thought: curiosity.

---Explication: one of the ways in which we communicate with ourselves is while reading stories, both fiction and nonfiction. In fact, we even communicate with ourselves while reading general academic and other materials. And for me (personally), my most common feeling is curiosity. E.g., are some of the characters going to die off, or "live happily ever after," etc.

---Application of first point/story farther above—i.e., the dubious gentleman—to improvement of learning methods: we can learn a lesson vis-à-vis the interactions we have with people—and the dialogues thereof, vs. your own internal dialogue/SC. In particular we can critically examine the way others think we think.

---For instance, your family and friends, or even strangers, might think that you lack self-confidence—similar to the incident with the above-mentioned gentleman. However, at the end of the day, only you can truly describe your true abilities. [Editorial note; caveat: unless you're afflicted by the Dunning-Kruger effect!] "You're the only one in your head."

---Technically, a self is composed of (a minimum of) two people; your primary self, and your secondary self. And it is your secondary self that discusses the senses, thoughts, feelings, actions, etc., of the primary self. [**Editorial note:** apparently, Gerald Edelman uses a similar term for consciousness: https://en.wikipedia.org/wiki/Secondary_consciousness] In any case, it is only the individual in question—e.g., myself—that can be absolutely certain of what/how that individual thinks.

---Thus, for purposes of self-actualization, we have to harness our own agency to rigorously review our mindsets for the relevant life-goals we have. In other words, if we know that they're wrong, we shouldn't allow people such as that dubious gentleman to convince us about what our mindsets are. "We deserve to have veto power in terms of what to do with our lives."

---Application of second point/story farther above to improvement of learning methods: first, luckily, I was able to ask myself that question, i.e., what is the remarkable thought/feeling/ etc. that I have while reading Grisham's and/or other books?

---Thus, we can/should use SC to better understand what we're learning. For instance, ask yourself whether or not you really understand what you're reading. If you think/feel you don't really understand it, Google it or use other references, etc.

---Further explication: example of Christopher Nolan's "Batman" trilogy. I've always had trouble understanding the plot of the final film of that trilogy. Thus, I have resolved to eventually read about the plot on Wikipedia to really understand it.

***

4) **_Voice-Note No.:_** <u>3.2.4</u>
   Date & Time Recorded: <u>Mon, March 22; 6:00 AM</u>
   Audio-Length: <u>22:41</u>
   Main Topic(s)/Theme(s)—Note: Add after listening to entire recording (!): <u>1)—"To-do": mentally or out-loud, saying what you need to get done; 2)—Rubber-ducky for "duck-debugger" method! 3)—Treating all learning similar to the way we learn our native tongues: prolonged engagement, consistency, practice (among other variables).</u>

Sub-Topic/Theme Summaries:

--- 1--"To-do" lists (TDLs): mentally or out-loud, saying what you need to get done. Explication: quick recap of research goal(s): general study of SC, and applying SC to improvement of learning methods.

   >> Regardless of if you're writing it down or saying it in your mind, TDLs are a form of us collecting our thoughts and taking stock of what we need to get done.

   ---The question: how can we use that to improve our learning methods? For one, we can treat learning as one big task, and we can then break it up into parts or sub-tasks, so to speak.

   >> Most or all of us, regardless of whether or not we realize it, use TDLs either in our minds or via writing.

   >> Using the example farther below of learning languages using prolonged engagement, consistency, practice (among other variables): supposing I want to learn how to be a good cook, via a variety of cuisines and cooking methods—frying, boiling, grilling, baking, etc., with a total of 10 cooking methods. I could say, "this week, I'm going to learn three methods." This can help you learn gradually.

--- 2--Rubber-ducky for "duck-debugger" method! Explication of debugging in computer-sci/ software-dev.: solving problems in a software program after you discover them via testing or other means.

   >> It helps to have a partner or someone you can talk to as you do this problem-solving, by talking out-loud about what you've written vs. what you're trying to write.

   >> If there's no one around, you can use a rubber-duck for that purpose—talk to the rubber duck the way you would a human partner, but with the same objective—realizing your mistake via verbally auditing the parts of your code.

---In my case, I don't have a rubber-ducky, but I have a tiny dinosaur that I stole from my niece Amber, and I named the dinosaur Amber, in honor of its real owner. And that for me satisfies the role of the rubber-ducky.

---But essentially, you're not really talking to the inanimate object; rather, you're talking to yourself. And when I came across that (rubber-duck debugger) method while studying soft-dev with the Ed-X/Harvard CS50 MOOC, I felt vindicated about the concept of SC.

--- 3--Treating all learning similar to the way we learn our native tongues: prolonged engagement, consistency, practice (among other variables techniques).

>> Question: how do we learn our mother tongues? Via the above methods. E.g., for prolonged engagement, as a baby, you're exposed to people speaking a given language all the time. E.g., telling you "no!"

>> E.g. my niece: used to call milk "*mik!*" And water, "*wooye!*" And fish, "*pssh!*"

---Application of above point to general SC and improvement of learning methods: basically, just use those same methods we use for language-learning, in the domain(s) of learning other topics/subjects/skills/etc.!

>> E.g., I have to use that strategy in own learning of computer science.

>> Self-critical question: are those techniques really a form of SC?

---Answer-attempt: I)--Per Doug Porpora: it's hard to think without using words. II)--Still, the question stands; are those techniques a form of SC?

>> Concession: apparently, via logic and close reflection, those methods don't seem to be a form of SC per se. Rather, they're a form of learning—i.e., "immersive learning," perhaps?

>> However, those methods are done via language, which is a mode of (inTER-personal) communication.

>>> E.g., compared to the rubber-duck-debugger method, which is clearly a form of SC.

***

5) *Voice-Note No.:* <u>3.2.5</u>
Date & Time Recorded: <u>Tues, March 23; 4:17 AM</u>
Audio-Length: <u>28:12</u>
Main Topic(s)/Theme(s)—Note: Add after listening to entire recording (!): <u>1)—"Lol" at "It belongs to God," pg. 141, "A Column of Fire"; 2)—Self-doubt about my belief that consistent exposure and practice result in mastery; 3)—What are the parts of the learning journey from 0 = Scared of the scary-looking formula, to Z = Master, able to rap about the topic/subject in your sleep? 4)—Related to above 2 points: my consumption of data science articles on LinkedIn and elsewhere, and my fascination—not fear—of the various concepts; 5)—Having to repeatedly look something up b4 fully internalizing it; 6)—Mastery for teaching.</u>

Sub-Topic/Theme Summaries:

--- 1--"Lol" at "It belongs to God," pg. 141 of "A Column of Fire" by Ken Follet. Explication: related to an earlier note I (believe I) made about the reactions I often have while reading fiction and nonfiction books or stories/literature. I.e., reactions I have when I come across parts of the stories/literature that are funny, sad, interesting, etc.

---In other words, "if it's funny, you'll laugh, … [if it's] sad, you [might] cry," etc. My argument: that is clearly a form of SC. The resultant question, re: the purposes of this study; how do you use that form of SC to improve your learning process?

---Ans.: it's possible to harness those reactions somehow—regardless of whatever it is you're reading, albeit in a narrower or more restricted fashion. For instance, you can keep checking your understanding of the material, e.g. with a biology or communication text. If after checking your understand, if you find that you don't understand the material, it is probably helpful for you to admit that to yourself, and to ask yourself why you're struggling to understand it.

---For instance, you can review your understanding of previous chapters; were you understanding those chapters well? And how come you're struggling with this chapter compared to those others?

--- 2--Self-doubt about my belief that consistent exposure and practice result in mastery.

>> First, a quick recap to review one of the previous days' points, re: treating all our learning the way we learn our native languages.

>> Thus, the context of the current point: I often have doubts about the effectiveness of that strategy.

---Further explication: a critical self-review of my learning of French.

>> Debate with self: is French a good example of evidence to support the non-effectiveness of my strategy?

>>> Negative answer: not really, seeing as I don't live in a French country/environment.

---Further explication:

>> Resolution of self-debate: perhaps the valid question isn't whether or not the strategy can work. Perhaps the valid question is **how exactly** does it work?

>> In any case, the entire debate above re: the efficacy of the "native-language learning strategy" is a prime example of SC.

---The debate as explicated above is felt/experienced as self-doubt or second-guessing myself; another good example of SC. And it is directly related to strategization of improving our own learning methods.

--- 3--What are the parts of the learning journey from 0 = Scared of the scary-looking formula, to Z = Master, able to rap about the topic/subject in your sleep?

>> That entire question is a question to myself, thus a prime example of SC.

>> It is also directly related to improving our own learning methods.

--- 4--Related to above 2 points: my consumption of data science articles on LinkedIn and elsewhere, and my fascination—not fear—of the various concepts, even though at the moment, I understand <2%, give or take

+ Specific example: stochastic processes!

---Explication: on LinkedIn, there is a page I follow titled "Towards Data Science," with a variety of topics about data-science.

>> I keep downloading/bookmarking/self-emailing the articles, even though I don't really understand them.

>> It might even have some scary-looking formulae, but those don't scare me anymore.

---E.g., my fascination with one particular topic, stochastic processes. I am pleasantly surprised by my ability to define stochastic processes, but it is a result of consistent exposure. If you keep engaging with a topic for a long period of time, you will eventually understand it.

---The above discussion is an expression of wonderment: I am fascinated by my fascination with those topics.

>> That is an example of me communicating with myself.

>> It's interesting how my notes are increasingly focused on learning. Relevant metaphor: if you're throwing darts at one of those boards with concentric circles, they might first start by landing in the outer circles. But with enough practice, they will eventually start falling in the inner circle(s).

>>> Thus, perhaps I am increasingly discussing learning processes because I have gradually conditioned my mind to the topic?

--- 5--Having to repeatedly look something up before fully internalizing it, e.g. benefits of shutting down vs. sleeping, and why re-starting machines helps repair their software-programs.

---Explication: I have looked up that topic a number of times. My educated-guess answer—i.e., to the question of why restarting computers fixes software issues: it's related to issues of RAM.

>> We can relate that issue to the way our brains cannot store long numbers—i.e., numbers with a big number of digits.

---That realization—i.e., that I have looked up that fact numerous times, is a form of SC.

>> As for relating it to improvement of learning methods: we can follow up such a realization with a technique to help ourselves—e.g., to write down the answer on a sticky note and put the sticky note somewhere on or close to our work-desk(s).

--- 6--Mastery for teaching: I believe many/most (?) profs regurgitate material, coz it's hard to paraphrase it, discuss it, etc.

>> This point/note is from personal experience as a college instructor. "Making the material come alive is hard, so you often find yourself regurgitating it the way it is written in the text.

---Further explication: on one hand, we have definitely have to help our students memorize the basic facts of the subjects/topics to help students pass quizzes or tests, etc. But on the other hand, we have to help the students via "making the material come alive" indeed. Thus, students can see how the different topics relate to each other, how the discipline relates to other disciplines, etc.

---I for one am an earnest believer in that goal, but I often "sacrifice it at the alter of helping students pass quizzes and tests, etc."

>> This was a longterm realization which came to me over a long period of time.

>> When I first started teaching, that point didn't quite click.

---After a while, you realize that you've heard these debates before—e.g., with that about the national K-12 curriculum, vis-à-vis too much testing, etc.

>> My own interpretation or opinion of the debate: what is the main purpose of your teaching—i.e., the K-12 students? Are you teaching them to pass tests, or to help them learn holistically?

---Again the above discussion is an example of SC, and is also directly related to improving our learning methods.

>> Further explication, re: improving our own learning methods: in my own learning of data-science, how do I do that appropriately? By memorizing facts, or by holistically learning the different topics in relation to themselves—i.e., within the discipline—as well as in relation to other subjects, etc.?

***

6) *Voice-Note No.:* <u>3.2.6</u>
    Date & Time Recorded: <u>Wed, March 24; 9:58 PM</u>
    Audio-Length: <u>22:17</u>
    Main Topic(s)/Theme(s)—Note: Add after listening to entire recording (!): <u>1)—Googling things and the sheer amount of information out there; 2)— Vindication for "rubber-duck debugger" method; implication: we don't want people to think we're crazy! 3)—Discuss: The messiness of all you have to do, combined; i.e., the often-unwieldy nature of "the TDL".</u>

Sub-Topic/Theme Summaries:

--- 1--Googling things and the sheer amount of information out there. Explication: we're very lucky to live in the age of Google. Back in the day, you had encyclopedias, dictionaries, and other reference resources in the library or elsewhere.

   ---Further explication via random-scenario example: you might be writing a paper, and you want to discuss the concept of intrapersonal communication. In my own case, I often hesitate to use words or concepts without being certain of their meaning or [general] context.

   >> Once you get to that point were you ask yourself—"hm, what do people mean by that (?)," you're indeed talking to yourself.

   >> This scenario is interesting, given the fact that by looking up a word or concept, I'm technically communicating indirectly with other people.

   ---Reason for above assertion: other people wrote and saved the meaning of the word or concept on Wikipedia or other sources retrievable online.

   >> In any case, the first necessary step is to ask yourself that question, "hm, I wonder what that means?"

   >> Improving our learning methods:
      >>> You can/should ask yourself: how often do I ask myself that question (*"hm, what do people mean by that (?)"*)? In my case, I have realized over time that I indeed do that—not only while writing papers, but even with information/message artefacts such as emails.

   ---Thus, if I know to expect that, I won't blame myself for Googling words, wrongly thinking I am wasting time, etc. I will in turn see that as part and parcel of my writing process, probably even budget time for it.

--- 2--Vindication for from "rubber-duck debugger" method; implication: we don't want people to think we're crazy!

   >> Quick recap of what the method is.

>> Explication of point: the method vindicates my advocacy for the effectiveness of intrapersonal-comm.

---Further explication: instead of using the rubber-duck, why not just talk directly to ourselves, saying e.g., "Seif, what am I doing wrong here?"

>> But, I guess it makes us feel better about our sanity, the fact that we're apparently talking to the rubber duck (and thus to ourselves indirectly), not ourselves directly.

>>> It's also comical that the method is publicized the way it is, which I believe is (at least partly) because again, we don't want people to think we're crazy!

---Recap of above point's relation to SC and learning-method improvement:

>> You talking to the rubber duck is you talking to yourself.

>> Learning-method improvement: by talking to the rubber duck more and more, you'll realize the types of errors you're prone to, either semantic or syntactic, or both, etc.

--- 3--Discuss: The messiness of all you have to do, combined; i.e., the often-unwieldy nature of "the TDL":

>> Imbuing the TDL with power over you and your id and ego.

---Explication: prioritization is tough when you have a lot of things to do.

>> For instance, supposing you have a number of items whose deadlines are far off into the future, and one item whose deadline is tomorrow. Obviously, you'll prioritize the item with tomorrow's deadline.

>> But what about the items whose deadlines are farther off into the future—supposing you have three of those [with the same deadline]?

>> The point I'm belaboring is: it's a messy process.

---Explication of subpoint, i.e., imbuing the TDL with power over you and your id and ego.:

>> First, a recap of what id, and ego are.

>>> Id = our basest instincts. Super-ego = the judgmental/moral part of you.

>>> Ego = the mediator between those two extremes.

---Further explication: apparently, we often imbue the TDL with power over us.

>> I for one have to "empower" the TDL, so to speak, for it to be effective.

---In fact, explicating the above point makes me think that essentially, that power—which we imbue into TDLs—is similar to the one we give written laws, constitutions, contracts, etc.

---Relation to SC and learning-method improvement:

>> As previously argued, a TDL is a form of SC.

>> Learning-method improvement: you cannot improve your learning methods without also [/first] improving your productivity as well. [**Editorial note:** I just struck through this point, as I disagree with it in hindsight.]

>> But we can indeed improve our learning-methods via improving our productivity first or in tandem. E.g., figuring out if you're a morning person or "night owl."

>> Based on your determination of which of the two is your habit, you can schedule your studying accordingly—i.e. either at night or in the morning, etc.

***

7)   *Voice-Note No.:* <u>3.2.7</u>
    Date & Time Recorded: <u>Fri, March 26; 1:26 PM</u>
    Audio-Length: <u>19:38</u>
    Main Topic(s)/Theme(s)—Note: Add after listening to entire recording (!): <u>1)—Discuss: Self-critique, e.g., looking at old writings including papers, emails, etc.; 2)—Discuss: Do I feel myself getting "smarter"? Maybe.</u>

Sub-Topic/Theme Summaries:

---Celebration of end of voice-note recordings!

---   1--Discuss: Self-critique, e.g., looking at old writings including papers, emails, etc. Explication: we often re-read materials we wrote in the past, and we react in a negative or positive way. E.g., reacting in a negative way, critiquing ourselves: "Oh my G_d, I wrote this?! I can't believe I wrote this!"

---And essentially, that self-critique or other communication (positive or negative) is you communicating with yourself. As for improving our own learning methods, the utility here is rather obvious. "By looking critically/closely at your own writings, and constantly keeping up with your [improving/deteriorating/unchanging] abilities," we can adjust our efforts vis-à-vis improving our craft(s).

---Specific example of the above phenomenon with my own life. My constant review of past emails or papers, etc., and my embarrassment vis-à-vis errors of judgment with sending the emails, or stylistic mannerisms, etc. E.g., my wrong usage of the semi-colon over the years, and my ensuing self-correction.

---   2--Discuss: Do I feel myself getting "smarter"? Maybe. But I still don't think I'm "smart"! The IQ-test debate is an interesting one. Explication: related to the above point. If I constantly self-critique my past works—i.e., in my own narrow context of professional scholarship work, then one naturally resultant question is this question, re: intelligence.

---"So what (?)"; why is that question important, what does it mean to be smart, and what do I want to do with that intelligence? That can be left as an open question. However, some tentative practical answers are:

>>   If I'm getting smart, great. If not, why not—as in, why am I not getting smarter [and how can I change that—i.e., how can I get smarter, going forward]?

>>   One way I define intelligence is vis-à-vis reasoning ability/capacity: for instance, has my reasoning ability improved since fifth grade? Thankfully, yes.
>>> What can I achieve with that improved ability? Better progress with my work, and general improvement of my wellbeing; dealing with people, ethics-wise, personal-planning—e.g., with my finances, etc.

---I for one don't quite believe in the validity of IQ tests, for I believe there are various types of intelligence. In any case, the discussion above just demonstrated the relevance of the point (#-2 above), re: the issue of if/how the question is a form of—or otherwise related to—SC, and how we can use it to improve our learning methods.

>>   Or rather, the discussion mostly answers the question of the efficacy of the that "am I getting smarter qn.," re: how we can use it to improve our learning methods.

>>   But the question is definitely a prime example of SC, it is a question I am posing directly to myself, and I answer it similar to the way I have above.

***

APPENDIX D

For Chapters 6, 8, and 9: Qualitative Meta-Synthesis

Part 1: Listing and categorization of all self-emailed articles related to learning processes and cognitive science.

I: Listing (grand-total: 42 articles)

   A—General/Other:

   • 12 Ways to Get Smarter in One Infographic; author: Jeff Desjardins

   • Creativity quiz from MindTools.com

   B—Data Science (Source: Data Science Central; https://www.datasciencecentral.com/**):**

   >> Batch 1:

   1. How to Create a Professional GitHub Data Science Repository; author: Ahilan Srivishnumohan

   2. Promising Careers for Data Scientist Besides the Data Science Industry; author: Yoey Thamas

   3. Top 7 Reasons Why You Need to Learn Python as a Data Scientist; author: Olha Bahaieva

   4. 7 Must-Know Visualizations for Better Data Analysis; A practical guide for ggplot2 package in R; author: Soner Yıldırım

   5. How to Learn to Code; 10 Tips to 10x Your Coding Skills; author: Eric Elliott

   6. The Brain's Most Precious Resource: The role of attention in neuroscience, deep learning, and everyday life. Author: Manuel Brenner

   7. Logic and Statistics; Joys and Sorrows of Markov Logic Networks—The best statistical thinking model; author: Eleonora Laurenza

   8. Free Book: Applied Stochastic Processes; author: Vincent Granville

   9. Single or Multiple Lines: What's Better For Python Code Readability? Author: Zoe Zbar

   10. Data Science Minimum: 10 Essential Skills You Need to Know to Start Doing Data Science; author: Benjamin Obi Tayo

   11. 29 Statistical Concepts Explained in Simple English—Part 13; author: Vincent Granville

   12. A Learning Path To Becoming a Data Scientist: The 10 steps roadmap to kickstarting your data science future; author: Sara A. Metwalli

   13. New Books and Resources for DSC Members; author: Vincent Granville

   14. Introduction to Probabilistic programming; author: Ajit Jaokar

   15. Orphaned Analytics: The Great Destroyers of Economic Value; author: Bill Schmarzo

   16. The Complete Guide to Time Series Analysis; author: Ram Tavva

   17. Probabilistic Machine Learning book – a great free reference for math of machine learning; author: Ajit Jaokar

   18. Hundreds of AI tools have been built to catch COVID. None of them helped. Author: Will Douglas Heaven

   19. Why AI tools Failed to Help With Detecting COVID; author: Ajit Jaokar

   >> Batch 2:

   1. A Step-By-Step Guide To AI Model Development; author: Jaimin Dave

   2. Not All Data is Useful. An Insight into Data Fitment Analysis; author: Raja Dev

3. Is Machine Learning an Art, a Science or Something Else? Author: Vincent Granville
4. 66 job interview questions for data scientists; author: Vincent Granville
5. Top Frameworks Every Python Developer Should Explore; author: INEXTURE Solutions LLP
6. PathQL: Intelligently finding knowledge as a path through a maze of facts; author: Peter Lawrence
7. Content rewriting techniques using NLP paraphrasers; author: Annie Moore
8. A Gentle Introduction to Non-Parametric Tests; author: Kushal Mukherjee
9. Future Proofing Your Career; author: Bill Schmarzo
10. Three Steps to Addressing Bias in Machine Learning; author: Vamshi Ambati
11. What is DevOps and How can it give a Boost to Software Development? Author: Varun Bhagat
12. Are Data Scientists Becoming Obsolete? Author: Vincent Granville
13. The Inverse Problem in Random Dynamical Systems; author: Vincent Granville
14. Orbits of Non-periodic Fourier Series: Simple Introduction, Cool Applications. Author: Vincent Granville
15. Understanding the Complexity of Meta classes and their Practical Applications. Author: Monika Sangwan
16. The Top Skills for a Career in Data Science in 2021. Author: Michael Kevin Spencer
17. How Artificial Intelligence is Revolutionizing Mental Healthcare; author: Aliha Tanveer
18. The Future of Pandemic Modeling; author: Stephanie Glen
19. What Skills to Look for When Hiring a Python Developer; author: INEXTURE Solutions LLP
20. A minimum viable learning framework for self-learning AI (machine learning and deep learning); author: Ajit Jaokar.
21. Topic Modeling: Algorithms, Techniques, and Application; author: Roger Max

II: Categorization of Data-Science & Related Articles

Subject	Articles By Topic/Theme				
	All articles are more-or-less about learning. In addition, you can categorize them under the following topics/themes:				
	1	**2**	**3**	**4**	**5**
	B-2.18	B-1.4	B-1.2	B-1.8	
	B-2.20	B-1.7	B-1.12	B-1.18	
		B-1.8	B-1.13	B-1.19	
		B-1.10	B-2.4	B-2.1.1	
		B-1.11	B-2.9	B-2.10	
Math, Statistics & Data-Science		B-1.15	B-2.12		
		B-1.16	B-2.16		
		B-1.17			
		B-2.2			
		B-2.3			
		B-2.6			
		B-2.7			
		B-2.8			
		B-2.13			
		B-2.14			
		B-2.21			
	All articles are more-or-less about learning. In addition, you can categorize them under the following topics/themes:				
	1	**2**	**3**	**4**	**5**
		B-1.1	B-1.3	B-1.5	
		B-1.14	B-2.19	B-1.9	
Software-Dev.		B-2.5			
		B-2.11			
		B-2.15			
		B-2.17			
Learning & Prof.-Dev	*All articles are more-or-less about learning. In addition, you can categorize them under the following topics/themes:*				
	1	**2**	**3**	**4**	**5**
	B-1.6				

Topics And Themes

    1--Learning

    2--Tools and Techniques

    3--Careers

    4--How-To

    5—Other

Partial List of Implications

- The articles reflect the breadth and depth of the tools and techniques available for the subjects.
- They also demonstrate that there is a focus—albeit not as intensive as one would wish—vis-à-vis guiding potential career-changers or early-career professionals.
- They also reflect the dearth of articles that provide step-by-step "how-to" instructions. However, there are plenty of online sources for such instructions, including MOOCs, YouTube, etc. And those other platforms are arguably better-suited for the purpose.

Part 2: Listing and thematic-analysis of all part 2 cognitive-science articles.

I: Listing

A—Cognitive Science Journal:

>> Batch 1:

1. The Relation Between Cognitive Abilities and the Distribution of Semantic Features Across Speech and Gesture in 4-year-olds; authors: Abramov et Al (2021).
2. Phonation Types Matter in Sound Symbolism; author: K. Akita (2021)
3. Compositionality in a Parallel Architecture for Language Processing; author: Baggio (2021)
4. Human Self-Domestication and the Evolution of Pragmatics; authors: Benítez-Burraco, Ferretti, and Progovacc (2021)
5. Do Humans and Deep Convolutional Neural Networks Use Visual Information Similarly for the Categorization of Natural Scenes? Authors: Cesarei, Cavicchi, Cristadoro, and Lippic (2021)
6. Simulating the Acquisition of Verb Inflection in Typically Developing Children and Children With Developmental Language Disorder in English and Spanish; authors: Freudenthal, Ramscar, Leonard, and Pine (2021).
7. Easier Said Than Done? Task Difficulty's Influence on Temporal Alignment, Semantic Similarity, and Complexity Matching Between Gestures and Speech; author: Jonge-Hoekstra, Cox, Van der Steen, and Dixon (2021).
8. Correlations Between Handshape and Movement in Sign Languages; authors: Napoli and Ferrarab (2021).
9. Attention Does Not Affect the Speed of Subjective Time, but Whether Temporal Information Guides Performance: A Large-Scale Study of Intrinsically Motivated Timers in a Real-Time Strategy Game; author: Van der Mijn and Van Rijn (2021).

>> Batch 2:

1. Semantic Similarity of Alternatives Fostered by Conversational Negation; authors: Capuano, Dudschig, Günther, and Kaup (2021).
2. Emergent Goal-Anticipatory Gaze in Infants via Event-Predictive Learning and Inference; authors: Gumbsch, Adam, Elsner, and Butz (2021).
3. Does Infant-Directed Speech Help Phonetic Learning? A Machine Learning Investigation; authors: Ludusan, Mazuka, and Dupoux (2021).
4. Robust Lexically Mediated Compensation for Coarticulation: Christmash Time Is Here Again; author: Luthra, Peraza-Santiago, Beeson, Saltzman, Crinnion, and Magnusona (2021).

5. Modeling Misretrieval and Feature Substitution in Agreement Attraction: A Computational Evaluation; authors: Paape, Avetisyan, Lago, and Vasishth (2021).

6. A Systematic Investigation of Gesture Kinematics in Evolving Manual Languages in the Lab; author: Pouw, Dingemanse, Motamedi, and Özyürek (2021).

7. Magnitude and Order are Both Relevant in SNARC and SNARC-like Effects: A Commentary on Casasanto and Pitt (2019); authors: Prpic, Mingolo, Agostini, Murgiab (2021).

8. Concept Appraisal; author: Sapphira R. Thorne, Jake Quilty-Dunn, Joulia Smortchkova, Nicholas Shea, James A. Hampton (2021).

9. Extensional Superposition and Its Relation to Compositionality in Language and Thought; author: Chris Thornton (2021).

>> Batch 3:

1. How Reliably Do Eye Parameters Indicate Internal Versus External Attentional Focus? Authors: Sonja Annerer-Walcher, Simon M. Ceh, Felix Putze, Marvin Kampen, Christof Körner, Mathias Benedeka (2021).

2. Monotone Quantifiers Emerge via Iterated Learning; authors: Fausto Carcassi, Shane Steinert-Threlkeld, Jakub Szymanika (2021).

3. Parsing as a Cue-Based Retrieval Model; author: Jakub Dotlacil (2021)

4. Extrafoveal Processing in Categorical Search for Geometric Shapes: General Tendencies and Individual Variations;
   authors: Dreneva, Anna Shvarts, Dmitry Chumachenko, Anatoly Krichevetsa (2021)

5. Are There Cross-Cultural Legal Principles? Modal Reasoning Uncovers Procedural Constraints on Law; authors: Hannikainen Et Al. (2021).

6. A Computational Evaluation of Two Models of Retrieval Processes in Sentence Processing in Aphasia; authors: Lisson (2021).

7. How Category Selection Impacts Inference Reliability: Inheritance Inference From an Ecological Perspective; authors: Paul D. Thorn & Gerhard Schurz (2021).

8. How Children Process Reduced Forms: A Computational Cognitive Modeling Approach to Pronoun Processing in Discourse; authors: Margreet Vogelzang, Maria Teresa Guasti, Hedderik vanRijn, Petra Hendrikse (2021).

B—Google-Scholar:

>> Logic:

1. The Logic of the Unified Model; author: Brian MacWhinney (2013).

2. Innovation, learning and industrial organisation; author: Bart Nooteboom (1999).

3. Logic and/in Psychology: The paradoxes of material implication and psychologism in the cognitive science of human reasoning.; author: Walter Schroyens (2010).

4. An Evaluation of Accelerated Learning in the CMU Open Learning Initiative Course "Logic & Proofs"; author: Christian D. Schunn & Mellisa M. Patchan (2009).

5. A little logic goes a long way: basing experiment on semantic theory in the cognitive science of conditional reasoning; author: Keith Stenning, Michiel van Lambalgen (2004).

>> Rules:

1. Learning General Phonological Rules From Distributional Information: A Computational Model; author: Shira Calamaro, Gaja Jarosz (2015).

2. The Knowledge-Learning-Instruction Framework: Bridging the Science-Practice Chasm to Enhance Robust Student Learning; author: Kenneth R. Koedinger, Albert T. Corbett, Charles Perfettic (2012).

3. Learning By Design: Iterations of Design Challenges for Better Learning of Science Skills; author: Janet L. Kolodner (2002).

4. Toward an Instructionally Oriented Theory of Example-Based Learning; author: Alexander Renkl (2014).

5. Formal Relational Rules of English Syntax for Cognitive Linguistics, Machine Learning, and Cognitive Computing; authors: Yingxu Wang and Robert C. Berwick (2013).

>> Concepts:

1. Simplicity: a unifying principle in cognitive science? Authors: Nick Chater and Paul Vitanyi (2003).

2. The Construction of Category Membership Judgments: Towards a Distributed Model; authors: Asher Koriat and Hila Sorka (2017).

3. Physically Distributed Learning: Adapting and Reinterpreting Physical Environments in the Development of Fraction Concepts; authors: Taylor Martin, Daniel L. Schwartz (2005).

4. Linking Cognitive Science to Education: Generation and Interleaving Effects; authors: Lindsey E. Richland, Robert A. Bjork, Jason R. Finley, Marcia C. Linn (2005).

5. On the Importance of a Rich Embodiment in the Grounding of Concepts: Perspectives From Embodied Cognitive Science and Computational Linguistics; authors: Serge Thill, Sebastian Pado, Tom Ziemkea (2014).

>> Analogies:

1. Analogical Reasoning; authors: Dedre Gentner and Francisco Maravilla (2018).

2. Cognitive science questions for cognitive development: the concepts of learning, analogy, and capacity; author: Graeme S. Halford and Julie McCredden (1998).

3. A new look in representations for mathematics and science learning; author: DAVID N. PERKINS & CHRIS UNGER (1994).

4. Cognitive Supports for Analogies in the Mathematics Classroom; authors: Lindsey E. Richland, Osnat Zur, Keith J. Holyoak (2007).

5. Analogy, higher order thinking, and education; authors: Lindsey Engle Richland and Nina Simms (2015).

>> Images:

1. Capturing human categorization of natural images by combining deep networks and cognitive models; authors: Ruairidh M. Battleday, Joshua C. Peterson, & Thomas L. Griffiths (2020).

2. Human Detection and Tracking for Video Surveillance: A Cognitive Science Approach; author: Vandit Gajjar, Ayesha Gurnani, Yash Khandhediya (2017).

3. Cognitive Theory of Multimedia Learning ; author: Richard E. Mayer (2005).

4. Learning to Decode Cognitive States from Brain Images; authors: Tom M. Mitchell, Rebecca Hutchinson, Radu S. Niculescu, Francisco Pereira, Xuerui Wang (2004).

5. Learning words from sights and sounds: a computational model; authors: Deb K. Roy, Alex P. Pentland (2002).

>> Connections:

1. Cognitive Science Connections Blog #3: Looking Back; author: Education Adds Up (Blog), 2013.
2. Implicit Learning of Form-Meaning Connections; authors: Janny Leung, John N. Williams (2006).
3. Supporting Students to Make Conceptual Connections; authors: Min Li, Ming-Chih Lan, Maria Araceli Ruiz-Primo, Michael Giamellaro, Ting Wang (2012).
4. Connections and symbols: closing the gap; author: Brian MacWhinney (1993).
5. Making Connections in Math: Activating a Prior Knowledge Analogue Matters for Learning; author: Pooja G. Sidney and Martha W. Alibali (2015).

II: Thematic Analysis

>>> Total Numbers of Articles And Mechanism For Choosing the 24 Articles:

>> Total numbers of articles by part and batch (grand-total of 113 cognitive-science articles):

>>+ Part 1 (Total Combined: 56):

>>++ "Cognitive Science" journal articles: 15

→ Batch 1: 8

→ Batch 2: 4

→ Batch 3: 3

>>++ Google-Scholar: 41

→ Logic: 8

→ Rules: 6

→ Concepts: 8

→ Analogies: 7

→ Images: 7

→ Connections: 5

>>+ Part 2 (Total Combined: 57):

>>++ "Cognitive Science" journal articles: 27

→ Batch 1: 10

→ Batch 2: 9

→ Batch 3: 8

>>++ Google-Scholar: 30

→ Logic: 5

→ Rules: 5

→ Concepts: 5

→ Analogies: 5

→ Images: 5

→ Connections: 5

>> *Mechanism For Choosing the 24 Articles*:

>>+ Part 1 (Total Combined: 12; purposive selection, based on helpfulness):

>>++ "Cognitive Science" journal articles: 6

→ Batch 1: 2 (i.e., the top two most helpful from chapter 1)

→ Batch 2: 2

→ Batch 3: 2

>>++ Google-Scholar: 6 (^ = one of the most helpful from chapter 1)

→ Logic: 1^

→ Rules: 1

→ Concepts: 1^

→ Analogies: 1^

→ Images: 1^

→ Connections: 1^

>>+ Part 2 (Total Combined: 12; random selection, using number generator):

>>++ "Cognitive Science" journal articles: 6

→ Batch 1: 2

→ Batch 2: 2

→ Batch 3: 2

>>++ Google-Scholar: 6

→ Logic: 1

→ Rules: 1

→ Concepts: 1

→ Analogies: 1

→ Images: 1

→ Connections: 1

Cog-Sci Article Basic Thematic Analysis

Article	Topic	Other Topics & Themes
1. <u>Selection Method: Most Helpful Or Clear.</u> (***Part 1; Cog-Sci Journal Batch 1***:) Divjak: Exploring and Exploiting Uncertainty	Statistical Learning Ability	• Linguistics • Language-Learning • Learning Via Patterns
2. <u>Selection Method: Most Helpful Or Clear.</u> (***Part 1; Cog-Sci Journal Batch 1***:) Hoerl: Temporal Binding, Causation & Agency	Temporal Binding, Causation, and Agency	• Philosophy • Agent-Causation vs. Event-Causation • Intellectual-Debates
3. <u>Selection Method: Most Helpful Or Clear.</u> (***Part 1; Cog-Sci Journal Batch 2***:) Atkinson, Blakey, & Caldwell: Inferring Behavior From Partial Social Information	Cultural Transmission of Adaptive Traits	• Human-Behavior • Learning-Methods • Research-Experiment Methodology
4. <u>Selection Method: Most Helpful Or Clear.</u> (***Part 1; Cog-Sci Journal Batch 2***:) Lee et. Al: Pain in the Past and Pleasure in the Future	Development of Past–Future Preferences for Hedonic Goods	• Pain-Pleasure Time-Location Preferences • Amount of Known Knowledge of Topic • Confirmation of Knowledge & Addition of Complexity
Arguable Random Common Theme 1: <u>*Knowledge*</u>		
5. <u>Selection Method: Most Helpful Or Clear.</u> (***Part 1; Cog-Sci Journal Batch 3***:) Dubey, Mehta, & Lombrozo: Curiosity Is Contagious	Contagiousness of Curiosity	• Socialization • Social-Cues'/Info's Influence on Curiosity & Behavior • Views on science, • Perception of Knowledge-Usefulness

continues on following page

Continued

Article	Topic	Other Topics & Themes
6. <u>Selection Method: Most Helpful Or Clear.</u> (***Part 1; Cog-Sci Journal Batch 3***:) Reins & Wiegman: Is Lying Bound to Commitment?	Ontology of Deception—i.e., Implicit Deception In Particular	• Ontology of lying • Experimentation • Experiment results: implications can be defined as lies
7. <u>Selection Method: Most Helpful Or Clear.</u> (***Part 1; Google-Scholar; Logic***:) Isaac et al.: Logic & Complexity in Cognitive Science	Use of Logic and Computational Complexity Theory in Cognitive Science	• Logic • Marr's three levels • Validity Via Empiricism
8. <u>Selection Method: Most Helpful Or Clear.</u> (***Part 1; Google-Scholar; Rules***:) Bunge: How we use rules to select actions	Rules: How rules are learned, stored in the brain, and retrieved and used.	• Cognitive dynamics of rules • Humans and non-human primates • Role of Ventrolateral prefrontal cortex (VLPFC) • Dorsolateral PFC, and other brain parts
Arguable Random Common Theme 2: *<u>Epistemology</u>*		
9. <u>Selection Method: Most Helpful Or Clear.</u> (***Part 1; Google-Scholar; Concepts***:) Weidman & Baker: The Cognitive-Science of Learning	From Abstract: "…*Cognitive load theory, <u>constructivism, and analogical transfer</u>* are concepts particularly beneficial to educators. An understanding of <u>*goal orientation, metacognition, retrieval, spaced learning, and deliberate practice*</u> will primarily benefit the learner. …"	• Physicians' education • Effective teaching • Study of learning in cog-sci • Cognitive load theory, constructivism, and analogical transfer • Goal orientation, metacognition, retrieval, spaced learning, and deliberate practice
10. <u>Selection Method: Most Helpful Or Clear.</u> (***Part 1; Google-Scholar; Analogies***:) Jee et al: Analogical Thinking in Geo-Science Education	Analogical Thinking	• Use of analogies in teaching • Interdisciplinary study of cog- and educational-sci • Definition of analogy • Effective analogies
11. <u>Selection Method: Most Helpful Or Clear.</u> (***Part 1; Google-Scholar; Images***:) Dowrick: Self Model Theory: Learning from the future	Use of Future-Self-Envisioning, Re: Self-Improvement of Learning	• Interdisciplinary study of learning • Ultra-rapid learning • Self-modelling video experiment • Neurological processes • Self-model theory
12. <u>Selection Method: Most Helpful Or Clear.</u> (***Part 1; Google-Scholar; Connections***:) Perconti & Plebe: Deep Learning & Cognitive-Science	Deep Learning & Cognitive-Science	• Deep-learning algorithms, • Deep-learning relevance to cog-sci, • Biological exaptation, • Argument for deep-learning in cog-sci, • Unanswered questions
Arguable Random Common Theme 3: *<u>Learning And Pedagogy</u>*		
13. <u>Selection Method:</u> Google Random-No.-Gen.: 3 (***Part 2; Cog-Sci Journal Batch 1***:) Baggio--Compositionality in a Parallel Architecture for Language Processing	Compositionality ("…*the nature and scope of the principle of compositionality (Partee, 1995) from the perspective of psycholinguistics and cognitive neuroscience. …*" —Abstract	• Linguistics and philosophy • Compositionality • Human language • Competence • Human brain • Language processing architecture
14. <u>Selection Method:</u> Google Random-No.-Gen.: 1 (***Part 2; Cog-Sci Journal Batch 1***:) Abramov Et. Al.: The Relation Between Cognitive Abilities and the Distribution of Semantic Features Across Speech and Gesture in 4-year-olds	Abstract: "…*how children at 4 years of age employ speech and iconic gestures to convey meaning in different kinds of spatial event descriptions, and how this relates to their cognitive abilities. …*"	• Non-verbal communication • Individual differences • Meaning-conveyance • Semantic-feature analysis • Use of non-verbal-comm = smarter (in kids)

continues on following page

Continued

Article	Topic	Other Topics & Themes
15. <u>Selection Method</u>: Google Random-No.-Gen.: 8 (***Part 2; Cog-Sci Journal Batch 2***:) Thorne Et. Al.: Concept Appraisal	Abstract: "*…the first empirical investigation of the hypothesis that epistemic appraisals form part of the structure of concepts. …*"	• Epistemic appraisals • Categorization • Thinking • Epistemology • Meta-epistemology • Psychology • Conceptual engineering
16. <u>Selection Method</u>: Google Random-No.-Gen.: 5 (***Part 2; Cog-Sci Journal Batch 2***:) Paape, Avetisyan & Vasishtha: Modeling Misretrieval and Feature Substitution in Agreement Attraction	Misretrieval and Feature Substitution in Agreement Attraction	• Number attraction effects, • K-fold cross-validation • Agreement attraction vs. retrieval-basis • Use of comprehension questions
Arguable Random Common Theme 4: <u>*Meaning*</u>		
17. <u>Selection Method</u>: Google Random-No.-Gen.: 6 (***Part 2; Cog-Sci Journal Batch 3***:) Lisson Et. Al.: A Computational Evaluation of Two Models of Retrieval Processes in Sentence Processing in Aphasia	Retrieval Processes in Sentence Processing in Aphasia	• Dependency completion processes • Empirical models • K-fold cross-validation • Intermittence; slowed syntax & LDA • Models of language-processing
18. <u>Selection Method</u>: Google Random-No.-Gen.: 7 (***Part 2; Cog-Sci Journal Batch 3***:) Thorn & Schurz: How Category Selection Impacts Inference Reliability: Inheritance Inference From an Ecological Perspective	Inheritance Inference	• Inheritance inference, • Reliability of Inheritance inference, • Statistical methods
19. <u>Selection Method</u>: Google Random-No.-Gen.: 5 (***Part 2; Google-Scholar; Logic***:) Stenning & Lambalgen: A little logic goes a long way	Semantic Analysis	• Interpretation and derivation • Importance of interpretative processes • Subjects' experiment-reactions • Semantic theory
20. <u>Selection Method</u>: Google Random-No.-Gen.: 2 (***Part 2; Google-Scholar; Rules***:) Koedinger, Corbett & Perfetti: The Knowledge-Learning-Instruction Framework	Knowledge-Learning-Instruction (KLI) Framework	• Intellectual debate • KLI applications • Coordinated taxonomies
Arguable Random Common Theme 5: <u>*Empirical Models*</u>		
21. <u>Selection Method</u>: Google Random-No.-Gen.: 3 (***Part 2; Google-Scholar; Concepts***:) Martin & Schwartz: Physically Distributed Learning	Physically Distributed Learning / Adapting and Reinterpreting Physical Environments For Better Learning	• Development of fraction concepts • Manipulating physical pieces • Environment-Repurposing
22. <u>Selection Method</u>: Google Random-No.-Gen.: 1 (***Part 2; Google-Scholar; Analogies***:) Gentner & Maravilla: Analogical Reasoning	Analogical Reasoning	• Analogical ability • Human cognition • Processes of analogical reasoning
23. <u>Selection Method</u>: Google Random-No.-Gen.: 4 (***Part 2; Google-Scholar; Images***:) Mitchell Et Al.: Learning to Decode Cognitive States from Brain Images	Use of machine-learning "*classifiers to distinguish cognitive states such as (1) whether the human subject is looking at a picture or a sentence, (2) whether the subject is reading an ambiguous or unambiguous sentence, and (3) whether the word the subject is viewing is a word describing food, people, buildings, etc.*"	• fMRI brain-research • Large amount of fMRI-sourced data • Research methodology • Brain state during single time-interval • ML (via classifiers)
24. <u>Selection Method</u>: Google Random-No.-Gen.: 5 (***Part 2; Google-Scholar; Connections***:) Sidney & Alibali: Making Connections in Math	Use of Prior Knowledge in Math-Learning	• Analogical transfer • Experimentation • Effects of prior-knowledge analogue • Mathematics instruction and curricular sequencing
Arguable Random Common Theme 6: <u>*Cognition and Experimentation*</u>		

APPENDIX E

Mazimhaka
("He Who Resolves Disputes")

Seif Sekalala
English 4817
Professor Connor
01/16/07
First Draft

Solomon Mazimhaka could vividly remember one of the worst days of his life.

He could vividly remember the day when he had realized that his life had been doomed from the very beginning, and there was nothing he could do about it. But how could he explain this to these Ivy League old money patrons?

In fact, almost each and every item in the imposing room, along with the panel of professors just made him more tense. It was one of the opulent conference rooms of the gothic styled main administration building, with a high ceiling, low-hanging chandeliers and meticulous stain glass windows. The walls had a layer of shiny polished mahogany wood in place of paint or wall paper, and floor was covered with white marble tiles. On the floor, between his desk and that of the professors lay a leopard skin. On the wall behind the professors, above the fireplace hung a huge portrait of a stern-looking Governor David Glover, founding benefactor of the university. It was as if the whole room had been arranged to purposefully remind the candidates of their place, albeit subtly.

"Mr. Mazimhaka?" Dr. Lweza's concerned voice startled Solomon. He had been trying to find the right words to answer the question, but words at this point seemed impotent.

He glanced at the cover of his dissertation manuscript. The bold words of the title seemed to be staring, even mocking him: *"Cursed: A Philosophical Reflection on Select Individuals' Perceived Bad Fate."*

The silence was deafening. Patiently, Lweza nudged: "Solomon, it's okay. Try to relax. Just try and remember whatever it was that inspired you to write on this subject."

He still sat motionless, as if in a trance. At first he had panicked, trying to quickly do some damage control. But as he tried more and more to find the words, his mind simply froze. It froze his consciousness and his entire person in the present, and instead vividly revisited another moment in his life, a moment so far away in time.

***

"Solley" (which rhymes with "sorry"), as all the boys called him wished he had never been born.

Why couldn't they understand? He wondered. Why did they all hate him so much? What had he ever done that had been so terrible for him to deserve such punishment?

The whole school, right from the primary sevens (7th graders) to the primary ones (1st graders) that whole week could not stop talking about the homo that brother Taylor had revealed on Monday morning prayer assembly after a tip-off. All the boys who were recounting the story kept cheekily mimicking brother Taylor's deep voice: "It has come to our attention that Solomon Mazimhaka of P.5 prefers to be kissed by fellow boys!" Anytime someone cut ahead of him in line at the mess, or anytime he dared to contest a dorm cleaning duty turn, Solley was quickly reminded that as a homo, he had no business saying anything to any other normal person. The only day the school had ever experienced such excitement was when the only girl in the school--Jill, the nurse's daughter had arrived.

The prestigious St. Francis Boys' Catholic Boarding Preparatory School at Kkobe Hill had an expansive tract of land overlooking the dead town of Kyengera and the swamps of river Mayanja on the outskirts of Kampala. Run by the Catholic Brothers of the order of St. Francis of Uganda, it was one of the select few elite schools that every old money club patron or politician felt compelled to take their heir. Rich in tradition and academic excellence, the school featured a grueling mix of tedious classroom instruction and the typical catholic co-curricular discipline. In the center of the plateau was the assembly ground of dusty gravel, in front of the flagpoles and the primary five-classroom block. A cluster of more red brick simple yet durable bungalows circled; classroom blocks, the staffroom and headmaster/brother provincial's office, general and mess halls and the dorms on the peripheral. Beyond the main complex lay the sports grounds; stadiums, courts and the pool, and even further, stretching as far as the eye could see was lush greenery dotted with mango trees in the school farm, that dreaded hell on earth where even the meanest of bullies were broken down during punishment-chores. Everyone agreed it was three times better for one to be suspended than to be placed in the care of brother Kizito, the farm manager.

But it is almost funny what happens to a place even as big as St. Francis once one gets into trouble. That whole tract of land shrinks and one feels as if the only place to hide is underneath it indeed. This is what Solley was thinking as he sat under the mango tree near the entrance of the farm. Every laugh, every yell, every sound that rhymed in the least bit similar to "homo" startled the 11-year-old primary five pupil. Nearby, a group of primary ones was playing "dool", a game a bit similar to pool, only played on the ground, without sticks and using rocks as balls. Solley had started concentrating on the game, temporarily forgetting his ordeal when something happened that rarely did.

Solley did not have that many friends. He was of below average height for his age and was not muscular, unlike the ones who played sports, had an oval symmetrical face with sleepy eyes and a long nose. This nose, along with his light complexion was also an easy target for the bullies; by now many of Solley's classmates had heard their parents complain about these stupid "Banyarwanda" (hardworking Rwandese refugees—usually long nosed and light skinned in complexion) who try to take all the jobs that truly belong to the proper citizens! At St. Francis, if the boys didn't know who your father was, thus assuming your family was neither rich nor powerful (as everybody else's), *and* you had a long nose and light complexion, just like the Banyarwanda, what did you expect?

This is why Solley was surprised that someone was giving him an "eye grab", especially at such a time as this (after he was known to be a homo)! An "eye grab" was a gesture used between best friends, and among loyal "camp" (clique) members. At a very unexpected moment, someone would swiftly sneak up from right behind you and cover your eyes, only letting go after you guessed who it was. Sometimes, one would feign ignorance by repeatedly guessing wrong on purpose just to prolong the pleasure of the game. He was in no mood for games, so he simply declared:

"Okay, I don't know who this is; just let go…"

"Man, you take the fun out of it!" replied a high-pitched hoarse voice. It was the voice of Elvis Baguma, one of Solley's seasonal friends. Actually, shortly before the whole homo business had begun, Solley had started hoping that Elvis would become his first real friend. In fact, the homo thing had started after someone saw Elvis jokingly trying to kiss him in the bathroom. Many gay boys had tried to kiss

Solley before because of his small and feminine-like stature; even some of the popular ones who were now mercilessly bullying him. Solley had never told anyone about the attempts. That day, he told Elvis about it, and Elvis couldn't stop laughing at the whole farce. He cordially teased him about it nonstop, even during shower-hour. This is when he had mockingly acted out what he thought the boys had done. Someone had seen them and tipped off brother Taylor, who had been on duty that week. But why was everyone making a big deal out of something many of them did? In fact, up until that very moment, Solley had not even realized that whoever had told the brothers about "the kiss" had for some reason left out Elvis.

"Hey Solley! Solley! Wake up, man!" Elvis' hoarse voice jolted Solley from his disturbed thoughts.

"Come on, Solley!" Elvis went on, "are you gonna let this ruin the rest of your life?" Solley shifted his attention from the game of "*dool*" he had been absentmindedly watching and took a close look at Elvis. Elvis had bushy eyebrows and a pair of small glaring eyes to go with them, a combination that made him look a little tough. That, and a very thin but visible early mustache made him score some points with the popular boys. Plus, Elvis was Solley's height but had some muscles.

"I don't know Vee…I don't think they're gonna forget this one…"

"Oh come on, Solley. Even Jill has a bunch of friends now," reassured Elvis. He put his arm around Solley's shoulders, then tried to cheer him up.

"Did you hear what happened to Jake?" Jake Twinomugisha was the section A class monitor, a position always given to one of the most popular. Grades were divided into A and B, each with about 25-40 pupils, and P.5A was Elvis' section.

"What happened?"

"Well, matron Liz found his sheets and mattress soaked and made him take them out. You know he's a seasonal bed-wetter--"

"Whoa! Him?"

"Yep…we were all shocked." The two sat silently for a couple of minutes. Solley got back to (absentmindedly) watching the primary ones' dool match, while Elvis tried to think of a way of cheering him up. Suddenly, he remembered.

"Are you sure there's nothing I can do to cheer you up?" Elvis mischievously asked Solley, in a tone of voice that told Solley exactly what he meant.

Normally, Solley would have been infuriated at the slightest hint of someone asking him to do something that would only add to his problems. He was not a goody two shoes of course and had no problem with the occasional sneak into the farm to steal mangoes or even taking formulae into a test. However, Solley was fast realizing that what had happened to him recently was no ordinary bout of trouble. Just as Joachim Pulkol or "Stinky Joe" had forever been damned thanks to that one incident in the first grade when he had peed all over himself, Solley was going to have to just deal with being "the fifth-grade homo". Since his good name could never be salvaged as far as the brothers were concerned, Solley figured he might as well use his tarnished reputation to his advantage and have some fun. Who knows, Solley thought, maybe he could even become one of the official "rebels", the cool "bad boys" of the school. After all, now he could never become one of the cool "goodies"—the prefects, neither could he join one of the many neutral "cool cliques"—like those of diplomats and central government ministers' kids, old money, the nouveau riche kids etc.

"Sure," Solley finally replied. "Meet at me at the usual spot after evening prayers."

At the agreed time and place of the rendezvous, Solley awaited his new friend. In fact, Elvis was, at least as of now his only friend. He really hoped this would become permanent; even the weirdest boys had at least one permanent friend, someone they could talk to every once in a while. Up to now, this had evaded Solley. He did not want a "best friend" or anything nearly as fancy. All he wanted was at least one person he could call a true friend. The more he thought about it, Solley realized that in the entire school, only a handful of kids, him included could fit the definition of "loner". Wow. If finally, after four years of primary school Solley was going to have a friend, then this whole homo business had not occurred in vain. A loud thump near one of the windows in the dark and dingy locker room jolted Solley out of his thoughts. It was Elvis, making his entrance through the window. Solley stumbled back and made a loud noise with his shoulders hitting the locker he had been leaning on, only to be shushed by Elvis.

"Where have you been man?" whispered Solley.

"I had to be careful on my way here", Elvis whispered back. "You know its brother Bwino on duty this week!" Brother Bwino was good at patrolling the school after lights out, they all knew better than to have "convos" or ghost story sessions as long as he was on duty. Yet here they were, about to find and shred the chore defaulters list from one of the prefects' lockers. It was the ultimate thrill; doing something so sweet, so taboo and all the while knowing that getting caught meant paying dearly indeed.

The two then continued with endless niceties and small talk and fidgeting, putting off the main course for as long as they could. Eventually, the small talk ran out, and the uneasy silence followed. This was always the most difficult part. Solley could feel the nausea, butterflies, and the sound of his own heavy breathing. Elvis told Solley to start picking the locker padlock as he staked out the balcony to make sure Bwino wasn't nearby.

The events were simultaneous. A flashlight beam focused directly onto his face, and a bunch of excited blubber broke out; around 5 to 7 voices. Solley singled out one of the voices as that of Jake. He confirmed that a half minute or so later, when Jake switched on the light and Solley stared at the Elvis' five classmates, along with two prefects, both primary sevens. He immediately realized what had just happened, and the answer to the question he had been battling with earlier was clearer than ever. The sly look on Elvis' face was more than enough confirmation. It had been a set up from the start. Elvis, the "true friend" he had been hoping for had set him up just so he could join one of the cliques, or perhaps become a prefect. Elvis had started the whole homo scandal.

***

In reality, it had only been a minute or less since Dr. Lweza had asked the question, but the tense silence made it seem like an eternity. The defense had been running smoothly for over thirty minutes. In fact, the candidate had been answering the panel's questions very brilliantly, and this is why they could not figure out why this particular question had made him react the way he did.

The panel had three professors. Dr. Lweza, the doctoral candidate (Solomon)'s main professor, Dr. Edward Taylor, the head of the philosophy department, and Dr. William Charneski, a senior professor. Solomon had first had Dr. Charneski as his main professor before joining Lweza. He had heard the rumors about him being unnecessarily stringent and unhelpful but had ignored them. He could handle it, he thought. Besides, Solomon didn't want to join Dr. Lweza because he thought Lweza would be too soft on him, what with the very close relationship they had. He was wrong on all counts. A very ambitious Charneski always had a gazillion meetings and seminars to attend and even more journal articles to write, yet he always ridiculed ideas. He would never edited or suggest ways to improve them; he simply shot them down. On the other hand, as soon as he joined him, Dr. Lweza provided Solomon with endless helpful lists. Of colleagues' contacts--experts on the subject--at other universities, publications and current research connected to the subject. Nevertheless, he always demanded the most stringent adherence to fact checking and close examination of these sources' research, among other high standards.

Lweza had also panicked at first after Solomon froze, before he realized what was going on.

It was not the first time he had seen that look on his face, and it would not be the last.

***

Glover Hall University in Virginia was picturesque. It had two small lakes, five ponds, and a river passing through it. Nearby, the forest reserve, one of the biggest in the United States provided a hunter's paradise with endless game to choose from, with dear and duck the easiest targets. Oaks were everywhere; along the redbrick footpaths, on the neat thick green grass lawns, even on the medians of the driveways.

It looked like a big cluster of antebellum plantations. Many buildings had white and red brick façades with roof-high Corinthian columns on the verandahs. The new classroom and office blocks, entertainment and other multi-purpose halls had a simple, yet clean cutting-edge look; see through glass and stainless steel. The inside of these buildings was a whole different story. "Functional Coziness" seemed to be the theme of the decor. Bright color paint, sofas, even lava lamps in the lounges and hallways! This interior design, which was a bit eccentric for a university, might have raised a few visitors' eyebrows every once in a while. But the students liked it, and the faculty did not complain.

Doctor Martin Ssekitto Lweza, a new professor in the Philosophy department was in one of these buildings, in the cafeteria of University Student Center. He had actually been admiring it that day before his mind wandered. He was sipping an espresso in the corner near the pool tables. Every once in a while, giggling sophomores would pass by and wave, and the young men would offer a pretentiously humble "Hey Doc!"

When would they ever have such buildings in universities in his native Uganda or anywhere else in Africa, he wondered; probably never. He had been asking himself questions such as these for most of his adult life, especially after he had moved to the United States all those years back. In fact, he realized, it had now been thirty years to be exact! Wow, he thought. When one's life makes a trip to hell and back, maybe time does fly by. One probably becomes oblivious to such concepts as menial as "time".

His grandfather, Sir Harold "Harry" Lweza was one of the Buganda kingdom chiefs that had signed the Buganda Agreement of 1900 with the Imperial British East African Company. They had been trying to ensure the continuity of the kingship while getting themselves a perk in the process, if millions of square miles of fertile lake-side land, half the entire kingdom can be called that (a "perk"). Yes, they had been a bit greedy, but they had saved the kingship. Henceforth, the British protectorate, later the independent republic of Uganda was born.

Martin had had an enchanting childhood. He attended St. Francis during the school term. In the holidays, especially the longest after the third term (shortly before and after Christmas), him and his siblings, as did many other chiefs' kids would take turns having sleep overs at the palace. The king loved children. When he was not attending to matters of the state, canoe-fishing, hunting and flying lesson trips occupied him and his children and their guests. Martin's maternal grandfather had been British, one of the few colonialists that had married the natives. These long holidays provided the best opportunity for the family to visit their connections in England, with whom they kept in close contact. Martin loved everything about London. The good British manners, the Marks and Spencer stores, the parks and sidewalks of West End by the river Thames.

Every time they visited, he wished they would not return to Uganda. Only two thoughts comforted him. That he would return the following year, and that the journey back would be fun. In those days, Uganda could not service big planes. Only medium sized seaplanes could land on lake Victoria at Entebbe. These would then go to Nairobi, from where they could fly directly to London Gatwick or Heathrow. The return journey was the same.

After finishing secondary school, Martin's father gave him three choices. He could go on to either the new London University College at Makerere in Uganda (the only university in the country at the time) or any other university in England and continue his studies, or he could get married immediately and go into managing the family's vast estates.

He chose neither. One of the princes, his best friend from their St. Francis days was planning an excursion into America. One beautiful April day in 1962, as their parents were negotiating for the country's independence in London, they set off for America. Martin could remember how he had felt as soon as they got off the plane in New York. He loved America, much more than he loved England. The people; black, white, Asian and Hispanic, all a bit suspicious of each other and yet somehow united, the expansive territory (50 states!), but most of all, the freedom. This country gave him a feeling of pure, unadulterated freedom.

Prince Simon Walugembe and Martin started a tours and travel agency in New York, targeting the then wave of "enlightened" blacks who wished to visit "the mother land".

After their success in America, Martin returned to Uganda to expand and co-ordinate their business there; Simon would stay and manage affairs in the states. Martin set up not only a branch of the Tours and Travel, but also the country's first five-star hotel in Jinja, near the Owen falls dam on the river Nile. His father had started warning him of tough times ahead when he was still in the states, even suggesting he should perhaps suspend their expansion plans and stay in the states for safety. Prime minister Milton Obote seemed intent on a stripping the kingdom of Buganda of its superior status among the other kingdoms of the country, and in fact, the ceremonial president, King Sir Edward Frederick "Freddie" William David Walugembe Mutebi Luwangula Mutesa II was mulling a possible "expulsion" of the central government administrative offices from his land.

But even as the political feuds grew worse, Martin, like other ordinary Ugandans stood his ground. He refused to go into exile. Obote started nationalizing all the investments in the country he could lay his hands on. By now, only a couple of plots of land remained of the thousands of hectares their family once owned. The hotel and travel agency was surviving, but Martin did not care; property could be replaced but human life couldn't. In 1969, king Freddie was assassinated, as were scores of his men with whom he had escaped to London. Among these was Harold Lweza II, Martin's father.

Idi Amin took over and the killing sprees continued with a vengeance. Shortly after, Martin's mum and two sisters disappeared. In those days, that only meant one thing. In fact, Ugandans knew better than to ask about anyone who disappeared, lest the same fate befell them. Men wearing sunglasses and driving Peugeots would just show up and stuff you into the car trunk, never to be heard of or from ever again. But the funny thing about this era too was that men's fortunes could change in the blink of an eye; a big "gift" for a tip to the army about rebels' hideouts or a bonus from the president on any day he felt happy. Martin was about to find out for himself.

General Amin used to be weary of all the usual spots frequented by the Kampala establishment. He knew that many of these patrons were Obote's cronies and didn't like him much, so he got into the habit of hoping around different clubs, lounges or restaurants.

One day, while on his way back to Kampala from commissioning a new air force academy in the Jinja town center, Amin stopped by Simon and Martin's hotel, the Jinja Owen Falls Hideout Inn . Martin had been helping out since the manager had disappeared the day before. As he was asking clients in the restaurant how their meals had been, he saw the entourage make their way into the lobby, where the president stopped to stare at the Picasso above the fireplace.

Martin had seen the man countless times on T.V but was nonetheless taken aback by the gigantic stature of the infamous leader. Idi Amin in the flesh! More than six feet tall with a very dark complexion, a round dimpled face, and a port belly. Feigning calmness, Martin approached the general.

"Mr. president, it's an honor". Amin's hand swallowed his, and the grip almost made him scream. Indeed not one to mince words, the general quipped:

"Who's the owner of this place? I've never heard of it. It's beautiful." His voice was neither deep nor soft, but had a consistent, smooth, authoritative and confident tone. Martin stammered:

"It's--it's mine sir…" Amin studied Martin with a steady gaze. Martin swallowed. Amin was a very paranoid man, and anything could happen now. Most likely, the general would "investigate and find out" the source of Martin's wealth (read accuse him of rebel collaboration), and that would be his end.

"Well do you have a name?" Timidly, Martin whispered:

"Martin--Lweza--"

"No! *The* Chief Lweza is your father? *Mungu yangu* [my God]!" General Amin broke into a hearty laugh, then grabbed Martin in an embrace. He revealed to Martin that he had actually met the chief once, and that he was sure his son was just as nice as him. Apparently, Martin could tell that the general had had no bad feelings towards the old man. He could also tell that indeed, as rumors had suggested, Amin was not half as bad as he seemed; many of his State Research Bureau (the internal intelligence organ) operatives (men driving Peugeots in sunglasses) were rogue and committed the atrocities on their own, usually after trying to extort money or while fulfilling personal vendettas. This must have been the way Martin's mum and two sisters had "disappeared".

But one of his many weak links was that of women. He was a very possessive man, and he assumed that any woman he liked him too, and thus was not to be touched by any other man. Unfortunate for him, Martin did not know this. Joyce, the Rwandese in-house masseuse had had a crash on him for long, but at the time, relationships were the farthest thing on his mind. But just as his bad luck would have it, the day he decided to take her up on her offer was the day one of the men in sunglasses was in the hotel. Apparently, she had caught the general's eye, and he had told "the men" to watch over her.

As Martin finished taking a shower after doing it, he heard her scream. Normally, of course, Martin's reflexes would have made him go into the room to try and help her but somehow, he knew that this was no ordinary stick up. With only a towel wrapped around his waist, Martin jumped out of the small bathroom window into the shamba behind the inn. One of the locals, a radical Buganda loyalist gave him some clothes and money and sneaked him across the border to Kenya in his car trunk.

In Nairobi, the capital, Martin managed to locate some exiles who helped him for a few weeks. He had already gotten in touch with Simon, and of course they knew what his only option was. Nairobi was not safe, neither was London. While Amin was a threat, the remnants of Obote's men were Martin and Simon's chief concern, and there were groups of them all over Africa and England. Assassinations among the exiles there were an everyday occurrence. These guys were determined to wipe out the entire Buganda royal clan. Besides, even if London had been safe, living with his British grandpa's connections there would have been too painful, a reminder of the happier times that once were. All Martin wanted to do was to get as far away as he could from all this—lost peace, lost African royalty, lost property and prestige, lost family…

The only place they could never suspect—it was too big and not yet uncharted for them anyway was America. Of the entire royal family, only Simon had survived the massacres and only a handful of people, including Solomon knew where he was. There was one thing Obote and Amin had not taken and could never take from Martin, and that was hope. For now at least, the embodiment of this hope was Simon. For as long as there was at least one potential heir to a restored Buganda throne, the mother land would regain sanity.

Martin returned to America a battered man. At JFK airport, Simon took one look at him and broke into an endless gush of tears. Martin cried too.

Simon had hoped that Martin would rejoin him in the business. He was surprised when instead, Martin embarked on a career in academia. He went to Columbia, then Yale and got his PhD in Philosophy and Economics. He could have become "famous" if he had wanted; journalists, authors, even presidents clamored for the astute scholar's opinion. Over the years, Simon's business ventures had also morphed into one of the top 10 fortune 500s, but he had listed it onto the NYSE. Both men had the ability to harness endless power in this new homeland of theirs, but they had seen what power can turn men into. They were self-effacing humble men; best friends who teased, joked, and counseled each other. They tried to avoid talking about the past. Each was afraid to ask or say anything to the other about it.

Surprisingly, one of the people from those days that Martin always thought about was Joyce the masseuse, a woman who should not have had that big of an impact on him. Perhaps it was because she had been one of the last people he had seen before he left the country…

Of course, both Simon and Martin knew that they would have to return to Uganda one day. Simon, perhaps to take back the throne and stay for good. And Martin, if only to get the closure he needed and only then finally return to the states for good. Democracy had been restored and the current president, Joel Saleh was in touch with Simon. He had requested him to return—

"Sir, I didn't understand the chapter; could you go over it for me some more?" The timid voice of one of his students in the upper-level Philosophy of Law course startled him. He quickly recomposed himself and cleared his throat:

"Why of course Solomon, have a seat." The first time he had seen Solomon, Martin had been taken aback by the body and facial resemblance they shared; around five foot eleven and unmuscular with a light complexion and an oval-symmetrical face, a long nose and a thin moustache and goatee. It was just like looking at himself in the mirror sans wrinkles and some gray hair. Martin always joked with Simon about the way black Americans looked just like Africans, as opposed to the fantasies they had had about them growing up in Uganda. "So, my friend, what exactly didn't you understand in the chapter," asked Martin, a bit sarcastically.

"Well…you mentioned that one can interpret Patrick Devlin's views as basically based on the fact that it is all about the principle?"

Solomon was one of the most active and insightful students Martin had ever come across in a class-room. Every once in a while he would start getting frustrated, thinking he had just wasted an entire hour of explaining a topic, only to be bombarded by a brilliant question from Solomon in the back of the lecture room, where he always sat. This would then trigger a collective "aha!", and a hot debate by the entire class, thus helping them to grasp the concepts. "But like I said Mr. Mazimhaka, his views were in support of legal moralism. With that doctrine, society's collective moral judgements rule, period."

Martin was very popular with students at Glover. The few interviews and television appearances he granted were always over-publicized to his chagrin, as this only intensified the unnecessary spotlight. He had been teaching at Duke for three years before being wooed by Glover, and shortly before his arrival, the campus had been buzzing with the news. Students always acted embarrassingly around him. This is why Martin was particularly fond of Solomon. He always seemed unfazed by Martin's "celebrity".

Solomon had kept quiet after Martin's response. He looked like he was trying to find the right words to phrase a rebuttal or another clever question. Meanwhile, Martin was smiling at him in a jokingly sly manner as if to say: "Gotcha!" Finally, in a somber tone, he spoke up.

"Dr. Lweza, I have to apologize. I didn't come to talk about philosophy today."

Martin was taken aback. "Hunh? Is everything alright?"

"Oh no, everything's fine," reassured Solomon. "I uh, I just wanted to share something with you."

"What is it?"

"Well, I read that you're from Uganda. Sorry about what happened by the way."

Oh no, Martin thought. Not another request for an article or blog interview! But he wasn't going to brush him off immediately, not least because he was one of his favorite students.

It was then that Martin first saw the look on Solomon's face. A look Simon had told him he always saw on *his* face, the kind of look that appears when people who have had a lot taken from them remember that which once was, or could have been.

Martin could tell that a revelation—a big revelation was about to made. "Mr. Mazimhaka?"

A startled Solomon quickly recomposed himself. "You see Dr. Lweza, I'm--I'm from Uganda too."

He had been expecting a bombshell, but not one nearly as big as this. "Wow. But then how come you--the accent!"

"I know. Everybody says that. The biography I read about you mentioned you attended St. Francis back there--is it true?" Martin nodded. "So did I."

"*Mungu yangu!*" exclaimed Martin, before whistling in awe. They both laughed heartily, after realizing how quickly a mother tongue returns after one is in a motherland state of mind. "So that's why you don't have the accent. Tell me, was brother Taylor still there when you attended?" Solomon nodded. "That man is built of steel!" They laughed again.

"It's actually funny. I was telling my friend Lugz the other day--"

"Wait a minute," interrupted an excited Solomon. "*The* Lugz?" By "Lugz", Solomon could tell that Martin meant the surviving prince of Buganda, Martin's best friend Simon Walugembe. An AP reporter had once heard Martin refer to him that way at a banquet, and from then on the western press had cheekily referred to him as Prince "Lugz", just as they had dubbed his father "Freddie".

Martin nodded, smiling shyly, but proudly. "Yep, that's him." He was starting to like the young man more and more by the minute. "So anyways, I was telling him about you. You know how some of these black Americans sometimes have these 'pan-African' names? Well, I thought you were one of them!" This prompted another bout of laughter from the new buddies.

"It's actually funny you should mention that." The somber tone quickly returned.

"Why?" Martin asked as he took a sip of the espresso. Solomon's face had gone back to the look of deep thought and puzzlement, and Martin was looking at him studiously in anticipation.

"Well, you know how the name-clan system is in Buganda?"

"Of course. Is your father Rwandese?" asked Martin. There are over 23 tribes in Uganda, but the Baganda were historically the strongest, most assimilated, and most sophisticated. The remaining twenty-two tribes had weaker kingdoms or were decentralized and only had chiefdoms. Each Bugandan tribe sir name is assigned a unique clan, and all kids have to be named according to the father's clan. Solomon's sir name, "Mazimhaka", which means "I've resolved all disagreements," was in Kinyarwanda, thus did not have an assigned clan, at least not a *Bugandan* one. Had it been uniquely Bugandan and thus *U*gandan for that matter, it would have been "Maze'mpaka." But in Buganda, there is no such name. Bugandan sir names were only composed up to around the 17th century. The vocabulary that was used to compose them had been unique to only before and up to that era. The language has since evolved, just as English evolved from Shakespeare's era to today.

"Exactly my point. Nope, pure Muganda, born and bred. I mean mum is," he chortled sarcastically, "but what's that got to do with it, right?"

"Precisely," agreed Martin, "if the child has to be named according to the father's clan, it's a no-brainer what tribe the name has to come from."

They were talking a bit bitterly, regretting their African traditions' established and archaic sexist system. Sons were prized and daughters caged, scolded, ridiculed and seen as possessions. In fact in Buganda, the women had it better than in many tribes all across Africa. Over the centuries, the kings and elders had started reforming many of the archaic customs. It was this willingness to embrace change and basic universal values that had convinced the British to grant Buganda semi-autonomy within the new country, have it as the trustee kingdom, and indeed name the country *U*ganda.

"Well, she probably just gave me the name to spite dad. They broke up after she became pregnant with me," concluded Solomon. "Ever heard of Gordon Mukasa?"

"Gordon's your dad?" Gordon Mukasa was one of the *Johnny-Come-Lately*s of the Ugandan city tycoons, the ones who always made sure their names appeared in the gossip columns.

"Unfortunately…" Solomon replied meekly.

"That bad huh," Martin teased. He could tell that this was not one of Solomon's best topics, so he steered away.

Their conversation then meandered through different subjects, mostly connected to catching up on the old country. The two were enjoying each other's company immensely. Solomon had had his eye on the professor for a long time, and indeed, good things come to those who wait. He was glad to be making such a deep connection with the professor, and he had a feeling that for his remaining tenure at Glover and for the rest of his life, Martin was destined to surpass the status of mentor, close friend and confidant.

For his part, Martin was also getting sucked in fast, like a vortex. He was intrigued by this young man who spoke and acted twenty years older. He could tell that just like himself and Simon, Solomon had had a tumultuous life in Uganda. He could tell from every wild gesture, facial expression and pain-ful puzzled look in the eyes—especially the painful puzzled look in the eyes, that just like himself and Simon, he believed that his fate had already been decided by forces beyond his control.

But there was something much deeper about Solomon that Martin could not quite place. Something much more than his charm, resourcefulness and poetic melancholia. Something that only *he* could put a finger on, but would evade him for a very long time. He had had this feeling before, the kind of feeling one has before getting an epiphany.

***

Between Lake Victoria and the middle of the Entebbe—Kampala highway lies a range of hills at Bwebajja, in Wakiso district. Their slopes have rich fertile soils; guava, mangoes and tomatoes grow unabated. In their valleys, rivers meander under the darkness of the thick palm tree leaf-shades. Snakes, beetles and snails abound in the swamps, and hundreds of bird-species feast on them there while monkeys tree-hop on the palms and fruit-trees. The highest of these hills is a peninsula that sits well into the lake, atop which one is kissed by a therapeutic breeze all year round. Here, on the western slopes of this hill sits the complex of buildings that house The Eleanor Academy of Uganda.

Solomon sat at a window in the dining room, absent-mindedly watching a group of monkeys playing in the valley. All that was left of the St. Francis years were vivid painful memories. He was now in his last year of high school, and it was crystal clear that all he had ever wanted was a father. All the identity and confidence issues he had had at St. Francis; not fitting into cliques and not knowing how to play soccer like all the other boys, having to explain to them why only his mother visited on the open days and why he had a Rwandese name, all these would not have existed if he had had a father.

As per the Ugandan education system, Solomon had completed his seventh-grade primary exams, and thus had had to move on to a secondary school of his choice. His mum assured him that he could go to any school he wanted. At the time, he had no idea that this generosity was not entirely altruistic. Dorothy Kanaani could see that her son was growing. He was only twelve years old at the time, but the teen-age years and all the deep mature questions that come with them had arrived. She knew that giving him a blank check for secondary schools, as opposed to the ones she would have preferred only meant one thing. Of course, he would choose a co-ed secular, *nouveau riche* one; the type preferred by western diplomats where they only have dress codes as opposed to uniforms, where they let them go home on weekends and feed them burgers and pizza, and let them kiss in the open. By letting him attend such a school, Dorothy would buy herself some precious years. He would be distracted by all the fun, and would thus have no chance to ask about the details of the failed relationship between her and his father. Those skeletons were better off left the way they were.

She was wrong. At first, the trick had worked. Eleanor was a far cry from the catholic primary school choke-hold of St. Francis, and Solomon loved every bit of it. Here, being different was praised and not frowned upon. He made lots of friends from different parts of the world, and the girls loved his shy demeanor. Sport participation was not enforced, but Solomon now enjoyed swimming and hiking in the hills. He was beginning to break from his shell. But every once in a while, Solomon was still reminded of his bastard status, sometimes subtly and other times boldly. By friends' questions about his family, by watching them go home with their fathers on the weekends, or by hearing them tell stories about their family escapades on exotic trips. In such moments, the St. Francis blues returned, and with them the melancholia he could never share with anyone. He would then isolate himself and reflect for hours. In fact, the particular dining room window at which he was sitting was one of his most favorite meditating spots at Eleanor.

"Solomon, what are you doing here, boy? I've been looking all over for you" The Texan drawl of Mr. Grier, the principal, startled Solomon.

Oh man! Now what? It was a well-known fact at Eleanor that for Principal Grier to personally look for you on campus, you must have *really* pissed him off, and if there is one thing you never want to do in your life, *never* piss off Mr. Allan Woolworth Grier! At six foot four, and weighing 300 pounds, Grier had graying black hair and a round chubby face. He wore thick-lenses and had a flat nose, and his belly poured over his belt.

At first, Solomon had expected him to start ranting about a missed homework deadline or class-room cleaning duty. But almost immediately, he could tell that something was terribly wrong. How else could he explain the kind and sad look on Grier's face? Everybody knew that the day Grier smiles, looks kind or sad will be the day Christ will return.

Nervous, Solomon asked: "Is everything alright, Mr. Grier?"

"No, I'm afraid son," replied the Texan. "Your mother's in the hospital. Come on, let me give you a ride."

In the intensive care unit of the International Hospital Kampala at Namuwongo, a teary-eyed Solomon looked at his dying mother. She looked like a hideous cyborg with all the wires on her arms and chest. These relayed up-to-the-second data to the annoying beeping monitors. In her left arm, an intravenous tube delivered medicine and glucose-water. Thankfully, the oxygen-ventilator, which had been making other annoying sounds—a whooshing and an alert beep from time to time had been switched off, and the tube had been removed from the nostrils of her beautiful long Rwandese nose. Her thick long hair, badly in need of a perm was held in a pony tail, and her skin was pale.

She opened her eyes and smiled at Solomon, and this made him sob even more.

"Hey," she whispered, "I'm not dead yet."

Vintage mum, thought Solomon. Even on her death bed, somehow, she tried to make light of the moment! "Oh mum…"

"Come on, son…you know you have to be strong for the both of us. Why didn't you call me over the weekend?"

Now how could he explain to this perfect woman how he hadn't called her, because he had been busy thinking about how terrible his life is, as if she hadn't done all she could to raise him comfortably.

Again, just like her classic self, she read his mind. "You've been thinking about him, haven't you?"

Solomon nodded, fresh tears streaming down his cheeks.

"You look just like him…"

He then asked her the question he had waited to ask his whole life, a question he knew she couldn't answer, but which he wanted to ask anyway. "Why did he have to abandon us mum? Why?"

"Oh Solomon…you look just like him…"

Solomon could tell she wasn't talking about Gordon. In all his life, he had never heard her talk fondly of him. In fact, Solomon had never met Gordon Mukasa in person. He just wired the money whenever she called him, and Solomon only saw him on T.V every once in a while, at a state dinner, a trade-show or a product launch.

Once, when he was twelve, Dorothy had explained to Solomon that while in exile, in London during the revolution, she and Gordon had had an affair, but she had refused to have an abortion. But if anything, that explanation had only helped create more questions than answers in his mind.

"Solomon, there's something you need to know." Her whispering voice startled him. "Gordon--Gordon is not your father."

By now, Solomon had stopped crying. Instead, shocked, he stared at his mother.

"When I arrived in the U.K, I was one month pregnant. Gordon and I had met a couple of times before at the hotel… I only told him he was your father so he could take care of--care of--" she coughed, then panted briefly. "So he could take care of us…"

"Mum, who is he?"

"Poor man--" she swallowed, then panted again, "Amin was going to kill him…" She paused and stared at him, then tenderly felt his right cheek. Her hand was cold. A tear trickled down; she wiped it off, then started crying herself.

"His name is--"

Suddenly, Dorothy started wheezing, then gasping loudly. Wide eyed, she clutched her neck with both her hands and sat up, and the annoying monitors went off frantically.

A doctor and two nurses rushed in:

"Please clear the room sir," barked one of them to a frozen Solomon. The third cardiac arrest was about to finish off Dorothy J. Kanaani.

The funeral was simple.

Dorothy's mum, a peasant, the late grandma Lydia had escaped the first Rwandese genocide with the then six months old Dorothy. Even then, Dorothy had climbed the Kampala social ladder fast, thanks to her good looks, street-smarts, and country-girl wisdom and humility. However, the U.K exile years had taken their toll on her, and the only socialites in attendance that day were the ones she had always considered her true friends. Indeed, they turned out to be genuine.

Solomon was numb throughout the whole affair. Dorothy's friends were full of kindness and encouragement, but to Solomon, all this was just a really bad nightmare. Eventually, he would wake up from it. His mum was not dead. *What are you doing, you fools? Stop! Get her out of there! She can't breathe! Mum is just taking a long afternoon nap.*

As soon as the men started lowering the mahogany gold-plated casket into the grave, Solomon let out a shrill, then passed out.

Two weeks after the funeral, Jonathan Muhwezi, Dorothy's attorney revealed to Solomon that he was the sole beneficiary of her entire estate; the ranch at Kabojja hill, 2.5 million dollars in mixed portfolio investments, jewelry, and the cash in her accounts. Also, since he had just turned 18, the full age of consent under Ugandan law, he could cash in or invest from the trust fund that she had established for him as a child.

But as soon as Jonathan finished reading the will, Solomon declared:

"Liquidate everything. All of it."

He just wanted to go away, as far away as he could. It did not matter where. He just wanted to get away from all the pain and the false hope. Now that his mum was gone, this place, this so-called *Uganda* was a cold, alien land. His father, who had abandoned him was a citizen here, but his beloved mum had been rejected to the end. To them, she was just another Rwandese refugee.

And him; citizen of the U.K by birth, citizen of nowhere in reality.

Absent-mindedly, Solomon whirled the spinning globe on Jonathan's desk.

His finger landed on the United States of America.

***

He was sitting at the very table where he had had the first conversation with Martin, in the cafeteria of the University Student Center. The dissertation defense had ended successfully. Martin had tactfully steered him on from the tense moment, and surprisingly, Taylor and Charneski had been gracious and supportive in the end anyway.

Now what? He had escaped to America to start afresh, and for the first time in six years, he felt he was ready. A Philosophy professor indeed! His mum would have been proud of him—

"*Doctor* Mazimhaka! May I join you?" teased Martin.

"Oh shut up, you old scoundrel," retorted Solomon.

Martin sat with his protégé while sipping from his espresso cup. "So…how does it feel?"

"Normal. I wish mum were here to see it. Oh, she'd be grinning from ear to ear, I tell you."

Martin stared long at Solomon, trying to gather his words.

Solomon could tell that something was bothering him. "Are you okay?" he asked.

Martin was startled. "Um…" he cleared his throat, "Solomon, what was your mum's middle name?"

Puzzled, Solomon answered: "Joyce. Why?"

"Did she ever work as a masseuse by any chance?"

Even more puzzled: "Yeah, before the exile years. She told me she used to work at the Jinja Owen Falls--"

"Hideout Inn."

Suddenly, Solomon understood. "Oh—my—God," he whispered.

It was an A-list affair. The who's who of politics, including both the American and Ugandan presidents, Hollywood stars and international royalty were all well-represented. Simon had also invited the press, but nobody knew why the shy exiled crown prince of Buganda was throwing the banquet. Of course he had already told Martin and Solomon, but they were tight-lipped.

In the corner of the ball-room, at the Marriot Hotel in Washington DC, Martin and Solomon continued their conversation. They had been talking since morning, but neither of them wanted to stop.

"Wow," said Solomon, "I still can't believe it. And yet…it explains so many things. You know just before she died, she talked about you for the first and last time…said Amin was going to kill you."

"Now you also know why she gave you a Rwandese name," added Martin. "It was the only neutral option she had. She didn't want to give you a clan-name of Gordon's…and of course she couldn't give you one of mine's either."

Simon took to the podium, and the whole room fell silent. The years had taken their toll on the prince, but the ladies still fell for him like flies. He looked just like his father, the late king "Freddie". Six feet in height, slender, with an oval face. He had a long nose, and he always combed his hair backwards; it had started graying. He spoke in a low baritone.

"Well, you all know I'm not good at making speeches, so I'll keep it brief. I'd like to thank everybody for coming; a special salute to my best friend—y'all know him; Dr. Martin Lweza and his son, the new *Doctor* Solomon Mazimhaka--"

A shocked Solomon looked at Martin, who was grinning "Did you put him up to this? I'll get you back--"

The whole room gave a round of applause, and Solomon was about to die from shyness. Simon went on to announce that he would return to Uganda to ceremoniously take back the throne, but pending parliament's approval, he would be permanently based in Washington, a compromise Martin had brokered between him and the Ugandan and U.S governments.

Solomon was elated. Every once in a while, a song he had heard at some point in his life would play in his head. That evening, the chorus of Phil Collin's *In The Air Tonight* was playing:

I can feel it coming in the air tonight, oh Lord
And I've been waiting for this moment for all my life, oh Lord
Can you feel it coming in the air tonight, oh Lord, oh Lord

He had incurred a lot of wounds growing up; all the bullying and not fitting in, not having a father figure, and the constant sense of anxiety, because of being raised by "a Rwandese refugee".

But that evening, for the first time in his life, he had a father. He was fitting in just fine with royalty, and he felt confident. He was ready to take on the world. He looked at Martin, and he had the same look in his eyes that Dorothy had often had. He knew what Solomon was thinking, and he agreed.

Related References

To continue our tradition of advancing information science and technology research, we have compiled a list of recommended IGI Global readings. These references will provide additional information and guidance to further enrich your knowledge and assist you with your own research and future publications.

Abir, J. I., & Shamim, T. F. (2020). What Compels Journalists to Take a Step Back?: Contextualizing the Media Laws and Policies of Bangladesh. In S. Jamil (Ed.), *Handbook of Research on Combating Threats to Media Freedom and Journalist Safety* (pp. 38–53). IGI Global. https://doi.org/10.4018/978-1-7998-1298-2.ch003

Adesina, K., Ganiu, O., & R., O. S. (2018). Television as Vehicle for Community Development: A Study of Lotunlotun Programme on (B.C.O.S.) Television, Nigeria. In A. Salawu, & T. Owolabi (Eds.), *Exploring Journalism Practice and Perception in Developing Countries* (pp. 60-84). Hershey, PA: IGI Global. https://doi.org/ doi:10.4018/978-1-5225-3376-4.ch004

Aggarwal, K., Singh, S. K., Chopra, M., & Kumar, S. (2022). Role of Social Media in the COVID-19 Pandemic: A Literature Review. In B. Gupta, D. Peraković, A. Abd El-Latif, & D. Gupta (Eds.), *Data Mining Approaches for Big Data and Sentiment Analysis in Social Media* (pp. 91–115). IGI Global. https://doi.org/10.4018/978-1-7998-8413-2.ch004

Ahmad, R. H., & Pathan, A. K. (2017). A Study on M2M (Machine to Machine) System and Communication: Its Security, Threats, and Intrusion Detection System. In M. Ferrag & A. Ahmim (Eds.), *Security Solutions and Applied Cryptography in Smart Grid Communications* (pp. 179–214). Hershey, PA: IGI Global. doi:10.4018/978-1-5225-1829-7.ch010

Akanni, T. M. (2018). In Search of Women-Supportive Media for Sustainable Development in Nigeria. In A. Salawu & T. Owolabi (Eds.), *Exploring Journalism Practice and Perception in Developing Countries* (pp. 126–149). Hershey, PA: IGI Global. doi:10.4018/978-1-5225-3376-4.ch007

Akçay, D. (2017). The Role of Social Media in Shaping Marketing Strategies in the Airline Industry. In V. Benson, R. Tuninga, & G. Saridakis (Eds.), *Analyzing the Strategic Role of Social Networking in Firm Growth and Productivity* (pp. 214–233). Hershey, PA: IGI Global. doi:10.4018/978-1-5225-0559-4.ch012

Akmese, Z. (2020). Media Literacy and Framing of Media Content. In N. Taskiran (Ed.), *Handbook of Research on Multidisciplinary Approaches to Literacy in the Digital Age* (pp. 73–87). IGI Global. doi:10.4018/978-1-7998-1534-1.ch005

Al-Jenaibi, B. (2021). Paradigms of Public Relations in an Age of Digitalization: Social Media Analytics in the UAE. In O. Yildiz (Ed.), *Recent Developments in Individual and Organizational Adoption of ICTs* (pp. 262–277). IGI Global. https://doi.org/10.4018/978-1-7998-3045-0.ch016

Al-Rabayah, W. A. (2017). Social Media as Social Customer Relationship Management Tool: Case of Jordan Medical Directory. In W. Al-Rabayah, R. Khasawneh, R. Abu-shamaa, & I. Alsmadi (Eds.), *Strategic Uses of Social Media for Improved Customer Retention* (pp. 108–123). Hershey, PA: IGI Global. doi:10.4018/978-1-5225-1686-6.ch006

Algül, A., & Akpınar, M. E. (2022). Hate Speech on Social Media: "Dunyaerkeklergunu" Hashtag on Twitter. In E. Öngün, N. Pembecioğlu, & U. Gündüz (Eds.), *Handbook of Research on Digital Citizenship and Management During Crises* (pp. 293–305). IGI Global. https://doi.org/10.4018/978-1-7998-8421-7.ch016

Almjeld, J. (2017). Getting "Girly" Online: The Case for Gendering Online Spaces. In E. Monske & K. Blair (Eds.), *Handbook of Research on Writing and Composing in the Age of MOOCs* (pp. 87–105). Hershey, PA: IGI Global. doi:10.4018/978-1-5225-1718-4.ch006

Alsalmi, J. M., & Shehata, A. M. (2022). Official Uses of Social Media. In M. Al-Suqri, O. Al-Shaqsi, & J. Alsalmi (Eds.), *Mass Communications and the Influence of Information During Times of Crises* (pp. 123–140). IGI Global. https://doi.org/10.4018/978-1-7998-7503-1.ch006

Altaş, A. (2017). Space as a Character in Narrative Advertising: A Qualitative Research on Country Promotion Works. In R. Yılmaz (Ed.), *Narrative Advertising Models and Conceptualization in the Digital Age* (pp. 303–319). Hershey, PA: IGI Global. doi:10.4018/978-1-5225-2373-4.ch017

Altıparmak, B. (2017). The Structural Transformation of Space in Turkish Television Commercials as a Narrative Component. In R. Yılmaz (Ed.), *Narrative Advertising Models and Conceptualization in the Digital Age* (pp. 153–166). Hershey, PA: IGI Global. doi:10.4018/978-1-5225-2373-4.ch009

Arda, Ö., & Akmeşe, Z. (2021). Media Ethics: Evaluation of Television News in the Context of the Media and Ethics Relationship. In M. Taskiran & F. Pinarbaşi (Eds.), *Multidisciplinary Approaches to Ethics in the Digital Era* (pp. 96–110). IGI Global. https://doi.org/10.4018/978-1-7998-4117-3.ch007

Arık, E. (2019). Popular Culture and Media Intellectuals: Relationship Between Popular Culture and Capitalism – The Characteristics of the Media Intellectuals. In O. Ozgen (Ed.), *Handbook of Research on Consumption, Media, and Popular Culture in the Global Age* (pp. 1–10). IGI Global. https://doi.org/10.4018/978-1-5225-8491-9.ch001

Aslan, F. (2021). Could There Be an Alternative Method of Media Literacy in Promoting Health in Children and Adolescents? Media Literacy and Health Promotion. In G. Sarı (Eds.), *Handbook of Research on Representing Health and Medicine in Modern Media* (pp. 191-199). IGI Global. https://doi.org/10.4018/978-1-7998-6825-5.ch013

Assay, B. E. (2018). Regulatory Compliance, Ethical Behaviour, and Sustainable Growth in Nigeria's Telecommunications Industry. In I. Oncioiu (Ed.), *Ethics and Decision-Making for Sustainable Business Practices* (pp. 90–108). Hershey, PA: IGI Global. doi:10.4018/978-1-5225-3773-1.ch006

Assensoh-Kodua, A. (2022). This Thing of Social Media!: Indeed a Platform for Running or Developing Business in the Financial Sector. In M. Ertz (Ed.), *Handbook of Research on the Platform Economy and the Evolution of E-Commerce* (pp. 389–414). IGI Global. https://doi.org/10.4018/978-1-7998-7545-1.ch017

Atar, Ö. G. (2019). Digital Media Literacy: In-Depth Interview With the Parents of the Students Who Use Digital Media. In G. Sarı (Eds.), *Handbook of Research on Children's Consumption of Digital Media* (pp. 139-155). IGI Global. https://doi.org/10.4018/978-1-5225-5733-3.ch011

Attié, E. A., Bouvet, A., & Guibert, J. (2022). The Stakes of Social Media: Analyzing User Sentiments. In B. Gupta, D. Peraković, A. Abd El-Latif, & D. Gupta (Eds.), *Data Mining Approaches for Big Data and Sentiment Analysis in Social Media* (pp. 196–222). IGI Global. https://doi.org/10.4018/978-1-7998-8413-2.ch009

Averweg, U. R., & Leaning, M. (2018). The Qualities and Potential of Social Media. In M. Khosrow-Pour, D.B.A. (Ed.), Encyclopedia of Information Science and Technology, Fourth Edition (pp. 7106-7115). Hershey, PA: IGI Global. doi:10.4018/978-1-5225-2255-3.ch617

Baarda, R. (2017). Digital Democracy in Authoritarian Russia: Opportunity for Participation, or Site of Kremlin Control? In R. Luppicini & R. Baarda (Eds.), *Digital Media Integration for Participatory Democracy* (pp. 87–100). Hershey, PA: IGI Global. doi:10.4018/978-1-5225-2463-2.ch005

Barbosa, C., & Pedro, L. (2019). Time Orientation and Media Use: The Rise of the Device and the Changing Nature of Our Time Perception. In L. Oliveira (Ed.), *Managing Screen Time in an Online Society* (pp. 78–98). IGI Global. https://doi.org/10.4018/978-1-5225-8163-5.ch004

Başal, B. (2017). Actor Effect: A Study on Historical Figures Who Have Shaped the Advertising Narration. In R. Yılmaz (Ed.), *Narrative Advertising Models and Conceptualization in the Digital Age* (pp. 34–60). Hershey, PA: IGI Global. doi:10.4018/978-1-5225-2373-4.ch003

Behjati, M., & Cosmas, J. (2017). Self-Organizing Network Solutions: A Principal Step Towards Real 4G and Beyond. In D. Singh (Ed.), *Routing Protocols and Architectural Solutions for Optimal Wireless Networks and Security* (pp. 241–253). Hershey, PA: IGI Global. doi:10.4018/978-1-5225-2342-0.ch011

Bekafigo, M., & Pingley, A. C. (2017). Do Campaigns "Go Negative" on Twitter? In Y. Ibrahim (Ed.), *Politics, Protest, and Empowerment in Digital Spaces* (pp. 178–191). Hershey, PA: IGI Global. doi:10.4018/978-1-5225-1862-4.ch011

Bekman, M. (2022). Interaction of Internet Addiction with FoMO: The Role of Digital Media. In E. Öngün, N. Pembecioğlu, & U. Gündüz (Eds.), *Handbook of Research on Digital Citizenship and Management During Crises* (pp. 116–133). IGI Global. https://doi.org/10.4018/978-1-7998-8421-7.ch007

Bishop, J. (2017). Developing and Validating the "This Is Why We Can't Have Nice Things Scale": Optimising Political Online Communities for Internet Trolling. In Y. Ibrahim (Ed.), *Politics, Protest, and Empowerment in Digital Spaces* (pp. 153–177). Hershey, PA: IGI Global. doi:10.4018/978-1-5225-1862-4.ch010

Bitrus-Ojiambo, U. A., & King'ori, M. E. (2020). Media and Child Rights in Africa: Narrative Analysis of Child Rights in Kenyan Media. In O. Oyero (Ed.), *Media and Its Role in Protecting the Rights of Children in Africa* (pp. 125–148). IGI Global. https://doi.org/10.4018/978-1-7998-0329-4.ch007

Black, S. (2019). Diversity and Inclusion: How to Avoid Bias and Social Media Blunders. In J. Joe & E. Knight (Eds.), *Social Media for Communication and Instruction in Academic Libraries* (pp. 100–118). IGI Global. https://doi.org/10.4018/978-1-5225-8097-3.ch007

Bolat, N. (2017). The Functions of the Narrator in Digital Advertising. In R. Yılmaz (Ed.), *Narrative Advertising Models and Conceptualization in the Digital Age* (pp. 184–201). Hershey, PA: IGI Global. doi:10.4018/978-1-5225-2373-4.ch011

Brown, M. A. Sr. (2017). SNIP: High Touch Approach to Communication. In *Solutions for High-Touch Communications in a High-Tech World* (pp. 71–88). Hershey, PA: IGI Global. doi:10.4018/978-1-5225-1897-6.ch004

Brown, M. A. Sr. (2017). Comparing FTF and Online Communication Knowledge. In *Solutions for High-Touch Communications in a High-Tech World* (pp. 103–113). Hershey, PA: IGI Global. doi:10.4018/978-1-5225-1897-6.ch006

Brown, M. A. Sr. (2017). Where Do We Go from Here? In *Solutions for High-Touch Communications in a High-Tech World* (pp. 137–159). Hershey, PA: IGI Global. doi:10.4018/978-1-5225-1897-6.ch008

Brown, M. A. Sr. (2017). Bridging the Communication Gap. In *Solutions for High-Touch Communications in a High-Tech World* (pp. 1–22). Hershey, PA: IGI Global. doi:10.4018/978-1-5225-1897-6.ch001

Brown, M. A. Sr. (2017). Key Strategies for Communication. In *Solutions for High-Touch Communications in a High-Tech World* (pp. 179–202). Hershey, PA: IGI Global. doi:10.4018/978-1-5225-1897-6.ch010

Bryant, K. N. (2017). WordUp!: Student Responses to Social Media in the Technical Writing Classroom. In K. Bryant (Ed.), *Engaging 21st Century Writers with Social Media* (pp. 231–245). Hershey, PA: IGI Global. doi:10.4018/978-1-5225-0562-4.ch014

Buck, E. H. (2017). Slacktivism, Supervision, and #Selfies: Illuminating Social Media Composition through Reception Theory. In K. Bryant (Ed.), *Engaging 21st Century Writers with Social Media* (pp. 163–178). Hershey, PA: IGI Global. doi:10.4018/978-1-5225-0562-4.ch010

Bull, R., & Pianosi, M. (2017). Social Media, Participation, and Citizenship: New Strategic Directions. In V. Benson, R. Tuninga, & G. Saridakis (Eds.), *Analyzing the Strategic Role of Social Networking in Firm Growth and Productivity* (pp. 76–94). Hershey, PA: IGI Global. doi:10.4018/978-1-5225-0559-4.ch005

Caldarola, G., D'Eredità, A., Falcone, A., Lo Blundo, M., & Mancini, M. (2020). Communicating Archaeology in a Social World: Social Media, Blogs, Websites, and Best Practices. In E. Proietti (Ed.), *Developing Effective Communication Skills in Archaeology* (pp. 259–284). IGI Global. https://doi.org/10.4018/978-1-7998-1059-9.ch013

Carbajal, D., & Ramirez, Q. A. (2022). Applying Theoretical Perspectives to Social Media Influencers: A Content Analysis on Social Media Influencers the LaBrant Family. In M. Al-Suqri, O. Al-Shaqsi, & J. Alsalmi (Eds.), *Mass Communications and the Influence of Information During Times of Crises* (pp. 69–98). IGI Global. https://doi.org/10.4018/978-1-7998-7503-1.ch004

Castellano, S., & Khelladi, I. (2017). Play It Like Beckham!: The Influence of Social Networks on E-Reputation – The Case of Sportspeople and Their Online Fan Base. In A. Mesquita (Ed.), *Research Paradigms and Contemporary Perspectives on Human-Technology Interaction* (pp. 43–61). Hershey, PA: IGI Global. doi:10.4018/978-1-5225-1868-6.ch003

Chepken, C. K. (2020). Mobile-Based Social Media, What Is Cutting?: Mobile-Based Social Media: Extensive Study Findings. In S. Kır (Ed.), *New Media and Visual Communication in Social Networks* (pp. 113–135). IGI Global. https://doi.org/10.4018/978-1-7998-1041-4.ch007

Chugh, R., & Joshi, M. (2017). Challenges of Knowledge Management amidst Rapidly Evolving Tools of Social Media. In R. Chugh (Ed.), *Harnessing Social Media as a Knowledge Management Tool* (pp. 299–314). Hershey, PA: IGI Global. doi:10.4018/978-1-5225-0495-5.ch014

Cole, A. W., & Salek, T. A. (2017). Adopting a Parasocial Connection to Overcome Professional Kakoethos in Online Health Information. In M. Folk & S. Apostel (Eds.), *Establishing and Evaluating Digital Ethos and Online Credibility* (pp. 104–120). Hershey, PA: IGI Global. doi:10.4018/978-1-5225-1072-7.ch006

Cossiavelou, V. (2017). ACTA as Media Gatekeeping Factor: The EU Role as Global Negotiator. *International Journal of Interdisciplinary Telecommunications and Networking*, 9(1), 26–37. doi:10.4018/IJITN.2017010103

Costanza, F. (2017). Social Media Marketing and Value Co-Creation: A System Dynamics Approach. In S. Rozenes & Y. Cohen (Eds.), *Handbook of Research on Strategic Alliances and Value Co-Creation in the Service Industry* (pp. 205–230). Hershey, PA: IGI Global. doi:10.4018/978-1-5225-2084-9.ch011

Cyrek, B. (2019). The User With a Thousand Faces: Campbell's "Monomyth" and Media Usage Practices. In J. Kreft, S. Kuczamer-Kłopotowska, & A. Kalinowska-Żeleźnik (Eds.), *Myth in Modern Media Management and Marketing* (pp. 50–68). IGI Global. https://doi.org/10.4018/978-1-5225-9100-9.ch003

Deniz, Ş. (2020). Is Somebody Spying on Us?: Social Media Users' Privacy Awareness. In S. Kır (Ed.), *New Media and Visual Communication in Social Networks* (pp. 156–172). IGI Global. https://doi.org/10.4018/978-1-7998-1041-4.ch009

Di Virgilio, F., & Antonelli, G. (2018). Consumer Behavior, Trust, and Electronic Word-of-Mouth Communication: Developing an Online Purchase Intention Model. In F. Di Virgilio (Ed.), *Social Media for Knowledge Management Applications in Modern Organizations* (pp. 58–80). Hershey, PA: IGI Global. doi:10.4018/978-1-5225-2897-5.ch003

Dolanbay, H. (2022). The Transformation of Literacy and Media Literacy. In C. Lane (Ed.), *Handbook of Research on Acquiring 21st Century Literacy Skills Through Game-Based Learning* (pp. 363–380). IGI Global. https://doi.org/10.4018/978-1-7998-7271-9.ch019

Dunn, R. A., & Herrmann, A. F. (2020). Comic Con Communion: Gender, Cosplay, and Media Fandom. In R. Dunn (Ed.), *Multidisciplinary Perspectives on Media Fandom* (pp. 37–52). IGI Global. https://doi.org/10.4018/978-1-7998-3323-9.ch003

DuQuette, J. L. (2017). Lessons from Cypris Chat: Revisiting Virtual Communities as Communities. In G. Panconesi & M. Guida (Eds.), *Handbook of Research on Collaborative Teaching Practice in Virtual Learning Environments* (pp. 299–316). Hershey, PA: IGI Global. doi:10.4018/978-1-5225-2426-7.ch016

Ekhlassi, A., Niknejhad Moghadam, M., & Adibi, A. (2018). The Concept of Social Media: The Functional Building Blocks. In *Building Brand Identity in the Age of Social Media: Emerging Research and Opportunities* (pp. 29–60). Hershey, PA: IGI Global. doi:10.4018/978-1-5225-5143-0.ch002

Ekhlassi, A., Niknejhad Moghadam, M., & Adibi, A. (2018). Social Media Branding Strategy: Social Media Marketing Approach. In *Building Brand Identity in the Age of Social Media: Emerging Research and Opportunities* (pp. 94–117). Hershey, PA: IGI Global. doi:10.4018/978-1-5225-5143-0.ch004

Ekhlassi, A., Niknejhad Moghadam, M., & Adibi, A. (2018). The Impact of Social Media on Brand Loyalty: Achieving "E-Trust" Through Engagement. In *Building Brand Identity in the Age of Social Media: Emerging Research and Opportunities* (pp. 155–168). Hershey, PA: IGI Global. doi:10.4018/978-1-5225-5143-0.ch007

El-Henawy, W. M. (2019). Media Literacy in EFL Teacher Education: A Necessity for 21st Century English Language Instruction. In M. Yildiz, M. Fazal, M. Ahn, R. Feirsen, & S. Ozdemir (Eds.), *Handbook of Research on Media Literacy Research and Applications Across Disciplines* (pp. 65–89). IGI Global. https://doi.org/10.4018/978-1-5225-9261-7.ch005

Elegbe, O. (2017). An Assessment of Media Contribution to Behaviour Change and HIV Prevention in Nigeria. In O. Nelson, B. Ojebuyi, & A. Salawu (Eds.), *Impacts of the Media on African Socio-Economic Development* (pp. 261–280). Hershey, PA: IGI Global. doi:10.4018/978-1-5225-1859-4.ch017

Endong, F. P. (2018). Hashtag Activism and the Transnationalization of Nigerian-Born Movements Against Terrorism: A Critical Appraisal of the #BringBackOurGirls Campaign. In F. Endong (Ed.), *Exploring the Role of Social Media in Transnational Advocacy* (pp. 36–54). Hershey, PA: IGI Global. doi:10.4018/978-1-5225-2854-8.ch003

Erkek, S. (2021). Health Communication and Social Media: A Study About Using Social Media in Medicine Companies. In G. Sarı (Eds.), *Handbook of Research on Representing Health and Medicine in Modern Media* (pp. 70-83). IGI Global. https://doi.org/10.4018/978-1-7998-6825-5.ch005

Erragcha, N. (2017). Using Social Media Tools in Marketing: Opportunities and Challenges. In M. Brown Sr., (Ed.), *Social Media Performance Evaluation and Success Measurements* (pp. 106–129). Hershey, PA: IGI Global. doi:10.4018/978-1-5225-1963-8.ch006

Ersoy, M. (2019). Social Media and Children. In G. Sarı (Ed.), *Handbook of Research on Children's Consumption of Digital Media* (pp. 11-23). IGI Global. https://doi.org/10.4018/978-1-5225-5733-3.ch002

Ezeh, N. C. (2018). Media Campaign on Exclusive Breastfeeding: Awareness, Perception, and Acceptability Among Mothers in Anambra State, Nigeria. In A. Salawu & T. Owolabi (Eds.), *Exploring Journalism Practice and Perception in Developing Countries* (pp. 172–193). Hershey, PA: IGI Global. doi:10.4018/978-1-5225-3376-4.ch009

Fawole, O. A., & Osho, O. A. (2017). Influence of Social Media on Dating Relationships of Emerging Adults in Nigerian Universities: Social Media and Dating in Nigeria. In M. Wright (Ed.), *Identity, Sexuality, and Relationships among Emerging Adults in the Digital Age* (pp. 168–177). Hershey, PA: IGI Global. doi:10.4018/978-1-5225-1856-3.ch011

Fayoyin, A. (2017). Electoral Polling and Reporting in Africa: Professional and Policy Implications for Media Practice and Political Communication in a Digital Age. In N. Mhiripiri & T. Chari (Eds.), *Media Law, Ethics, and Policy in the Digital Age* (pp. 164–181). Hershey, PA: IGI Global. doi:10.4018/978-1-5225-2095-5.ch009

Fayoyin, A. (2018). Rethinking Media Engagement Strategies for Social Change in Africa: Context, Approaches, and Implications for Development Communication. In A. Salawu & T. Owolabi (Eds.), *Exploring Journalism Practice and Perception in Developing Countries* (pp. 257–280). Hershey, PA: IGI Global. doi:10.4018/978-1-5225-3376-4.ch013

Fechine, Y., & Rêgo, S. C. (2018). Transmedia Television Journalism in Brazil: Jornal da Record News as Reference. In R. Gambarato & G. Alzamora (Eds.), *Exploring Transmedia Journalism in the Digital Age* (pp. 253–265). Hershey, PA: IGI Global. doi:10.4018/978-1-5225-3781-6.ch015

Fener, E. (2021). Social Media and Health Communication. In G. Sarı (Ed.), *Handbook of Research on Representing Health and Medicine in Modern Media* (pp. 16-32). IGI Global. https://doi.org/10.4018/978-1-7998-6825-5.ch002

Fernandes dos Santos, N. (2020). The Use of Twitter During the 2013 Protests in Brazil: Mainstream Media at Stake. In A. Solo (Ed.), *Handbook of Research on Politics in the Computer Age* (pp. 181–202). IGI Global. https://doi.org/10.4018/978-1-7998-0377-5.ch011

Fiore, C. (2017). The Blogging Method: Improving Traditional Student Writing Practices. In K. Bryant (Ed.), *Engaging 21st Century Writers with Social Media* (pp. 179–198). Hershey, PA: IGI Global. doi:10.4018/978-1-5225-0562-4.ch011

Friesem, E., & Friesem, Y. (2019). Media Literacy Education in the Era of Post-Truth: Paradigm Crisis. In M. Yildiz, M. Fazal, M. Ahn, R. Feirsen, & S. Ozdemir (Eds.), *Handbook of Research on Media Literacy Research and Applications Across Disciplines* (pp. 119–134). IGI Global. https://doi.org/10.4018/978-1-5225-9261-7.ch008

Fung, Y., Lee, L., Chui, K. T., Cheung, G. H., Tang, C., & Wong, S. (2022). Sentiment Analysis and Summarization of Facebook Posts on News Media. In B. Gupta, D. Peraković, A. Abd El-Latif, & D. Gupta (Eds.), *Data Mining Approaches for Big Data and Sentiment Analysis in Social Media* (pp. 142–154). IGI Global. https://doi.org/10.4018/978-1-7998-8413-2.ch006

Gambarato, R. R., Alzamora, G. C., & Tárcia, L. P. (2018). 2016 Rio Summer Olympics and the Transmedia Journalism of Planned Events. In R. Gambarato & G. Alzamora (Eds.), *Exploring Transmedia Journalism in the Digital Age* (pp. 126–146). Hershey, PA: IGI Global. doi:10.4018/978-1-5225-3781-6.ch008

Ganguin, S., Gemkow, J., & Haubold, R. (2017). Information Overload as a Challenge and Changing Point for Educational Media Literacies. In R. Marques & J. Batista (Eds.), *Information and Communication Overload in the Digital Age* (pp. 302–328). Hershey, PA: IGI Global. doi:10.4018/978-1-5225-2061-0.ch013

Gardner, G. C. (2017). The Lived Experience of Smartphone Use in a Unit of the United States Army. In F. Topor (Ed.), *Handbook of Research on Individualism and Identity in the Globalized Digital Age* (pp. 88–117). Hershey, PA: IGI Global. doi:10.4018/978-1-5225-0522-8.ch005

Garg, P., & Pahuja, S. (2020). Social Media: Concept, Role, Categories, Trends, Social Media and AI, Impact on Youth, Careers, Recommendations. In S. Alavi & V. Ahuja (Eds.), *Managing Social Media Practices in the Digital Economy* (pp. 172–192). IGI Global. https://doi.org/10.4018/978-1-7998-2185-4.ch008

Golightly, D., & Houghton, R. J. (2018). Social Media as a Tool to Understand Behaviour on the Railways. In S. Kohli, A. Kumar, J. Easton, & C. Roberts (Eds.), *Innovative Applications of Big Data in the Railway Industry* (pp. 224–239). Hershey, PA: IGI Global. doi:10.4018/978-1-5225-3176-0.ch010

Gouveia, P. (2020). The New Media vs. Old Media Trap: How Contemporary Arts Became Playful Transmedia Environments. In C. Soares & E. Simão (Eds.), *Multidisciplinary Perspectives on New Media Art* (pp. 25–46). IGI Global. https://doi.org/10.4018/978-1-7998-3669-8.ch002

Gundogan, M. B. (2017). In Search for a "Good Fit" Between Augmented Reality and Mobile Learning Ecosystem. In G. Kurubacak & H. Altinpulluk (Eds.), *Mobile Technologies and Augmented Reality in Open Education* (pp. 135–153). Hershey, PA: IGI Global. doi:10.4018/978-1-5225-2110-5.ch007

Gupta, H. (2018). Impact of Digital Communication on Consumer Behaviour Processes in Luxury Branding Segment: A Study of Apparel Industry. In S. Dasgupta, S. Biswal, & M. Ramesh (Eds.), *Holistic Approaches to Brand Culture and Communication Across Industries* (pp. 132–157). Hershey, PA: IGI Global. doi:10.4018/978-1-5225-3150-0.ch008

Guzman-Garcia, P. A., Orozco-Quintana, E., Sepulveda-Gonzalez, D., Cooley-Magallanes, A., Salas-Velazquez, D., Lopez-Garcia, C., Ramírez-Treviño, A., Espinoza-Moran, A. L., Lopez, M., & Segura-Azuara, N. D. (2022). Ending Health Promotion Lethargy: A Social Media Awareness Campaign to Face Hypothyroidism. In M. Lopez (Ed.), *Advancing Health Education With Telemedicine* (pp. 165–182). IGI Global. https://doi.org/10.4018/978-1-7998-8783-6.ch009

Hafeez, E., & Zahid, L. (2021). Sexism and Gender Discrimination in Pakistan's Mainstream News Media. In S. Jamil, B. Çoban, B. Ataman, & G. Appiah-Adjei (Eds.), *Handbook of Research on Discrimination, Gender Disparity, and Safety Risks in Journalism* (pp. 60–89). IGI Global. https://doi.org/10.4018/978-1-7998-6686-2.ch005

Hai-Jew, S. (2017). Creating "(Social) Network Art" with NodeXL. In S. Hai-Jew (Ed.), *Social Media Data Extraction and Content Analysis* (pp. 342–393). Hershey, PA: IGI Global. doi:10.4018/978-1-5225-0648-5.ch011

Hai-Jew, S. (2017). Employing the Sentiment Analysis Tool in NVivo 11 Plus on Social Media Data: Eight Initial Case Types. In N. Rao (Ed.), *Social Media Listening and Monitoring for Business Applications* (pp. 175–244). Hershey, PA: IGI Global. doi:10.4018/978-1-5225-0846-5.ch010

Hai-Jew, S. (2017). Conducting Sentiment Analysis and Post-Sentiment Data Exploration through Automated Means. In S. Hai-Jew (Ed.), *Social Media Data Extraction and Content Analysis* (pp. 202–240). Hershey, PA: IGI Global. doi:10.4018/978-1-5225-0648-5.ch008

Hai-Jew, S. (2017). Applied Analytical "Distant Reading" using NVivo 11 Plus. In S. Hai-Jew (Ed.), *Social Media Data Extraction and Content Analysis* (pp. 159–201). Hershey, PA: IGI Global. doi:10.4018/978-1-5225-0648-5.ch007

Hai-Jew, S. (2017). Flickering Emotions: Feeling-Based Associations from Related Tags Networks on Flickr. In S. Hai-Jew (Ed.), *Social Media Data Extraction and Content Analysis* (pp. 296–341). Hershey, PA: IGI Global. doi:10.4018/978-1-5225-0648-5.ch010

Hai-Jew, S. (2017). Manually Profiling Egos and Entities across Social Media Platforms: Evaluating Shared Messaging and Contents, User Networks, and Metadata. In V. Benson, R. Tuninga, & G. Saridakis (Eds.), *Analyzing the Strategic Role of Social Networking in Firm Growth and Productivity* (pp. 352–405). Hershey, PA: IGI Global. doi:10.4018/978-1-5225-0559-4.ch019

Hai-Jew, S. (2017). Exploring "User," "Video," and (Pseudo) Multi-Mode Networks on YouTube with NodeXL. In S. Hai-Jew (Ed.), *Social Media Data Extraction and Content Analysis* (pp. 242–295). Hershey, PA: IGI Global. doi:10.4018/978-1-5225-0648-5.ch009

Hai-Jew, S. (2018). Exploring "Mass Surveillance" Through Computational Linguistic Analysis of Five Text Corpora: Academic, Mainstream Journalism, Microblogging Hashtag Conversation, Wikipedia Articles, and Leaked Government Data. In *Techniques for Coding Imagery and Multimedia: Emerging Research and Opportunities* (pp. 212–286). Hershey, PA: IGI Global. doi:10.4018/978-1-5225-2679-7.ch004

Hai-Jew, S. (2018). Exploring Identity-Based Humor in a #Selfies #Humor Image Set From Instagram. In *Techniques for Coding Imagery and Multimedia: Emerging Research and Opportunities* (pp. 1–90). Hershey, PA: IGI Global. doi:10.4018/978-1-5225-2679-7.ch001

Hai-Jew, S. (2018). See Ya!: Exploring American Renunciation of Citizenship Through Targeted and Sparse Social Media Data Sets and a Custom Spatial-Based Linguistic Analysis Dictionary. In *Techniques for Coding Imagery and Multimedia: Emerging Research and Opportunities* (pp. 287–393). Hershey, PA: IGI Global. doi:10.4018/978-1-5225-2679-7.ch005

Hasan, H., & Linger, H. (2017). Connected Living for Positive Ageing. In S. Gordon (Ed.), *Online Communities as Agents of Change and Social Movements* (pp. 203–223). Hershey, PA: IGI Global. doi:10.4018/978-1-5225-2495-3.ch008

Hersey, L. N. (2017). CHOICES: Measuring Return on Investment in a Nonprofit Organization. In M. Brown Sr., (Ed.), *Social Media Performance Evaluation and Success Measurements* (pp. 157–179). Hershey, PA: IGI Global. doi:10.4018/978-1-5225-1963-8.ch008

Heuva, W. E. (2017). Deferring Citizens' "Right to Know" in an Information Age: The Information Deficit in Namibia. In N. Mhiripiri & T. Chari (Eds.), *Media Law, Ethics, and Policy in the Digital Age* (pp. 245–267). Hershey, PA: IGI Global. doi:10.4018/978-1-5225-2095-5.ch014

Hopwood, M., & McLean, H. (2017). Social Media in Crisis Communication: The Lance Armstrong Saga. In V. Benson, R. Tuninga, & G. Saridakis (Eds.), *Analyzing the Strategic Role of Social Networking in Firm Growth and Productivity* (pp. 45–58). Hershey, PA: IGI Global. doi:10.4018/978-1-5225-0559-4.ch003

Horst, S., & Murschetz, P. C. (2019). Strategic Media Entrepreneurship: Theory Development and Problematization. *Journal of Media Management and Entrepreneurship, 1*(1), 1–26. https://doi.org/10.4018/JMME.2019010101

Hotur, S. K. (2018). Indian Approaches to E-Diplomacy: An Overview. In S. Bute (Ed.), *Media Diplomacy and Its Evolving Role in the Current Geopolitical Climate* (pp. 27–35). Hershey, PA: IGI Global. doi:10.4018/978-1-5225-3859-2.ch002

Inder, S. (2021). Social Media, Crowdsourcing, and Marketing. In A. Singh (Ed.), *Big Data Analytics for Improved Accuracy, Efficiency, and Decision Making in Digital Marketing* (pp. 64–73). IGI Global. https://doi.org/10.4018/978-1-7998-7231-3.ch005

Işık, T. (2021). Media and Health Communication Campaigns. In G. Sarı (Ed.), *Handbook of Research on Representing Health and Medicine in Modern Media* (pp. 1-15). IGI Global. https://doi.org/10.4018/978-1-7998-6825-5.ch001

Iwasaki, Y. (2017). Youth Engagement in the Era of New Media. In M. Adria & Y. Mao (Eds.), *Handbook of Research on Citizen Engagement and Public Participation in the Era of New Media* (pp. 90–105). Hershey, PA: IGI Global. doi:10.4018/978-1-5225-1081-9.ch006

Jamieson, H. V. (2017). We have a Situation!: Cyberformance and Civic Engagement in Post-Democracy. In R. Shin (Ed.), *Convergence of Contemporary Art, Visual Culture, and Global Civic Engagement* (pp. 297–317). Hershey, PA: IGI Global. doi:10.4018/978-1-5225-1665-1.ch017

Jimoh, J., & Kayode, J. (2018). Imperative of Peace and Conflict-Sensitive Journalism in Development. In A. Salawu & T. Owolabi (Eds.), *Exploring Journalism Practice and Perception in Developing Countries* (pp. 150–171). Hershey, PA: IGI Global. doi:10.4018/978-1-5225-3376-4.ch008

Joseph, J. J., & Florea, D. (2020). Clinical Topics in Social Media: The Role of Self-Disclosing on Social Media for Friendship and Identity in Specialized Populations. In M. Desjarlais (Ed.), *The Psychology and Dynamics Behind Social Media Interactions* (pp. 28–56). IGI Global. https://doi.org/10.4018/978-1-5225-9412-3.ch002

Kaale, K. B., & Mgeta, M. B. (2020). Photojournalism Ethics: Portraying Children's Photos in Tanzanian Media. In O. Oyero (Ed.), *Media and Its Role in Protecting the Rights of Children in Africa* (pp. 149–168). IGI Global. https://doi.org/10.4018/978-1-7998-0329-4.ch008

Kanellopoulos, D. N. (2018). Group Synchronization for Multimedia Systems. In M. Khosrow-Pour, D.B.A. (Ed.), Encyclopedia of Information Science and Technology, Fourth Edition (pp. 6435-6446). Hershey, PA: IGI Global. doi:10.4018/978-1-5225-2255-3.ch559

Kapepo, M. I., & Mayisela, T. (2017). Integrating Digital Literacies Into an Undergraduate Course: Inclusiveness Through Use of ICTs. In C. Ayo & V. Mbarika (Eds.), *Sustainable ICT Adoption and Integration for Socio-Economic Development* (pp. 152–173). Hershey, PA: IGI Global. doi:10.4018/978-1-5225-2565-3.ch007

Karahoca, A., & Yengin, İ. (2018). Understanding the Potentials of Social Media in Collaborative Learning. In M. Khosrow-Pour, D.B.A. (Ed.), Encyclopedia of Information Science and Technology, Fourth Edition (pp. 7168-7180). Hershey, PA: IGI Global. doi:10.4018/978-1-5225-2255-3.ch623

Kasemsap, K. (2017). Professional and Business Applications of Social Media Platforms. In V. Benson, R. Tuninga, & G. Saridakis (Eds.), *Analyzing the Strategic Role of Social Networking in Firm Growth and Productivity* (pp. 427–450). Hershey, PA: IGI Global. doi:10.4018/978-1-5225-0559-4.ch021

Kasemsap, K. (2017). Mastering Social Media in the Modern Business World. In N. Rao (Ed.), *Social Media Listening and Monitoring for Business Applications* (pp. 18–44). Hershey, PA: IGI Global. doi:10.4018/978-1-5225-0846-5.ch002

Kaufmann, H. R., & Manarioti, A. (2017). Consumer Engagement in Social Media Platforms. In *Encouraging Participative Consumerism Through Evolutionary Digital Marketing: Emerging Research and Opportunities* (pp. 95–123). Hershey, PA: IGI Global. doi:10.4018/978-1-68318-012-8.ch004

Kavak, B., Özdemir, N., & Erol-Boyacı, G. (2020). A Literature Review of Social Media for Marketing: Social Media Use in B2C and B2B Contexts. In S. Alavi & V. Ahuja (Eds.), *Managing Social Media Practices in the Digital Economy* (pp. 67–96). IGI Global. https://doi.org/10.4018/978-1-7998-2185-4.ch004

Kavoura, A., & Kefallonitis, E. (2018). The Effect of Social Media Networking in the Travel Industry. In M. Khosrow-Pour, D.B.A. (Ed.), Encyclopedia of Information Science and Technology, Fourth Edition (pp. 4052-4063). Hershey, PA: IGI Global. doi:10.4018/978-1-5225-2255-3.ch351

Kawamura, Y. (2018). Practice and Modeling of Advertising Communication Strategy: Sender-Driven and Receiver-Driven. In T. Ogata & S. Asakawa (Eds.), *Content Generation Through Narrative Communication and Simulation* (pp. 358–379). Hershey, PA: IGI Global. doi:10.4018/978-1-5225-4775-4.ch013

Kaya, A., & Mantar, O. B. (2021). Social Media and Health Communication: Vaccine Refusal/Hesitancy. In G. Sarı (Ed.), *Handbook of Research on Representing Health and Medicine in Modern Media* (pp. 33-53). IGI Global. https://doi.org/10.4018/978-1-7998-6825-5.ch003

Kaya, A. Y., & Ata, F. (2022). New Media and Digital Paranoia: Extreme Skepticism in Digital Communication. In H. Aker & M. Aiken (Eds.), *Handbook of Research on Cyberchondria, Health Literacy, and the Role of Media in Society's Perception of Medical Information* (pp. 330–343). IGI Global. https://doi.org/10.4018/978-1-7998-8630-3.ch018

Kell, C., & Czerniewicz, L. (2017). Visibility of Scholarly Research and Changing Research Communication Practices: A Case Study from Namibia. In A. Esposito (Ed.), *Research 2.0 and the Impact of Digital Technologies on Scholarly Inquiry* (pp. 97–116). Hershey, PA: IGI Global. doi:10.4018/978-1-5225-0830-4.ch006

Kharade, S. S. (2022). An Adverse Effect of Social, Gaming, and Entertainment Media on Overall Development of Adolescents. In S. Malik, R. Bansal, & A. Tyagi (Eds.), *Impact and Role of Digital Technologies in Adolescent Lives* (pp. 26–34). IGI Global. https://doi.org/10.4018/978-1-7998-8318-0.ch003

Kılınç, U. (2017). Create It! Extend It!: Evolution of Comics Through Narrative Advertising. In R. Yılmaz (Ed.), *Narrative Advertising Models and Conceptualization in the Digital Age* (pp. 117–132). Hershey, PA: IGI Global. doi:10.4018/978-1-5225-2373-4.ch007

Kocakoç, I. D., & Özkan, P. (2022). Clubhouse Experience: Sentiment Analysis of an Alternative Platform From the Eyes of Classic Social Media Users. In B. Gupta, D. Peraković, A. Abd El-Latif, & D. Gupta (Eds.), *Data Mining Approaches for Big Data and Sentiment Analysis in Social Media* (pp. 244–264). IGI Global. https://doi.org/10.4018/978-1-7998-8413-2.ch011

Kreft, J. (2019). A Myth and Media Management: The Facade Rhetoric and Business Objectives. In J. Kreft, S. Kuczamer-Kłopotowska, & A. Kalinowska-Żeleźnik (Eds.), *Myth in Modern Media Management and Marketing* (pp. 118–141). IGI Global. https://doi.org/10.4018/978-1-5225-9100-9.ch006

Krishnamurthy, R. (2019). Social Media as a Marketing Tool. In P. Mishra & S. Dham (Eds.), *Application of Gaming in New Media Marketing* (pp. 181–201). IGI Global. https://doi.org/10.4018/978-1-5225-6064-7.ch011

Kumar, D., & Gupta, P. (2021). Communicating in Media Dark Areas. In R. Jackson & A. Reboulet (Eds.), *Effective Strategies for Communicating Insights in Business* (pp. 141-156). IGI Global. https://doi.org/10.4018/978-1-7998-3964-4.ch009

Kumar, P., & Sinha, A. (2018). Business-Oriented Analytics With Social Network of Things. In H. Bansal, G. Shrivastava, G. Nguyen, & L. Stanciu (Eds.), *Social Network Analytics for Contemporary Business Organizations* (pp. 166–187). Hershey, PA: IGI Global. doi:10.4018/978-1-5225-5097-6.ch009

Kunock, A. I. (2017). Boko Haram Insurgency in Cameroon: Role of Mass Media in Conflict Management. In N. Mhiripiri & T. Chari (Eds.), *Media Law, Ethics, and Policy in the Digital Age* (pp. 226–244). Hershey, PA: IGI Global. doi:10.4018/978-1-5225-2095-5.ch013

Labadie, J. A. (2018). Digitally Mediated Art Inspired by Technology Integration: A Personal Journey. In A. Ursyn (Ed.), *Visual Approaches to Cognitive Education With Technology Integration* (pp. 121–162). Hershey, PA: IGI Global. doi:10.4018/978-1-5225-5332-8.ch008

Lantz, E. (2020). Immersion Domes: Next-Generation Arts and Entertainment Venues. In J. Morie & K. McCallum (Eds.), *Handbook of Research on the Global Impacts and Roles of Immersive Media* (pp. 314–346). IGI Global. https://doi.org/10.4018/978-1-7998-2433-6.ch016

Lasisi, M. I., Adebiyi, R. A., & Ajetunmobi, U. O. (2020). Predicting Migration to Developed Countries: The Place of Media Attention. In N. Okorie, B. Ojebuyi, & J. Macharia (Eds.), *Handbook of Research on the Global Impact of Media on Migration Issues* (pp. 293–311). IGI Global. https://doi.org/10.4018/978-1-7998-0210-5.ch017

Lefkowith, S. (2017). Credibility and Crisis in Pseudonymous Communities. In M. Folk & S. Apostel (Eds.), *Establishing and Evaluating Digital Ethos and Online Credibility* (pp. 190–236). Hershey, PA: IGI Global. doi:10.4018/978-1-5225-1072-7.ch010

Lekic-Subasic, Z. (2021). Women and Media: What Public Service Media Can Do to Ensure Gender Equality. In S. Jamil, B. Çoban, B. Ataman, & G. Appiah-Adjei (Eds.), *Handbook of Research on Discrimination, Gender Disparity, and Safety Risks in Journalism* (pp. 8–23). IGI Global. https://doi.org/10.4018/978-1-7998-6686-2.ch002

Luppicini, R. (2017). Technoethics and Digital Democracy for Future Citizens. In R. Luppicini & R. Baarda (Eds.), *Digital Media Integration for Participatory Democracy* (pp. 1–21). Hershey, PA: IGI Global. doi:10.4018/978-1-5225-2463-2.ch001

Maher, D. (2018). Supporting Pre-Service Teachers' Understanding and Use of Mobile Devices. In J. Keengwe (Ed.), *Handbook of Research on Mobile Technology, Constructivism, and Meaningful Learning* (pp. 160–177). Hershey, PA: IGI Global. doi:10.4018/978-1-5225-3949-0.ch009

Makhwanya, A. (2018). Barriers to Social Media Advocacy: Lessons Learnt From the Project "Tell Them We Are From Here". In F. Endong (Ed.), *Exploring the Role of Social Media in Transnational Advocacy* (pp. 55–72). Hershey, PA: IGI Global. doi:10.4018/978-1-5225-2854-8.ch004

Malicki-Sanchez, K. (2020). Out of Our Minds: Ontology and Embodied Media in a Post-Human Paradigm. In J. Morie & K. McCallum (Eds.), *Handbook of Research on the Global Impacts and Roles of Immersive Media* (pp. 10–36). IGI Global. https://doi.org/10.4018/978-1-7998-2433-6.ch002

Manli, G., & Rezaei, S. (2017). Value and Risk: Dual Pillars of Apps Usefulness. In S. Rezaei (Ed.), *Apps Management and E-Commerce Transactions in Real-Time* (pp. 274–292). Hershey, PA: IGI Global. doi:10.4018/978-1-5225-2449-6.ch013

Manrique, C. G., & Manrique, G. G. (2017). Social Media's Role in Alleviating Political Corruption and Scandals: The Philippines during and after the Marcos Regime. In K. Demirhan & D. Çakır-Demirhan (Eds.), *Political Scandal, Corruption, and Legitimacy in the Age of Social Media* (pp. 205–222). Hershey, PA: IGI Global. doi:10.4018/978-1-5225-2019-1.ch009

Marjerison, R. K., Lin, Y., & Kennedyd, S. I. (2019). An Examination of Motivation and Media Type: Sharing Content on Chinese Social Media. *International Journal of Social Media and Online Communities*, *11*(1), 15–34. https://doi.org/10.4018/IJSMOC.2019010102

Marovitz, M. (2017). Social Networking Engagement and Crisis Communication Considerations. In M. Brown Sr., (Ed.), *Social Media Performance Evaluation and Success Measurements* (pp. 130–155). Hershey, PA: IGI Global. doi:10.4018/978-1-5225-1963-8.ch007

Martin, P. M., & Onampally, J. J. (2019). Patterns of Deceptive Communication of Social and Religious Issues in Social Media: Representation of Social Issues in Social Media. In I. Chiluwa & S. Samoilenko (Eds.), *Handbook of Research on Deception, Fake News, and Misinformation Online* (pp. 490–502). IGI Global. https://doi.org/10.4018/978-1-5225-8535-0.ch026

Masterson, J. R. (2020). Chinese Citizenry Social Media Pressures and Public Official Responses: The Double-Edged Sword of Social Media in China. In S. Edwards III & D. Santos (Eds.), *Digital Transformation and Its Role in Progressing the Relationship Between States and Their Citizens* (pp. 139-181). IGI Global. https://doi.org/10.4018/978-1-7998-3152-5.ch007

Maulana, I. (2018). Spontaneous Taking and Posting Selfie: Reclaiming the Lost Trust. In S. Hai-Jew (Ed.), *Selfies as a Mode of Social Media and Work Space Research* (pp. 28–50). Hershey, PA: IGI Global. doi:10.4018/978-1-5225-3373-3.ch002

Mayo, S. (2018). A Collective Consciousness Model in a Post-Media Society. In M. Khosrow-Pour (Ed.), *Enhancing Art, Culture, and Design With Technological Integration* (pp. 25–49). Hershey, PA: IGI Global. doi:10.4018/978-1-5225-5023-5.ch002

Mazur, E., Signorella, M. L., & Hough, M. (2018). The Internet Behavior of Older Adults. In M. Khosrow-Pour, D.B.A. (Ed.), Encyclopedia of Information Science and Technology, Fourth Edition (pp. 7026-7035). Hershey, PA: IGI Global. doi:10.4018/978-1-5225-2255-3.ch609

McCallum, K. M. (2020). Immersive Experience: Convergence, Storyworlds, and the Power for Social Impact. In J. Morie & K. McCallum (Eds.), *Handbook of Research on the Global Impacts and Roles of Immersive Media* (pp. 453–484). IGI Global. https://doi.org/10.4018/978-1-7998-2433-6.ch022

McGuire, M. (2017). Reblogging as Writing: The Role of Tumblr in the Writing Classroom. In K. Bryant (Ed.), *Engaging 21st Century Writers with Social Media* (pp. 116–131). Hershey, PA: IGI Global. doi:10.4018/978-1-5225-0562-4.ch007

McKee, J. (2018). Architecture as a Tool to Solve Business Planning Problems. In M. Khosrow-Pour, D.B.A. (Ed.), Encyclopedia of Information Science and Technology, Fourth Edition (pp. 573-586). Hershey, PA: IGI Global. doi:10.4018/978-1-5225-2255-3.ch050

McMahon, D. (2017). With a Little Help from My Friends: The Irish Radio Industry's Strategic Appropriation of Facebook for Commercial Growth. In V. Benson, R. Tuninga, & G. Saridakis (Eds.), *Analyzing the Strategic Role of Social Networking in Firm Growth and Productivity* (pp. 157–171). Hershey, PA: IGI Global. doi:10.4018/978-1-5225-0559-4.ch009

McPherson, M. J., & Lemon, N. (2017). The Hook, Woo, and Spin: Academics Creating Relations on Social Media. In A. Esposito (Ed.), *Research 2.0 and the Impact of Digital Technologies on Scholarly Inquiry* (pp. 167–187). Hershey, PA: IGI Global. doi:10.4018/978-1-5225-0830-4.ch009

Melro, A., & Oliveira, L. (2018). Screen Culture. In M. Khosrow-Pour, D.B.A. (Ed.), Encyclopedia of Information Science and Technology, Fourth Edition (pp. 4255-4266). Hershey, PA: IGI Global. doi:10.4018/978-1-5225-2255-3.ch369

Meral, K. Z. (2021). Social Media Ethics and Children in the Digital Era: Social Media Risks and Precautions. In M. Taskiran & F. Pinarbaşi (Eds.), *Multidisciplinary Approaches to Ethics in the Digital Era* (pp. 166–182). IGI Global. https://doi.org/10.4018/978-1-7998-4117-3.ch011

Meral, Y., & Özbay, D. E. (2020). Electronic Trading, Electronic Advertising, and Social Media Literacy: Using Local Turkish Influencers in Social Media for International Trade Products Marketing. In N. Taskiran (Ed.), *Handbook of Research on Multidisciplinary Approaches to Literacy in the Digital Age* (pp. 224–261). IGI Global. https://doi.org/10.4018/978-1-7998-1534-1.ch012

Mhiripiri, N. A., & Chikakano, J. (2017). Criminal Defamation, the Criminalisation of Expression, Media and Information Dissemination in the Digital Age: A Legal and Ethical Perspective. In N. Mhiripiri & T. Chari (Eds.), *Media Law, Ethics, and Policy in the Digital Age* (pp. 1–24). Hershey, PA: IGI Global. doi:10.4018/978-1-5225-2095-5.ch001

Miliopoulou, G., & Cossiavelou, V. (2019). Brand Management and Media Gatekeeping: Exploring the Professionals' Practices and Perspectives in the Social Media. In N. Meghanathan (Ed.), *Strategic Innovations and Interdisciplinary Perspectives in Telecommunications and Networking* (pp. 56–82). IGI Global. https://doi.org/10.4018/978-1-5225-8188-8.ch004

Miranda, S. L., & Antunes, A. C. (2021). Golden Years in Social Media World: Examining Behavior and Motivations. In P. Wamuyu (Ed.), *Analyzing Global Social Media Consumption* (pp. 261–276). IGI Global. https://doi.org/10.4018/978-1-7998-4718-2.ch014

Miron, E., Palmor, A., Ravid, G., Sharon, A., Tikotsky, A., & Zirkel, Y. (2017). Principles and Good Practices for Using Wikis within Organizations. In R. Chugh (Ed.), *Harnessing Social Media as a Knowledge Management Tool* (pp. 143–176). Hershey, PA: IGI Global. doi:10.4018/978-1-5225-0495-5.ch008

Moeller, C. L. (2018). Sharing Your Personal Medical Experience Online: Is It an Irresponsible Act or Patient Empowerment? In S. Sekalala & B. Niezgoda (Eds.), *Global Perspectives on Health Communication in the Age of Social Media* (pp. 185–209). Hershey, PA: IGI Global. doi:10.4018/978-1-5225-3716-8.ch007

Mosanako, S. (2017). Broadcasting Policy in Botswana: The Case of Botswana Television. In O. Nelson, B. Ojebuyi, & A. Salawu (Eds.), *Impacts of the Media on African Socio-Economic Development* (pp. 217–230). Hershey, PA: IGI Global. doi:10.4018/978-1-5225-1859-4.ch014

Mukherjee Das, M. (2020). Harnessing the "Crowd" and the Rise of "Prosumers" in Filmmaking in India. In S. Biswal, K. Kusuma, & S. Mohanty (Eds.), *Handbook of Research on Social and Cultural Dynamics in Indian Cinema* (pp. 350–359). IGI Global. https://doi.org/10.4018/978-1-7998-3511-0.ch029

Musemburi, D., & Nhendo, C. (2019). Media Information Literacy: The Answer to 21st Century Inclusive Information and Knowledge-Based Society Challenges. In C. Chisita & A. Rusero (Eds.), *Exploring the Relationship Between Media, Libraries, and Archives* (pp. 102–135). IGI Global. https://doi.org/10.4018/978-1-5225-5840-8.ch007

Noor, R. (2017). Citizen Journalism: News Gathering by Amateurs. In M. Adria & Y. Mao (Eds.), *Handbook of Research on Citizen Engagement and Public Participation in the Era of New Media* (pp. 194–229). Hershey, PA: IGI Global. doi:10.4018/978-1-5225-1081-9.ch012

Obermayer, N., Csepregi, A., & Kővári, E. (2017). Knowledge Sharing Relation to Competence, Emotional Intelligence, and Social Media Regarding Generations. In A. Bencsik (Ed.), *Knowledge Management Initiatives and Strategies in Small and Medium Enterprises* (pp. 269–290). Hershey, PA: IGI Global. doi:10.4018/978-1-5225-1642-2.ch013

Obermayer, N., Gaál, Z., Szabó, L., & Csepregi, A. (2017). Leveraging Knowledge Sharing over Social Media Tools. In R. Chugh (Ed.), *Harnessing Social Media as a Knowledge Management Tool* (pp. 1–24). Hershey, PA: IGI Global. doi:10.4018/978-1-5225-0495-5.ch001

Odebiyi, S. D., & Elegbe, O. (2020). Human Rights Abuses Against Internally Displaced Persons (IDPs) in Nigeria: Investigating Media Reportage. In N. Okorie, B. Ojebuyi, & J. Macharia (Eds.), *Handbook of Research on the Global Impact of Media on Migration Issues* (pp. 180–200). IGI Global. https://doi.org/10.4018/978-1-7998-0210-5.ch011

Okoroafor, O. E. (2018). New Media Technology and Development Journalism in Nigeria. In A. Salawu & T. Owolabi (Eds.), *Exploring Journalism Practice and Perception in Developing Countries* (pp. 105–125). Hershey, PA: IGI Global. doi:10.4018/978-1-5225-3376-4.ch006

Okpara, S. N. (2020). Child Protection and Development in Nigeria: Towards a More Functional Media Intervention. In O. Oyero (Ed.), *Media and Its Role in Protecting the Rights of Children in Africa* (pp. 57–79). IGI Global. https://doi.org/10.4018/978-1-7998-0329-4.ch004

Olaleye, S. A., Sanusi, I. T., & Ukpabi, D. C. (2018). Assessment of Mobile Money Enablers in Nigeria. In F. Mtenzi, G. Oreku, D. Lupiana, & J. Yonazi (Eds.), *Mobile Technologies and Socio-Economic Development in Emerging Nations* (pp. 129–155). Hershey, PA: IGI Global. doi:10.4018/978-1-5225-4029-8.ch007

Pacchiega, C. (2017). An Informal Methodology for Teaching Through Virtual Worlds: Using Internet Tools and Virtual Worlds in a Coordinated Pattern to Teach Various Subjects. In G. Panconesi & M. Guida (Eds.), *Handbook of Research on Collaborative Teaching Practice in Virtual Learning Environments* (pp. 163–180). Hershey, PA: IGI Global. doi:10.4018/978-1-5225-2426-7.ch009

Pant, L. D. (2021). Gender Mainstreaming in the Media: The Issue of Professional and Workplace Safety of Women Journalists in Nepal. In S. Jamil, B. Çoban, B. Ataman, & G. Appiah-Adjei (Eds.), *Handbook of Research on Discrimination, Gender Disparity, and Safety Risks in Journalism* (pp. 194–210). IGI Global. https://doi.org/10.4018/978-1-7998-6686-2.ch011

Pase, A. F., Goss, B. M., & Tietzmann, R. (2018). A Matter of Time: Transmedia Journalism Challenges. In R. Gambarato & G. Alzamora (Eds.), *Exploring Transmedia Journalism in the Digital Age* (pp. 49–66). Hershey, PA: IGI Global. doi:10.4018/978-1-5225-3781-6.ch004

Patkin, T. T. (2017). Social Media and Knowledge Management in a Crisis Context: Barriers and Opportunities. In R. Chugh (Ed.), *Harnessing Social Media as a Knowledge Management Tool* (pp. 125–142). Hershey, PA: IGI Global. doi:10.4018/978-1-5225-0495-5.ch007

Pavlíček, A. (2017). Social Media and Creativity: How to Engage Users and Tourists. In A. Kiráľová (Ed.), *Driving Tourism through Creative Destinations and Activities* (pp. 181–202). Hershey, PA: IGI Global. doi:10.4018/978-1-5225-2016-0.ch009

Pérez-Gómez, M. Á. (2020). Augmented Reality and Franchising: The Evolution of Media Mix Through Invizimals. In V. Hernández-Santaolalla & M. Barrientos-Bueno (Eds.), *Handbook of Research on Transmedia Storytelling, Audience Engagement, and Business Strategies* (pp. 90–102). IGI Global. https://doi.org/10.4018/978-1-7998-3119-8.ch007

Phiri, S., & Mokorosi, L. (2020). Of Elephants and Men: Understanding Gender-Based Hate Speech in Zambia's Social Media Platforms. In J. Kurebwa (Ed.), *Understanding Gender in the African Context* (pp. 105–125). IGI Global. https://doi.org/10.4018/978-1-7998-2815-0.ch006

Pillai, A. P. (2019). Nuances of Media Planning in New Media Age. In P. Mishra & S. Dham (Eds.), *Application of Gaming in New Media Marketing* (pp. 151–170). IGI Global. https://doi.org/10.4018/978-1-5225-6064-7.ch009

Pillay, K., & Maharaj, M. (2017). The Business of Advocacy: A Case Study of Greenpeace. In V. Benson, R. Tuninga, & G. Saridakis (Eds.), *Analyzing the Strategic Role of Social Networking in Firm Growth and Productivity* (pp. 59–75). Hershey, PA: IGI Global. doi:10.4018/978-1-5225-0559-4.ch004

Piven, I. P., & Breazeale, M. (2017). Desperately Seeking Customer Engagement: The Five-Sources Model of Brand Value on Social Media. In V. Benson, R. Tuninga, & G. Saridakis (Eds.), *Analyzing the Strategic Role of Social Networking in Firm Growth and Productivity* (pp. 283–313). Hershey, PA: IGI Global. doi:10.4018/978-1-5225-0559-4.ch016

Pokharel, R. (2017). New Media and Technology: How Do They Change the Notions of the Rhetorical Situations? In B. Gurung & M. Limbu (Eds.), *Integration of Cloud Technologies in Digitally Networked Classrooms and Learning Communities* (pp. 120–148). Hershey, PA: IGI Global. doi:10.4018/978-1-5225-1650-7.ch008

Porlezza, C., Benecchi, E., & Colapinto, C. (2018). The Transmedia Revitalization of Investigative Journalism: Opportunities and Challenges of the Serial Podcast. In R. Gambarato & G. Alzamora (Eds.), *Exploring Transmedia Journalism in the Digital Age* (pp. 183–201). Hershey, PA: IGI Global. doi:10.4018/978-1-5225-3781-6.ch011

Ramluckan, T., Ally, S. E., & van Niekerk, B. (2017). Twitter Use in Student Protests: The Case of South Africa's #FeesMustFall Campaign. In M. Korstanje (Ed.), *Threat Mitigation and Detection of Cyber Warfare and Terrorism Activities* (pp. 220–253). Hershey, PA: IGI Global. doi:10.4018/978-1-5225-1938-6.ch010

Rao, N. R. (2017). Social Media: An Enabler for Governance. In N. Rao (Ed.), *Social Media Listening and Monitoring for Business Applications* (pp. 151–164). Hershey, PA: IGI Global. doi:10.4018/978-1-5225-0846-5.ch008

Redi, F. (2017). Enhancing Coopetition Among Small Tourism Destinations by Creativity. In A. Kiráľová (Ed.), *Driving Tourism through Creative Destinations and Activities* (pp. 223–244). Hershey, PA: IGI Global. doi:10.4018/978-1-5225-2016-0.ch011

Resuloğlu, F., & Yılmaz, R. (2017). A Model for Interactive Advertising Narration. In R. Yılmaz (Ed.), *Narrative Advertising Models and Conceptualization in the Digital Age* (pp. 1–20). Hershey, PA: IGI Global. doi:10.4018/978-1-5225-2373-4.ch001

Richards, M. B. (2022). Media and Parental Communication: Effects on Millennials' Value Formation. In S. Malik, R. Bansal, & A. Tyagi (Eds.), *Impact and Role of Digital Technologies in Adolescent Lives* (pp. 64–82). IGI Global. https://doi.org/10.4018/978-1-7998-8318-0.ch006

Robinson, W. R. (2021). The Intellectual Soul Food Lunch Buffet: The Classroom to Student Media Entrepreneurship. In L. Byrd (Ed.), *Cultivating Entrepreneurial Changemakers Through Digital Media Education* (pp. 108–121). IGI Global. https://doi.org/10.4018/978-1-7998-5808-9.ch007

Ross, D. B., Eleno-Orama, M., & Salah, E. V. (2018). The Aging and Technological Society: Learning Our Way Through the Decades. In V. Bryan, A. Musgrove, & J. Powers (Eds.), *Handbook of Research on Human Development in the Digital Age* (pp. 205–234). Hershey, PA: IGI Global. doi:10.4018/978-1-5225-2838-8.ch010

Rusko, R., & Merenheimo, P. (2017). Co-Creating the Christmas Story: Digitalizing as a Shared Resource for a Shared Brand. In I. Oncioiu (Ed.), *Driving Innovation and Business Success in the Digital Economy* (pp. 137–157). Hershey, PA: IGI Global. doi:10.4018/978-1-5225-1779-5.ch010

Sabao, C., & Chikara, T. O. (2018). Social Media as Alternative Public Sphere for Citizen Participation and Protest in National Politics in Zimbabwe: The Case of #thisflag. In F. Endong (Ed.), *Exploring the Role of Social Media in Transnational Advocacy* (pp. 17–35). Hershey, PA: IGI Global. doi:10.4018/978-1-5225-2854-8.ch002

Saçak, B. (2019). Media Literacy in a Digital Age: Multimodal Social Semiotics and Reading Media. In M. Yildiz, M. Fazal, M. Ahn, R. Feirsen, & S. Ozdemir (Eds.), *Handbook of Research on Media Literacy Research and Applications Across Disciplines* (pp. 13–26). IGI Global. https://doi.org/10.4018/978-1-5225-9261-7.ch002

Samarthya-Howard, A., & Rogers, D. (2018). Scaling Mobile Technologies to Maximize Reach and Impact: Partnering With Mobile Network Operators and Governments. In S. Takavarasha Jr & C. Adams (Eds.), *Affordability Issues Surrounding the Use of ICT for Development and Poverty Reduction* (pp. 193–211). Hershey, PA: IGI Global. doi:10.4018/978-1-5225-3179-1.ch009

Sandoval-Almazan, R. (2017). Political Messaging in Digital Spaces: The Case of Twitter in Mexico's Presidential Campaign. In Y. Ibrahim (Ed.), *Politics, Protest, and Empowerment in Digital Spaces* (pp. 72–90). Hershey, PA: IGI Global. doi:10.4018/978-1-5225-1862-4.ch005

Schultz, C. D., & Dellnitz, A. (2018). Attribution Modeling in Online Advertising. In K. Yang (Ed.), *Multi-Platform Advertising Strategies in the Global Marketplace* (pp. 226–249). Hershey, PA: IGI Global. doi:10.4018/978-1-5225-3114-2.ch009

Schultz, C. D., & Holsing, C. (2018). Differences Across Device Usage in Search Engine Advertising. In K. Yang (Ed.), *Multi-Platform Advertising Strategies in the Global Marketplace* (pp. 250–279). Hershey, PA: IGI Global. doi:10.4018/978-1-5225-3114-2.ch010

Seçkin, G. (2020). The Integration of the Media With the Power in Turkey (2002-2019): Native, National Media Conception. In S. Karlidag & S. Bulut (Eds.), *Handbook of Research on the Political Economy of Communications and Media* (pp. 206–226). IGI Global. https://doi.org/10.4018/978-1-7998-3270-6.ch011

Senadheera, V., Warren, M., Leitch, S., & Pye, G. (2017). Facebook Content Analysis: A Study into Australian Banks' Social Media Community Engagement. In S. Hai-Jew (Ed.), *Social Media Data Extraction and Content Analysis* (pp. 412–432). Hershey, PA: IGI Global. doi:10.4018/978-1-5225-0648-5.ch013

Sharma, A. R. (2018). Promoting Global Competencies in India: Media and Information Literacy as Stepping Stone. In M. Yildiz, S. Funk, & B. De Abreu (Eds.), *Promoting Global Competencies Through Media Literacy* (pp. 160–174). Hershey, PA: IGI Global. doi:10.4018/978-1-5225-3082-4.ch010

Sharma, D., & Bhattacharya, S. (2022). Complexity of Digital Media Crowning the Mental Health of Adolescents. In S. Malik, R. Bansal, & A. Tyagi (Eds.), *Impact and Role of Digital Technologies in Adolescent Lives* (pp. 100–117). IGI Global. https://doi.org/10.4018/978-1-7998-8318-0.ch008

Sillah, A. (2017). Nonprofit Organizations and Social Media Use: An Analysis of Nonprofit Organizations' Effective Use of Social Media Tools. In M. Brown Sr., (Ed.), *Social Media Performance Evaluation and Success Measurements* (pp. 180–195). Hershey, PA: IGI Global. doi:10.4018/978-1-5225-1963-8.ch009

Silva, H., & Simão, E. (2019). Thinking Art in the Technological World: An Approach to Digital Media Art Creation. In E. Simão & C. Soares (Eds.), *Trends, Experiences, and Perspectives in Immersive Multimedia and Augmented Reality* (pp. 102–121). IGI Global. https://doi.org/10.4018/978-1-5225-5696-1.ch005

Škorić, M. (2017). Adaptation of Winlink 2000 Emergency Amateur Radio Email Network to a VHF Packet Radio Infrastructure. In A. El Oualkadi & J. Zbitou (Eds.), *Handbook of Research on Advanced Trends in Microwave and Communication Engineering* (pp. 498–528). Hershey, PA: IGI Global. doi:10.4018/978-1-5225-0773-4.ch016

Soares, C., & Simão, E. (2020). Software-Based Media Art: From the Artistic Exhibition to the Conservation Models. In C. Soares & E. Simão (Eds.), *Multidisciplinary Perspectives on New Media Art* (pp. 47–63). IGI Global. https://doi.org/10.4018/978-1-7998-3669-8.ch003

Sonnenberg, C. (2020). Mobile Media Usability: Evaluation of Methods for Adaptation and User Engagement. *Journal of Media Management and Entrepreneurship*, 2(1), 86–107. https://doi.org/10.4018/JMME.2020010106

Sood, T. (2017). Services Marketing: A Sector of the Current Millennium. In T. Sood (Ed.), *Strategic Marketing Management and Tactics in the Service Industry* (pp. 15–42). Hershey, PA: IGI Global. doi:10.4018/978-1-5225-2475-5.ch002

Sudarsanam, S. K. (2017). Social Media Metrics. In N. Rao (Ed.), *Social Media Listening and Monitoring for Business Applications* (pp. 131–149). Hershey, PA: IGI Global. doi:10.4018/978-1-5225-0846-5.ch007

Swiatek, L. (2017). Accessing the Finest Minds: Insights into Creativity from Esteemed Media Professionals. In N. Silton (Ed.), *Exploring the Benefits of Creativity in Education, Media, and the Arts* (pp. 240–263). Hershey, PA: IGI Global. doi:10.4018/978-1-5225-0504-4.ch012

Teurlings, J. (2017). What Critical Media Studies Should Not Take from Actor-Network Theory. In M. Spöhrer & B. Ochsner (Eds.), *Applying the Actor-Network Theory in Media Studies* (pp. 66–78). Hershey, PA: IGI Global. doi:10.4018/978-1-5225-0616-4.ch005

Tilwankar, V., Rai, S., & Bajpai, S. P. (2019). Role of Social Media in Environment Awareness: Social Media and Environment. In S. Narula, S. Rai, & A. Sharma (Eds.), *Environmental Awareness and the Role of Social Media* (pp. 117–139). IGI Global. https://doi.org/10.4018/978-1-5225-5291-8.ch006

Tokbaeva, D. (2019). Media Entrepreneurs and Market Dynamics: Case of Russian Media Markets. *Journal of Media Management and Entrepreneurship*, 1(1), 40–56. https://doi.org/10.4018/JMME.2019010103

Tomé, V. (2018). Assessing Media Literacy in Teacher Education. In M. Yildiz, S. Funk, & B. De Abreu (Eds.), *Promoting Global Competencies Through Media Literacy* (pp. 1–19). Hershey, PA: IGI Global. doi:10.4018/978-1-5225-3082-4.ch001

Topçu, Ç. (2022). Social Media and the Knowledge Gap: Research on the Appearance of COVID-19 in Turkey and the Knowledge Level of Users. In H. Aker & M. Aiken (Eds.), *Handbook of Research on Cyberchondria, Health Literacy, and the Role of Media in Society's Perception of Medical Information* (pp. 344–361). IGI Global. https://doi.org/10.4018/978-1-7998-8630-3.ch019

Toscano, J. P. (2017). Social Media and Public Participation: Opportunities, Barriers, and a New Framework. In M. Adria & Y. Mao (Eds.), *Handbook of Research on Citizen Engagement and Public Participation in the Era of New Media* (pp. 73–89). Hershey, PA: IGI Global. doi:10.4018/978-1-5225-1081-9.ch005

Trauth, E. (2017). Creating Meaning for Millennials: Bakhtin, Rosenblatt, and the Use of Social Media in the Composition Classroom. In K. Bryant (Ed.), *Engaging 21st Century Writers with Social Media* (pp. 151–162). Hershey, PA: IGI Global. doi:10.4018/978-1-5225-0562-4.ch009

Trucks, E. (2019). Making Social Media More Social: A Literature Review of Academic Libraries' Engagement and Connections Through Social Media Platforms. In J. Joe & E. Knight (Eds.), *Social Media for Communication and Instruction in Academic Libraries* (pp. 1–16). IGI Global. https://doi.org/10.4018/978-1-5225-8097-3.ch001

Udenze, S. (2021). Social Media and Nigeria's Politics. In S. Aririguzoh (Ed.), *Global Perspectives on the Impact of Mass Media on Electoral Processes* (pp. 83–96). IGI Global. https://doi.org/10.4018/978-1-7998-4820-2.ch005

Uprety, S. (2018). Print Media's Role in Securitization: National Security and Diplomacy Discourses in Nepal. In S. Bute (Ed.), *Media Diplomacy and Its Evolving Role in the Current Geopolitical Climate* (pp. 56–82). Hershey, PA: IGI Global. doi:10.4018/978-1-5225-3859-2.ch004

Uprety, S., & Chand, O. B. (2021). Trump's Declaration of the Global Gag Rule: Understanding Socio-Political Discourses Through Media. In E. Hancı-Azizoglu & M. Alawdat (Eds.), *Rhetoric and Sociolinguistics in Times of Global Crisis* (pp. 277–294). IGI Global. https://doi.org/10.4018/978-1-7998-6732-6.ch015

van der Vyver, A. G. (2018). A Model for Economic Development With Telecentres and the Social Media: Overcoming Affordability Constraints. In S. Takavarasha Jr & C. Adams (Eds.), *Affordability Issues Surrounding the Use of ICT for Development and Poverty Reduction* (pp. 112–140). Hershey, PA: IGI Global. doi:10.4018/978-1-5225-3179-1.ch006

van Niekerk, B. (2018). Social Media Activism From an Information Warfare and Security Perspective. In F. Endong (Ed.), *Exploring the Role of Social Media in Transnational Advocacy* (pp. 1–16). Hershey, PA: IGI Global. doi:10.4018/978-1-5225-2854-8.ch001

Varnali, K., & Gorgulu, V. (2017). Determinants of Brand Recall in Social Networking Sites. In W. Al-Rabayah, R. Khasawneh, R. Abu-shamaa, & I. Alsmadi (Eds.), *Strategic Uses of Social Media for Improved Customer Retention* (pp. 124–153). Hershey, PA: IGI Global. doi:10.4018/978-1-5225-1686-6.ch007

Varty, C. T., O'Neill, T. A., & Hambley, L. A. (2017). Leading Anywhere Workers: A Scientific and Practical Framework. In Y. Blount & M. Gloet (Eds.), *Anywhere Working and the New Era of Telecommuting* (pp. 47–88). Hershey, PA: IGI Global. doi:10.4018/978-1-5225-2328-4.ch003

Velikovsky, J. T. (2018). The Holon/Parton Structure of the Meme, or The Unit of Culture. In M. Khosrow-Pour, D.B.A. (Ed.), Encyclopedia of Information Science and Technology, Fourth Edition (pp. 4666-4678). Hershey, PA: IGI Global. https://doi.org/ doi:10.4018/978-1-5225-2255-3.ch405

Venkatesh, R., & Jayasingh, S. (2017). Transformation of Business through Social Media. In N. Rao (Ed.), *Social Media Listening and Monitoring for Business Applications* (pp. 1–17). Hershey, PA: IGI Global. doi:10.4018/978-1-5225-0846-5.ch001

Vijayakumar, D. S., M., S., Thangaraju, J., & V., S. (2021). Social Media Content Analysis: Machine Learning. In V. Sathiyamoorthi, & A. Elci (Eds.), *Challenges and Applications of Data Analytics in Social Perspectives* (pp. 156-174). IGI Global. https://doi.org/10.4018/978-1-7998-2566-1.ch009

Virkar, S. (2017). Trolls Just Want to Have Fun: Electronic Aggression within the Context of E-Participation and Other Online Political Behaviour in the United Kingdom. In M. Korstanje (Ed.), *Threat Mitigation and Detection of Cyber Warfare and Terrorism Activities* (pp. 111–162). Hershey, PA: IGI Global. doi:10.4018/978-1-5225-1938-6.ch006

Wakabi, W. (2017). When Citizens in Authoritarian States Use Facebook for Social Ties but Not Political Participation. In Y. Ibrahim (Ed.), *Politics, Protest, and Empowerment in Digital Spaces* (pp. 192–214). Hershey, PA: IGI Global. doi:10.4018/978-1-5225-1862-4.ch012

Wamuyu, P. K. (2021). Social Media Consumption Among Kenyans: Trends and Practices. In P. Wamuyu (Ed.), *Analyzing Global Social Media Consumption* (pp. 88–120). IGI Global. https://doi.org/10.4018/978-1-7998-4718-2.ch006

Wright, K. (2018). "Show Me What You Are Saying": Visual Literacy in the Composition Classroom. In A. August (Ed.), *Visual Imagery, Metadata, and Multimodal Literacies Across the Curriculum* (pp. 24–49). Hershey, PA: IGI Global. doi:10.4018/978-1-5225-2808-1.ch002

Wright, M. F. (2020). Cyberbullying: Negative Interaction Through Social Media. In M. Desjarlais (Ed.), *The Psychology and Dynamics Behind Social Media Interactions* (pp. 107–135). IGI Global. https://doi.org/10.4018/978-1-5225-9412-3.ch005

Yang, K. C. (2018). Understanding How Mexican and U.S. Consumers Decide to Use Mobile Social Media: A Cross-National Qualitative Study. In K. Yang (Ed.), *Multi-Platform Advertising Strategies in the Global Marketplace* (pp. 168–198). Hershey, PA: IGI Global. doi:10.4018/978-1-5225-3114-2.ch007

Yarchi, M., Wolfsfeld, G., Samuel-Azran, T., & Segev, E. (2017). Invest, Engage, and Win: Online Campaigns and Their Outcomes in an Israeli Election. In M. Brown Sr., (Ed.), *Social Media Performance Evaluation and Success Measurements* (pp. 225–248). Hershey, PA: IGI Global. doi:10.4018/978-1-5225-1963-8.ch011

Yeboah-Banin, A. A., & Amoakohene, M. I. (2018). The Dark Side of Multi-Platform Advertising in an Emerging Economy Context. In K. Yang (Ed.), *Multi-Platform Advertising Strategies in the Global Marketplace* (pp. 30–53). Hershey, PA: IGI Global. doi:10.4018/978-1-5225-3114-2.ch002

Yılmaz, R., Çakır, A., & Resuloğlu, F. (2017). Historical Transformation of the Advertising Narration in Turkey: From Stereotype to Digital Media. In R. Yılmaz (Ed.), *Narrative Advertising Models and Conceptualization in the Digital Age* (pp. 133–152). Hershey, PA: IGI Global. doi:10.4018/978-1-5225-2373-4.ch008

Yusuf, S., Hassan, M. S., & Ibrahim, A. M. (2018). Cyberbullying Among Malaysian Children Based on Research Evidence. In M. Khosrow-Pour, D.B.A. (Ed.), Encyclopedia of Information Science and Technology, Fourth Edition (pp. 1704-1722). Hershey, PA: IGI Global. doi:10.4018/978-1-5225-2255-3.ch149

Zbinden, B. (2019). Restricted Communication: Social Relationships and the Media Use of Prisoners. In L. Oliveira & D. Graça (Eds.), *Infocommunication Skills as a Rehabilitation and Social Reintegration Tool for Inmates* (pp. 238–267). IGI Global. doi:10.4018/978-1-5225-5975-7.ch011

Zhou, M., Matsika, C., Zhou, T. G., & Chawarura, W. I. (2022). Harnessing Social Media to Improve Educational Performance of Adolescent Freshmen in Universities. In S. Malik, R. Bansal, & A. Tyagi (Eds.), *Impact and Role of Digital Technologies in Adolescent Lives* (pp. 51–63). IGI Global. https://doi.org/10.4018/978-1-7998-8318-0.ch005

About the Author

Seif Sekalala is a communication studies and social-science researcher and instructor, project manager, and mental health activist. For over 13 years, he has taught in various colleges and universities in the United States and China (including courses such as communication research methods, business-professional communication, intercultural communication, and sociology, among others). Dr. Sekalala has also published several research articles and books with foci on intercultural communication; health communication and peace and conflict studies—e.g., about sensemaking and resilience among Rwandan genocide survivors; and theorization and implementation of best practices for effective listening among healthcare workers. He is passionate about the amelioration of mental illness in the general population in the USA and around the world, and among US minorities and Black men in particular. Dr. Sekalala is also an avid self-taught software-developer, and is particularly keen about the potential contribution of open-source software projects vis-à-vis sustainability.

Index

A

AI-Bias 147
Artificial Intelligence 117
autoethnography 1, 9-10, 27-28, 30, 62, 77, 95, 168

B

Barack Obama 12, 15, 64-65, 77-79, 91, 97, 108, 138
blogging 30-32, 51, 53, 60, 77

C

Charlotte Linde 155, 166
cognitive science 4, 18-19, 27-28, 75, 83, 94, 116-117, 122-124, 126, 128-129, 133-134, 154, 161, 167-168
Coherence System 160, 166
crucible moments 84, 167, 171

D

deliberate SC 148, 155, 157, 166, 168
Discourse Analysis 117
Distress 11, 135-137, 139, 141, 146

E

emotions 11, 14-15, 17-18, 36, 40-41, 46-48, 64, 66-67, 69-71, 80, 83, 85, 92, 99, 108, 128, 135-136, 139-140, 143, 148, 152, 157-158
enhanced equilibrium and distress 135-136, 146
equilibrium 84, 135-137, 146, 171
evolution 8, 21, 30, 32, 131

G

Google Scholar 5, 117, 122, 124-127, 129
guilt 15, 36, 47, 69, 83, 135-136, 139

H

Human Communication 5, 28, 147

I

ICT 8, 147-154, 156, 161, 167
interpersonal communication 1, 4, 23, 26-28, 147-148, 150, 156, 161
Intrapersonal(Self) Communication (SC) 1-5, 7-11, 13, 15-16, 20, 23, 26-28, 77, 79, 117, 135, 147, 155, 158, 166-167

J

journaling 15-16, 23-24, 26, 31-32, 34, 36, 38, 56, 59-60, 69, 74-75, 77, 83, 85, 93, 96, 98, 110-111, 116, 132, 136, 145, 148, 154, 157, 166, 173

M

Machine Learning 127, 147
Margaret Archer 29, 155
Mass Media 155
Matthew McConaughey 155, 163
meditation 3, 15, 51, 69-71, 135-137, 149
mental-health 8, 33, 40, 135-136, 145
Metacognition 3, 15, 21, 30, 32, 60, 62-63, 69-71, 74-75, 77, 83, 86-88, 94-96, 98, 105-107, 109, 112, 116, 130, 133-134, 152, 155, 157, 161, 166, 168
mindfulness 56, 73-74, 107, 135, 141, 143, 145, 150-152, 156

O

Oprah Winfrey 135

P

Prince Harry 52, 135

R

Reality and Perception-Processing I; Without Metacognition 75, 94, 116
Reality and Perception-Processing II; With Metacognition 75, 94, 116, 134
reflexivity 4-8, 21, 27, 29, 62-63, 95-96, 156, 166

S

Secondary Intra-Interpersonal Cognition and Communication Spectrum 28
self-doubt 15, 67, 97, 100, 102-104, 115, 135-136, 138-141
self-help 155-156, 163-164
sensemaking 77, 117, 119, 127, 156, 161, 170
Sherry Turkle 147
Significance Analysis 21, 122-124, 126, 134
Social Reflexivity 6, 29
Social Science 10, 29
socialization 4, 7-8, 21, 63, 77, 86, 156, 158-159, 161
social-reflexivity 6, 26

social-science 1, 6, 9, 11, 97, 170
software-development 12, 14, 24, 65, 67, 99, 102, 105, 110-111, 117-118, 122, 138, 141, 143

T

Thinking-Triggered 135-136, 146
tip of the tongue 15-16, 65, 68, 70, 77, 79, 96, 137, 149
to-do lists 24, 32, 70, 72, 77
Tom Hanks 167

U

Unintentional SC 166

W

writing 4, 23, 25, 27, 30, 33-36, 38-39, 41-43, 49, 54, 58, 60, 65-66, 72, 74, 77, 79, 81, 86-89, 92-93, 95, 98, 113, 126, 131, 142, 147, 151, 159-160, 162, 164, 168

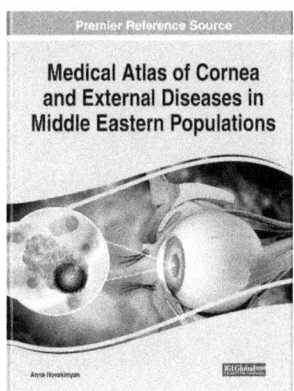

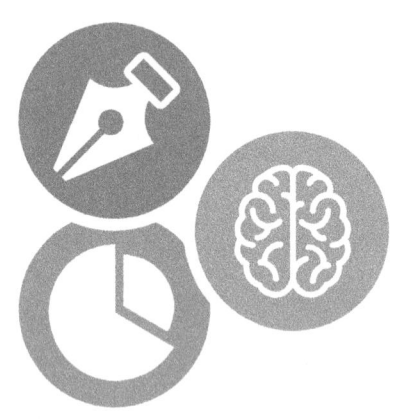

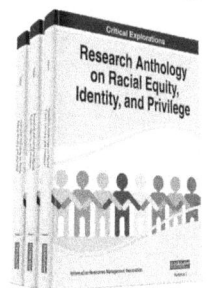